Place-Names of Carmarthenshire

Richard Morgan

Caerfyrddin/Carmarthen. A line engraving after Henry Gastineau c.1835.

Place-Names of Carmarthenshire

Richard Morgan

welsh academic press
Cardiff

Published in Wales by Welsh Academic Press, an imprint of

Ashley Drake Publishing Ltd
PO Box 733
Cardiff
CF14 7ZY

www.welsh-academic-press.wales

First Edition - 2022
Paperback - 978 1 86057 1572
eBook - 978 1 86057 1589

© Ashley Drake Publishing Ltd 2022
Text © Richard Morgan 2022

The right of Richard Morgan to be identified as the author of this work has been asserted in accordance with the Copyright Design and Patents Act of 1988.

Every effort has been made to contact copyright holders. However, the publishers will be glad to rectify in future editions any inadvertent omissions brought to their attention.

Ashley Drake Publishing Ltd hereby exclude all liability to the extent permitted by law for any errors or omissions in this book and for any loss, damage or expense (whether direct or indirect) suffered by a third party relying on any information contained in this book.

All rights reserved. No part of this publication may be reproduced, stored in a retrieval system, or transmitted, in any form or by any means without the prior permission of the publishers.

British Library Cataloguing-in-Publication Data.
A CIP catalogue for this book is available from the British Library.

Typeset by Prepress Plus, India (www.prepressplus.in)
Cover designed by Books Council of Wales

CONTENTS

Acknowledgements	vii
List of illustrations	viii
Preface	xi
Introduction	xii
Carmarthenshire Place-Names: study and survey	xii
Carmarthenshire Place-Names: research and analysis	xiv
Map 1 Medieval Carmarthenshire	xix
Map 2 English influence before 1500	xix
Selection of names	xx
Map 3: County boundary 1536-1974	xxi
Map 4: Unitary authorites from 2003	xxii
Editorial method	xxiii
Guide to the International Phonetic Alphabet	xxvi
Abbreviations and Bibliography	xxix
Online Databases and Reference Resources	xxxix
Glossary of Place-Name Elements, Personal Names and River-Names	xlii
Common place-name elements	xliii
List of personal names and surnames	lxxvi
List of river-names	lxxxi
A Aberarad to Ashfield	1
B Babel to Bynea	6
C Caeo to Cywyn	16
D Dafen to Dynevor	53
E East Marsh to Esgob	59
F Faenor to Furnace	63
G Ganol to Gwynfe	70
H Halfpenny Furze to Horeb	84
I Iddole to Is-morlais	89
J Johnstown	91
K Kidwelly to Kingsland	92
L Lacques to Loughor	93
M Mabelfyw to Myrtle Hill	114
N Nant Aeron to Newton	134
P Pantarfon to Pysgotwr	141
R Ram to Roche Castle	155
S Salem to Sylgen	160
T Tachlouan to Tywi	164
U Uwchcoed Morris to Uwch Sawdde	177
W Waun Baglam to Wysg	178
Y Ydw to Ystumgwili	181
List of Subscribers	184

For Tim Morgan

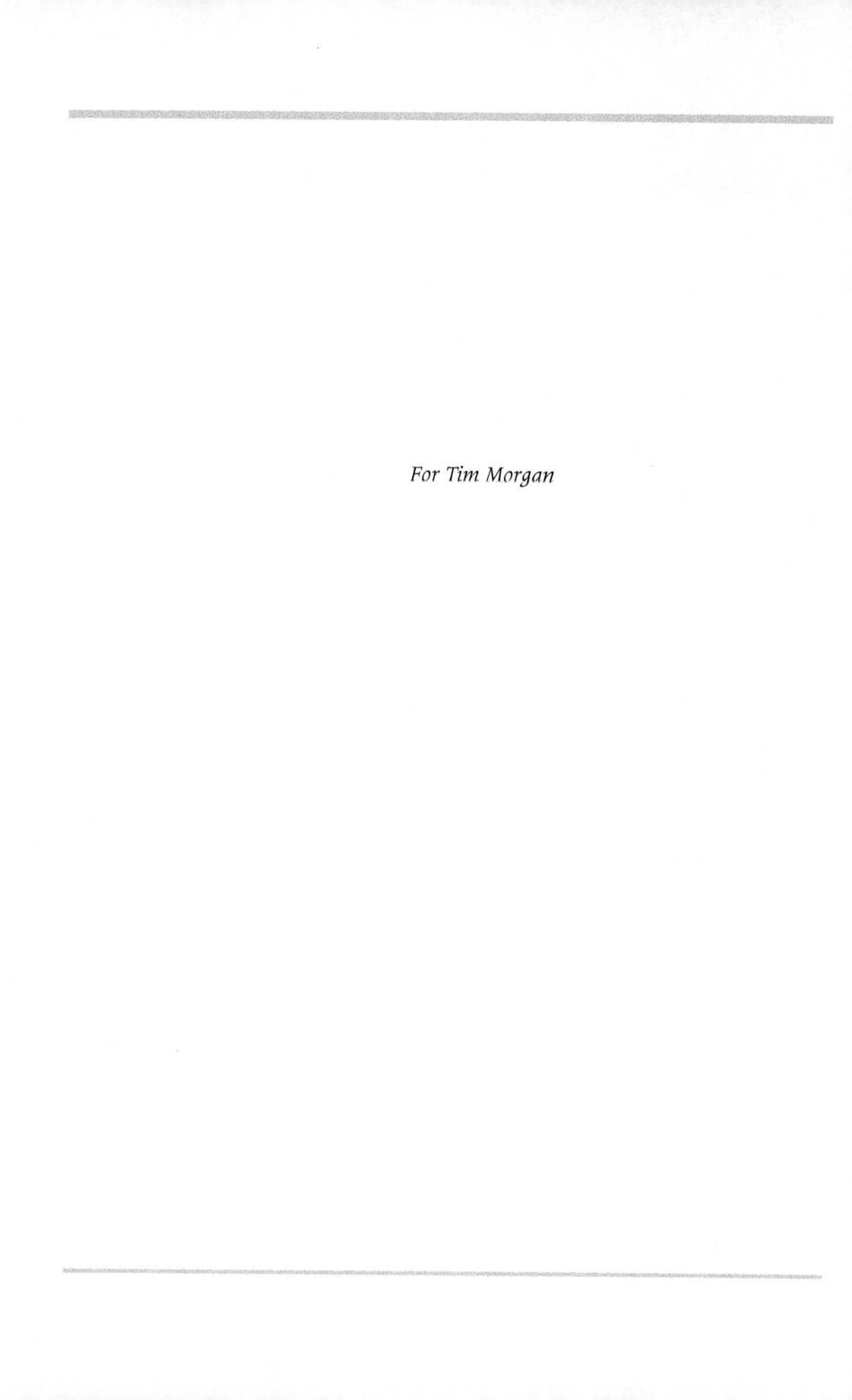

ACKNOWLEDGEMENTS

I am especially indebted to Dorothy Bere for her assistance in extracting place-names from Ordnance Survey maps. Advice in identifying historical sources and/or critical commentaries have been made by Gareth Beavan, Hugh Brodie, Richard Coates, Rhiannon Comeau, Bruce Coplestone-Crow, Thomas Owen Clancy, Byron Duckfield, Dylan Foster Evans, Dai Hawkins, Alan G. James, Heather James, Deric John, Lyn John, Ann Parry Owen, Hywel Wyn Owen, Oliver J. Padel, Gwynedd O. Pierce †, Eiluned Rees, Guto Rhys, Alan Richards, Sara Elin Roberts, Patrick Sims-Williams, David H. Williams and David Thorne. Terry Wells kindly provided the map of Carmarthenshire for the book cover and three of the illustrations taken from Thomas Kitchin's Accurate Map of Carmarthen Shire c.1762, and the copy of the engraving of the town of Carmarthen c.1835, on behalf of Carmarthenshire Archives. I am grateful to staff of Cardiff University's Arts and Humanities Library and Special Collections and to my former colleagues in Glamorgan Archives for assisting with my enquiries.

LIST OF ILLUSTRATIONS

Abergorlech
The former Wheaten Sheaf public house c.1935.

Brechfa
Forest Arms Hotel and St Teilo's church c.1930.

Cydweli/Kidwelly
Castle Street near former school looking towards the castle c.1910.

Dinefwr Castle
Viewed from the south, 1740.

Egrmwnt/Egremont
Taken from Thomas Kitchin's 'Accurate Map of Carmarthen Shire' c.1762.

Fforest, Llanedi
Taken from Thomas Kitchin's 'Accurate Map of Carmarthen Shire' c.1762.

Gwynfe
Pont Glan-rhyd, over Afon Clydach, and Capel Jerusalem c.1900

Hendy-gwyn/Whitland
Llan-gan Road looking towards Market Street c.1915.

Iddole
The former Capel Seion.

Johnstown/Tre Ioan
Looking west towards the Toll House with Tafarn y Cyfeillion/The Friends Arms behind, 2021.

Kingsland
Kingsland hamlet, in Llanboidy. Taken from Thomas Kitchin's 'Accurate Map of Carmarthen Shire' c.1762.

Llanelli
Station Road looking north towards the Town Hall c.1910.

List of illustrations

Meidrum
Viewed from the south, 1905.

Nantgaredig
Looking north from Heol yr Orsaf/Station Road c.1920.

Pen-dein / Pendine
Viewed westwards along seashore c.1910.

Rhydaman/Ammanford
Looking southwestwards down Wind Street c.1910.

Sanclêr/St Clears
Looking northwards up High Street from Butchers' Arms towards Capel Mair c.1900.

Trimsaran
Looking northwestwards from bridge over Afon Morlais to Sardis Independent chapel and Bryncaerau c.1930.

Uwch Sawdde area between Llangadog and Llanddeusant
Taken from Thomas Kitchin's 'Accurate Map of Carmarthen Shire' c.1762.

Waungilwen, Drefach Felindre 1932
Looking northwestwards to houses on the road from Dre-fach towards Pentrecagal c.1910.

Ystrad Tywi near Tŷ-gwyn-mawr, Llandeilo
Looking northwestwards over Afon Tywi c.1920.

This volume has been published with the generous financial support of

MARC FITCH FUND

PREFACE

The background to *Place-Names of Carmarthenshire* is set out in its companion volume *Place-Names of Glamorgan* (Welsh Academic Press 2018) and there is little point in repeating it save to add that publication of the Glamorgan volume has highlighted the need for similar publications in other parts of Wales. Ideally, Carmarthenshire deserves the close detailed attention given to other historic counties such as Ceredigion and Pembrokeshire but this would demand academic and financial resources which are not yet available. *Place-Names of Carmarthenshire* is instead a selection of 923 place-names – names of individual towns, villages and historic divisions – with the addition of a selection of topographical features such as major rivers. Names of individual fields, houses and most topographical names have had to be omitted because that is a task which is best tackled by an historical society with dozens of volunteers guided by those with specialist knowledge of language, historical sources, archaeology and topography. My hope is that *Place-Names of Carmarthenshire* will encourage this sort of research at a local level as first steps to a closer survey. A great deal of research underlies every place-name entry but this is not comprehensive and it follows that any analysis and interpretation of individual names will need to be re-examined at some point. That is the fundamental nature of place-name studies. Some of the evidence presented in *Place-Names of Carmarthenshire* was gathered before 2018 during preparation of the Glamorgan volume but the net had to be cast further in order to collect information relating specifically to Carmarthenshire. This extended research draws on archives held in Carmarthenshire Archives, Pembrokeshire Record Office and The National Archives, Kew, together with evidence accessible online from The National Library of Wales, Aberystwyth. A large number of publications relating specifically to Carmarthenshire were accessed in Cardiff University's Arts and Social Studies Library and in Special Collections. Some amendments have been made to the format set out in *Place-Names of Glamorgan*, notably by expansion of the Glossary in order to cross-refer place-name elements to place-name entries.

INTRODUCTION

Carmarthenshire Place-Names: study and survey

Everyone interested in history and language would accept the importance of place-name research. To quote the late Professor Melville Richards in translation, 'Tracing the history of place-names sheds a ray of light on the ways in which our ancestors lived and how they thought about the visible world around them' ('Mae olrhain hanes enwau lleoedd yn taflu ffrwd o oleuni ar y modd y byddai'n cyndadau yn byw, a sut y meddylient am y byd gweledig o'u cwmpas': *Enwau Tir a Gwlad* 1998). Yet place-name research in Wales has until recent years been slow for reasons which are summarised in the *Dictionary of the Place-Names of Wales* (2007) and *Place-Names of Glamorgan* (2018). Carmarthenshire historians may take comfort in the fact that they were among the first in Wales to understand the importance of place-names, however. In 1908 Morgan Hugh Jones (CAS IV, 28-29) observed that the Congress of Archaeological Studies, with which the Carmarthen County Antiquary Society was then affiliated, had published notes on the systematic study of English place-names, and noted that the Guild of Graduates of the University of Wales had appointed a committee to organise a similar study of Welsh place-names. Jones suggested that the society and readers of the newspaper 'The Welshman' should outline a study of Carmarthenshire place-names with the object of discouraging 'the popular but generally worthless fanciful etymologies and substitute a more scientific process by recording actual facts as to Names.' Jones set out a method of approach which was clear and practical but ultimately came to little. As if to underline his concerns, a short series of articles by J. Lloyd James on place-names and antiquities in the Taf valley – published in the same volume of the *Transactions of the Carmarthenshire Antiquary Society* – was riddled with the 'fanciful etymologies' detested by Jones. The poor scholarship of the articles by J. Lloyd James earned him deserved rebukes from H.E.H. James of Haverfordwest and D. Cledlyn Evans, fellow members of the Carmarthenshire Antiquarian Society (CAS IV, 78-85; V, 7, 16-17). M.H. Jones posted several times further in the *Transactions* urging the society in 1911 to collect material for 'an exhaustive study of the place-names of Carmarthenshire' (CAS VII, 8-9). He compiled a list of ecclesiastical place-names in the county (AC 1915, 321-332, 395-404) and personally submitted work to the *Transactions* compiled by E. Aman Jones on place-names of the Aman valley and Evan E. Morgan on field-names of Pen-bre/Pembrey to John Rhŷs (1840-1915), Professor of Celtic at Jesus College, Oxford. Subsequent articles in the *Transactions* continued this work, laying a useful foundation for further place-name study. It is difficult to avoid

Introduction

the conclusion, however, that the scale of the task was too daunting for most of his colleagues and successors.

Decades passed with little progress until 1990 when the Carmarthenshire Place-name Survey was set up by the Carmarthenshire Antiquarian Society. A joint meeting with representatives of the Powysland Club (for Montgomeryshire) and Clwyd Place-name Council led to the formation of a sub-committee 'to organise the collection and study of place-names'. Meetings followed at Gregynog which in turn led to the setting-up of the Place-Names Survey of Wales under the directorship of Professor Emeritus Gwynedd O. Pierce and to the publication of a report entitled 'Place-Name Surveys of Wales: A brief guide to the collection and recording of place-name forms', published by the Board of Celtic Studies, that same year. The Carmarthenshire Place-Name Survey began at a day school in March 1990 with the declared aim of transcribing all names which appear on the Ordnance Survey 1:2,500, 1:10,560 and 1:63,360 maps and plans in addition to the Ordnance Survey drawings. Members involved in the Survey were fully aware of the scale of the task and estimated that the OS 1:2,500 plans alone would amount to about 50,000 names (Terrence James, 'The Carmarthenshire Place-Name Survey', CAS 26, 91-4, and a summary compiled by James in November 2005 <http://www.carmants.org.uk/placenamessearch.html>). Terry James designed and commenced development on a computerised programme in which place-name data was entered with the aim of making this available 'for all branches of local study and to dovetail with moves in different parts of Wales to computerise place-names' (Peter Wihl, co-ordinator, 'Carmarthenshire Place-Name Survey', CAS 32, 129-130; with updates by Terrence James, ~ 34, 128-9, and Peter Wihl, ~ 37 (2001), 129-30). Despite the best efforts of its co-ordinators, the work was not completed – largely owing to the limited capacity of technological data storage and retrieval at that time. The database may be accessed from Cymdeithas Enwau Lleoedd Cymru/Welsh Place-Name Society. As with Montgomeryshire, the work of the Survey came to rest heavily on the shoulders of just a small number of members of the Carmarthenshire Antiquarian Society and the aims set out in 1990 were largely unfulfilled. By 1996, the survey had entered over 36,000 names in the database but the number of volunteers had dwindled to ten. The Survey was dealt a particularly bad blow by the passing of Terrence James in 2007 (obituary in CAS 42, 174).

Fortunately, the past twelve years have seen significant progress in place-names studies and in the gathering of evidence and the compilation of databases. The work of the Carmarthenshire Place-Names Survey and other research in Wales is being supplemented by the digital List of Historic Place Names compiled by the Royal Commission on Ancient and Historic Monuments of Wales (https://historicplacenames.rcahmw.gov.uk). Described as 'an index of names for geographical locations gathered from a variety of historical sources by a number of different projects', this aims to raise public awareness of the rich legacy of historic place names in Wales and to encourage continuing use of these important elements of our nation's heritage. The incorporation of data from other sources, such as evidence from tithe plans and apportionments (Cynefin: https://places.library.wales) and data contributed by Cymdeithas Enwau Lleoedd Cymru/Welsh Place-Name Society, local history groups and private individuals, is gradually producing a large resource which can provide both a starting-point and a source of reference for place-name studies. Contributory data from Cynefin and the Historic Environment Record now enable place-name studies to make important links to archaeological and historical databases.

PLACE-NAMES OF CARMARTHENSHIRE

Other online databases should be examined. Key sources now include historic Ordnance Survey maps and plans at The National Library of Scotland (www.maps.nls.uk) supplemented by Ordnance Survey drawings in The British Library (http://www.bl.uk/onlinegallery/onlineex/ordsurvdraw). Ordnance Survey first edition maps on the scale 1:63,360 are both online (http://www.vision of britain.org.uk/maps) and in print from the publishers David & Charles. Online sources of particular value to toponymists and historians, notably probate records, newspaper holdings (down to 1920) and many journals have been digitised by The National Library of Wales, Aberystwyth.

For toponymists – place-name specialists – the single most valuable resource at present is that of the late Professor Melville Richards in Archif Melville Richards (http://www.e-gymraeg.co.uk) at Bangor University. Digitisation of Richards' collection in 2005 by the university led the way in providing online access to place-name evidence. Historic forms and references on the database should nonetheless be checked before they are cited in publications. It is important to emphasise that a number of archives repositories have changed their names or moved site since Richards died in 1973 and some of the archives collections mentioned on his database have been moved to other repositories. Research remains challenging and can prove slow and expensive when archives are poorly served by public transport. The recent Covid-19 pandemic has added to these difficulties compelling the temporary closure of archives repositories and major libraries and/or restricted access to collections. This has been a heavy blow to archives repositories in particular; many have for many years lacked sufficient staff and financial resources to provide detailed lists of individual collections. Researchers must expect to be disappointed to find that some archives collections are listed only down to series level or, worse still, under a general summary description. Examination of individual items within these collections may involve weeks of research. These difficulties will in large part explain the paucity of evidence for some place-names examined in *Place-Names of Carmarthenshire*.

Carmarthenshire Place-Names: research and analysis

Place-Names of Carmarthenshire covers 923 place-names in the historic county – sufficient I think for those looking for a general and introductory survey. Hopefully, this and publications cited in the Bibliography will encourage further research. Sources of all historic forms cited in this volume are recorded in an online database accessible at Cymdeithas Enwau Lleoedd/Welsh Place-Name Society (see Selection of Names below). Perhaps this will lead others on to the sort of detailed analysis found in, for example, *Place-Names of Pembrokeshire* by B.G. Charles (PNPemb). This will require taking the research down to the names of individual houses, streets and fields, i.e. local names in contrast to the broader selection used in *Place-Names of Carmarthenshire* and *Place-Names of Glamorgan*. That will also demand closer examination of the vocabulary, phonology and meaning of individual names – both individually and collectively. At that point it should be possible to relate names to social structure, topography, chronological sequence and geographical distribution. That is clearly beyond the scope of *Place-Names of Carmarthenshire* though a few general comments are set out below and within individual entries. It is also worth emphasising that compiling place-name publications, whatever their scope and

Introduction

remit, is time-consuming and exceptionally demanding for even the most efficient and knowledgeable historian. All too often, it can lead to publications which are beyond the financial means of many readers and which fail to spread the historical message. Digitisation will solve many of these difficulties but at present toponymists have few up-to-date reservoirs of historical evidence on which to draw and reliable comparative studies are uncommon. Discussion has to be concise and guarded – given the present position – and every place-name entry in every county survey will at some time need reconsideration as historical and language research progress.

A few general matters can be looked at. Firstly, the great majority of place-names covered in *Place-Names of Carmarthenshire* are undoubtedly Welsh (with its predecessor British) or, in a very small number of cases, Irish. The latter is included with Welsh place-names because they are so intertwined with Welsh place-names and have so many etymological overlaps that they are sometimes difficult to distinguish. The paucity of early evidence, coupled with the limited research so far written on the languages of early Wales, often prevents clear separation. One particular matter, namely the relationship between the suffixes *-ach* and *-og*, found in many river-names, illustrates this problem well and a concise discussion can be found in the first entry for Clydach. A fuller understanding of this relationship will eventually be found when all the place-names of Carmarthenshire – and Wales in general – are examined and related to other facets of history, language and archaeology. Secondly, it is not an exaggeration to say that the Welsh language is the bedrock on which the earlier history of Wales lies. In some measure, this is better appreciated by examining the relationship between Welsh, its 'sibling languages' Cornish, Breton and Cumbrian (once spoken in north-west England and Scotland), its 'cousin language' Irish, and its 'distant relatives' English and Old Scandinavian. The last has been omitted from *Place-Names of Carmarthenshire* because no definite examples of place-names in this language have so far been identified – in contrast to the adjoining counties of Pembroke and Glamorgan.

Welsh (and Irish) place-names account for about 95% of the names examined in *Place-Names of Carmarthenshire* – making allowance of 2% or 3% variation in determining the language of individual place-names. Counter-intuitively, the clearest way of appreciating the dominance of Welsh place-names is by looking at the chronological and geographical distribution of English place-names and the effect which the English language had upon Welsh place-names. For convenience, these can be subdivided into (1) English-only place-names (with one Anglo-French example) (2) English place-names which have been affected in written form, pattern and phonology by Welsh-speakers (cymricisations) (3) Welsh place-names influenced in the same manner by English-speakers (anglicisations) and (4) dual names, i.e. places which bear both Welsh and English names. The 'dual' category should be regarded as the least easily defined since it overlaps with the other categories and includes place-names which are partly related semantically such as Porth Tywyn/Burry Port. Out of 923 place-names just 40 can be regarded as English and cymricisations. Making due allowance for uncertain names they amount to just 4.33% of the total figure.

An attempt has been made in Map 2 to show the distribution of the first three categories before 1500 in order to show areas where Anglo-Norman settlement took place. Only six names are English-only (0.65%) and even if we expand the figure to include cymricisations, anglicisations and dual names the total rises to only 22 names (2.38%). It has to be stressed, however, that a number of place-names recorded soon after 1500 may have to be added if earlier evidence is found. Names such as Egrmwnt/Egremont (recorded 1513) and Lacques (1616), for example, certainly have an 'old appearance' – if that is a permissible description. The map distribution with all its limitations is interesting since the majority of these names are within the lordship of Talacharn[1]/Laugharne and the lowland part of Sanclêr/St Clears with a very small scatter between Caerfyrddin/ Carmarthen and Cydweli/Kidwelly and isolated examples at Castellnewydd Emlyn/ Newcastle Emlyn, Pinged and Berwig (Llanelli). The westernmost concentration – south of Afon Taf and extending from Talacharn[2]/Laugharne to the Pembrokeshire border – forms a continuance with the far more numerous English place-names in the southern part of Pembrokeshire. The English language has an unbroken history here since the twelfth century. All of these concentrations correspond with those areas first conquered by Anglo-Normans. This intrusion was accompanied by English settlement along the coast and around newly-built castles.

This distribution is generally confirmed by evidence for the period after 1500. English settlement seems to have made little progress beyond the areas described above. It is important to emphasise, however, that we should make allowance for the possibility of English settlement elsewhere in Carmarthenshire where early historical evidence is so far lacking. There are a number of English place-names – particularly names of houses and fields – mentioned in later sources in the area between the rivers Taf and Tywi in Llanfihangel Abercywyn and Llansteffan, and further to the east around Cydweli. A useful, if somewhat outdated list of English place-names, appears in *Non-Celtic Place-Names in Wales* by B.G. Charles in 1938 (NCPN 111) and we can supplement this with more evidence for street-names in boroughs founded by the Anglo-Normans, notably Talacharn[2]/Laugharne, Cydweli/Kidwelly and especially Caerfyrddin/Carmarthen. The last has at least thirteen examples in this category for the period before 1600 including *Alremill* 1300, Bridge Street 1575, Canon Hill 1358, Cock Mill 1300, Dam Mill 1275, Gaol Street 1568, King Street 1575, Priory Street 1594, Quay Street 1509, St Mary Street 1560, Spilman Street 1405, unlocated *Sutternistret* 1356, and Water Street 1560. English street-names in Talacharn[2]/Laugharne for the same period include Gosport Street 1598 and it lays claim to Court Lane, Frog Street, Market Street, Towns End and Upton Street before 1700. Cydweli's street-names include Bower Street, Ditch Street, St Mary Street and Long Street c.1500, Causey Street 1575, Frogmoor Street 1345, Monksford Street and Water Street 1609, and Shoe Lane Street 1622. Numbers can be increased substantially if we were to add names of fields and individual buildings to the total. Such place-names would be commonplace in England and have parallels in other parts of south Wales and the border areas. Evidence for other boroughs in Carmarthenshire is much scarcer than those already mentioned but it is difficult to avoid the general conclusion that English-speakers here made up just a small part of the townspeople. It is important to qualify these observations since most of our early evidence is English. Some of these boroughs were probably much more Welsh than incidental references imply. English street-names in Llanymddyfri/Llandovery, for example, also bore Welsh names: Orchard Street and Stone Street were otherwise known as Heol y Berllan and Heol Gerrig. More extensive research will undoubtedly add to the evidence and amend these observations.

No attempt has been made to plot English place-names after 1500 systematically because that is beyond the scope of *Place-Names of Carmarthenshire*. A handful of general comments will have to suffice. Firstly, we know comparatively little about the relationship of place-names and language and secondly, our sources are largely English owing to the

Introduction

weaker legal status of the Welsh language as reflected in conveyances of property and financial accounts as examples. Welsh evidence is generally later and often anecdotal. In the areas first conquered by Anglo-Normans, minor names and cymricisations suggest that the Welsh language may have later gained ground, particularly in the Llansteffan area and around Cydweli and Llanelli before about 1700 – effectively burying some of the English place-names noted above. A few English names appear before about the middle of the eighteenth century but these include those of mansions and reflect fashionability rather than widespread language change. Golden Grove, for example, is first recorded in 1578 and its Welsh partner Gelli-aur (q.v.) in 1596 though both are likely to have been coined in 1560 when the first mansion was constructed. The houses Alltycadno and Nantyrarian, both in Llangyndeyrn, are recorded respectively as Foxgrove from c.1662 down to c.1730 and Silver Grove in 1666, but both adoptions subsequently disappear from the record. Dolgrogws (*dôl, crocws* nm. 'crocus, saffron'), in Llanfihangel-ar-arth, is Saffron Mead in 1741 (NLW Probate SD/1741/80). This phenomenon is found elsewhere in Wales, notably Crosswood, co. Cardigan, which has now reverted to Trawsgoed.

The most radical changes in the nature and geographical spread of place-names were sparked by religious nonconformity, commercial travel and industrialisation. The first category covers Biblical chapel-names such as Bethlehem and Hebron (ten examples or less than 2%) but there is a bias here because *Place-Names of Carmarthenshire* covers only chapel-names which rose to the level of villages and hamlets. Place-names in this category were often in rural areas where chapels might encourage further settlement. These have been included with Welsh place-names because the services of the great majority were Welsh by language and their names frequently match the regularity of Welsh spelling. The second category is names associated with travel such as Glanyfferi/Ferryside, connecting Llanismel/St Ishmael parish with Llansteffan parish, and inns. The latter total sixteen if we include Cross Inn (recorded in 1807) which was the nucleus of Rhydaman/Ammanford and eighteen with Pedair-heol and Pump-heol where proof of inns postdate their first historical references (1.95%). The earliest is New Inn[1], in Llandeilo parish, in 1678. Inn-names in the Welsh language are far less common, partly reflecting the bias in historical sources and partly because of their evident unfashionability among long-distance travellers. Two instances may be mentioned, viz. Tafarn Ddiflas (*tafarn, diflas*), in Llan-dawg, recorded in 1811 but a ruinous cottage by 1907 and Tafarn-sbeit/Tavernspite (*tafarn, sbeit*), recorded in 1763, just over the border in Pembrokeshire. Both inn-names are derogatory and bring to mind Tafarn y Maidd Sur (SO205115), cos. Brecon and Monmouth, 'the sour-whey tavern' (*tafarn, y, maidd, sur*).

Industrialisation – notably tinplate and coal – and commercial trade have wrought the greatest changes in Carmarthenshire but the effect on major place-names has been far less profound than in the neighbouring county of Glamorgan. *Place-Names of Carmarthenshire* includes only Ffwrnes[1]/Furnace, in Llanelli, and Furnace[2], in Pen-bre, which take their names from industrial works though Porth Tywyn/Burry Port recalls the industrial port developed in association with the coal and metal trades. Local family-names made much the same impact. Parc Howard takes its name from local landowners and benefactors, Pemberton recalls a family of coal pioneers, Trehopcyn/Hopkinstown a farming family, and Tre Ioan/Johnstown is named from a mayor of Caerfyrddin/Carmarthen. Urbanisation made a greater impact on place-names than all of these categories, introducing new names such Glan-môr/Seaside, Sandy (1803), Marble Hall (1837), Mount Pleasant (from a house), Myrtle Hill (from a house), and the curious adoption Swiss Valley (1847) recently translated as Glyn y Swisdir, all in Llanelli, and most notably Rhydaman/Ammanford, a new name for the area which developed around the Cross Inn. Urbanisation coupled with anglicisation made its greatest impression on names which have had to be left out of *Place-Names of Carmarthenshire*. A mere glance at a gazetteer or town plan is sufficient witness to the adoption – over the past two-hundred years – of English house-names and street-names: the familiar catalogue of imperial heroes and battles, monarchs, and politicians.

Several matters must be mentioned in conclusion. Welsh historians and language specialists have long recognised the importance of recording local vocabulary and dialect (note the sound archive in the Museum of Welsh Life: museum.wales/curatorial/social-cultural-history/archives) and defining the geographical spread of language and dialect (WDS). More locally, D. Trevor Williams discusses this in a short article 'Linguistic divides in south Wales' (AC 1935: 260-3) but only B.G. Charles – best known for *The Place-Names of Pembrokeshire* (PNPemb) and *Non-Celtic Place-Names in Wales* (NCPN) – seems to have been fully conscious of the importance of studying the relationship between Welsh and English and the subject of linguistic change through the perspective of place-names. In particular he notes the apparent revival of the Welsh language in the Talacharn[2]/ Laugharne area in *Carmarthenshire Studies* (CStudies). The relevance of place-name evidence in studying dialect and language change, however, was often overlooked or underappreciated. Even M.H. Jones who investigated the Welsh 'Demetian dialect' in articles published by the Carmarthenshire Antiquarian Society (CAS I: 99-100; II: 121-2, 125-6, 128) and Leslie Baker-Jones who researched Welsh vocabulary in the Llangeler and Penboyr area in the north-west of the county in 1997-1998 (CAS 33, 34 and 37) found little space for it. English dialect was just as neglected: T. Witton Davies (AC 1920: 183-7) noted some dialect words of south-western Carmarthenshire and similarities with language in Pembrokeshire and Gower but no one took the subject further – at least in print. It was not until the appearance of the Carmarthenshire Place-Name Survey that anyone paid particular attention to individual place-name elements and vocabulary. Examples in the Glossary (below) such as **bac, parc, cae, clos** and **syddyn** have interesting regional distributions *within* Carmarthenshire but they need to be compared with county-wide surveys in other parts of Wales.

The link between place-names and archaeology has been similarly underrated. Terrence James, who was so active in promoting place-names studies in Carmarthenshire, drew attention to this matter as long ago as 1998 in a paper entitled 'Place-name Distributions and Field Archaeology in South-west Wales' (online at web.archive.org/web/20070721214657/http:/www.terra-demetarum.org.uk) read at the founding conference of the Scottish Place-Name Society. The paper was published in *The Uses of Place-names* edited by Simon Taylor. If anyone doubts the importance of the connection between place-names and archaeology, they need only look at the entry for Caerfyrddin/Carmarthen.

Introduction

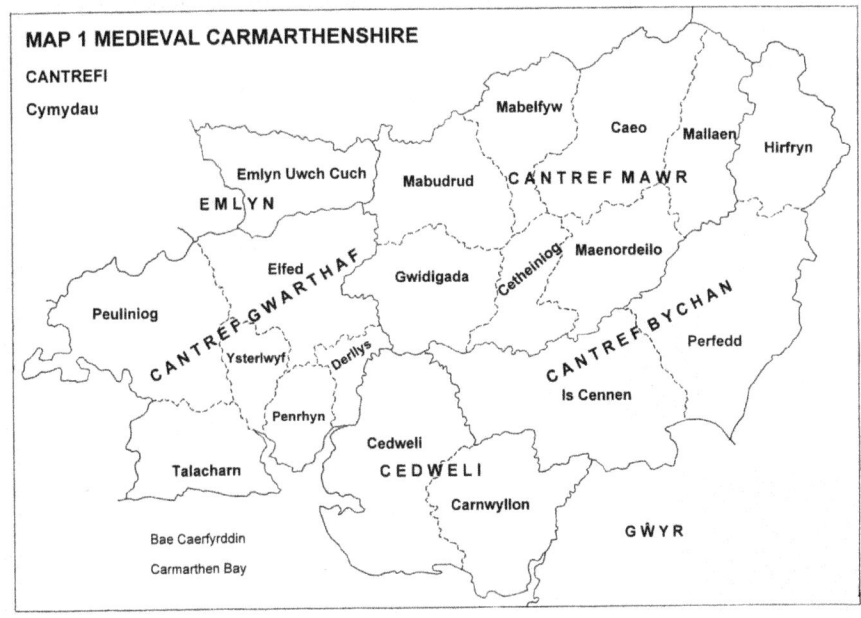

Map 1 Medieval Carmarthenshire

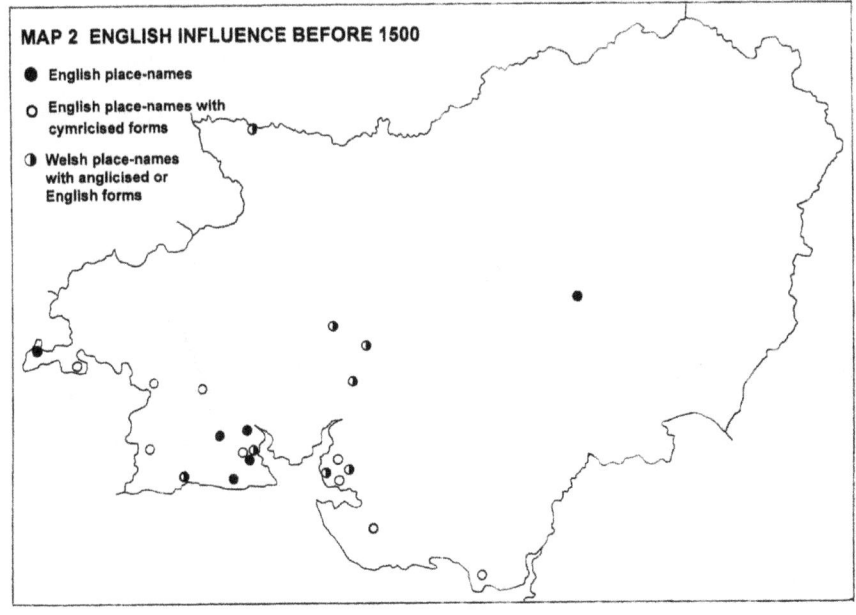

Map 2 English influence before 1500

PLACE-NAMES OF CARMARTHENSHIRE

Selection of names

The choice of placenames follows that described in *Place-Names of Glamorgan* with some amendments. Since the great majority of place-names in Carmarthenshire are undoubtedly Welsh, it was felt more appropriate to arrange entries according to standard Welsh form arranged according to the international alphabet rather than the Welsh alphabet. More topographical names – particularly those of rivers and streams (shown on Ordnance Survey Landranger 1:50,000 maps) – have been added to the text and the glossary has been expanded in order to cite relevant place-names in the main text. Some compensation for excluded names will be found within individual entries which frequently refer to topographical and comparative place-name evidence in Carmarthenshire as well as other parts of Wales. River-names such as Afon Gwydderig and Afon Tywi appear in alphabetical sequence under Gwydderig and Tywi (rather than Afon as in the Gazetter of Welsh Place-Names) with the exception of river-names where Afon is qualified by an adjective as in the case of Afon Fawr (***mawr***). Stream-names by contrast appear under Nant (***nant***). No attempt has been made to examine the names of houses, fields and minor topographical names except in passing; this is a far greater task which needs to be undertaken by an historical society or Cymdeithas Enwau Lleoedd Cymru/ Welsh Place-Name Society.

Selected place-names are drawn from:
(1) The Ordnance Survey map *Wales & West Midlands* (1:250 000) 1997.
(2) Larger-case place-names on the Ordnance Survey 1:25,000 Explorer
(3) Historic parishes, lordships and manors from *Welsh Administrative and Territorial Units* (WATU) with a small number of exceptions where reliable evidence is lacking
(4) Larger case place-names on current Ordnance Survey Landranger maps. Examples on individual maps include Pont-tyweli (sheet 146), Bancyfelin (159). Distinguishing these from other printed names has nonetheless proved difficult and some omissions are possible. One name Harford (SN 636430), in Caeo, found on the OS Landranger (2016) has been omitted because this properly applies to Harford House (recorded in 1888) named from the Harford family of Peterwell and Falcondale, co. Cardigan.

The area chosen is that of the present Carmarthenshire Council (formed in 1996) which includes the old civil parishes of Llanfallteg West and Llan-gan West which formerly lay in the historic county of Pembroke (1536-1974). *Place-Names of Carmarthenshire* also includes the whole of the former civil parishes of Llandysilio East, Egrmwnt/Egremont, and Castelldwyran which make up the greater part of the present Clunderwen Community transferred from Carmarthenshire to Pembrokeshire Council in 2003 (see David Rees, 'County Boundary Changes', *Carmarthen Antiquarian Society Transactions* CAS 39, 166-8, with map). These are omitted in B.G. Charles' survey of the place-names of the county of Pembroke (PNPemb) and it was felt that their omission would have left an inconvenient vacuum. Places which lay in historic Pembrokeshire are identified as 'Pemb' prefixed to grid references.

Introduction

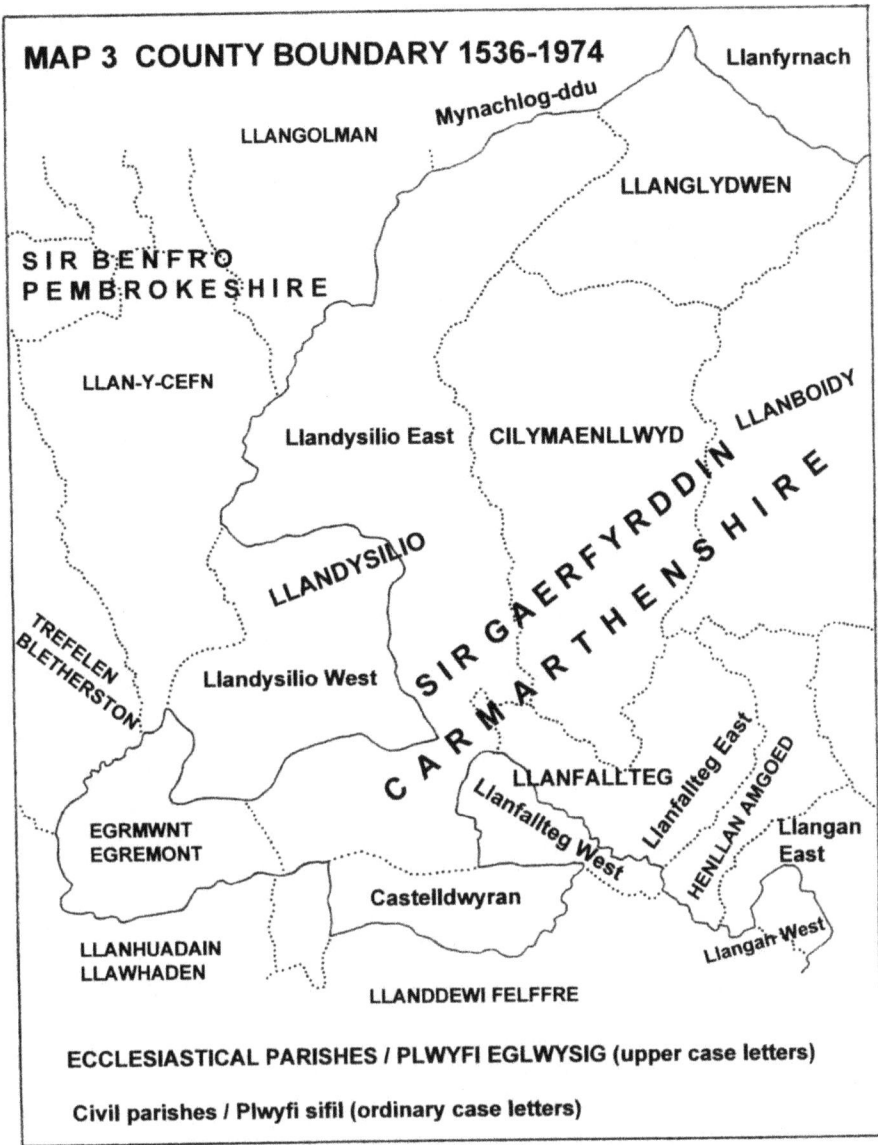

Map 3: County boundary 1536-1974

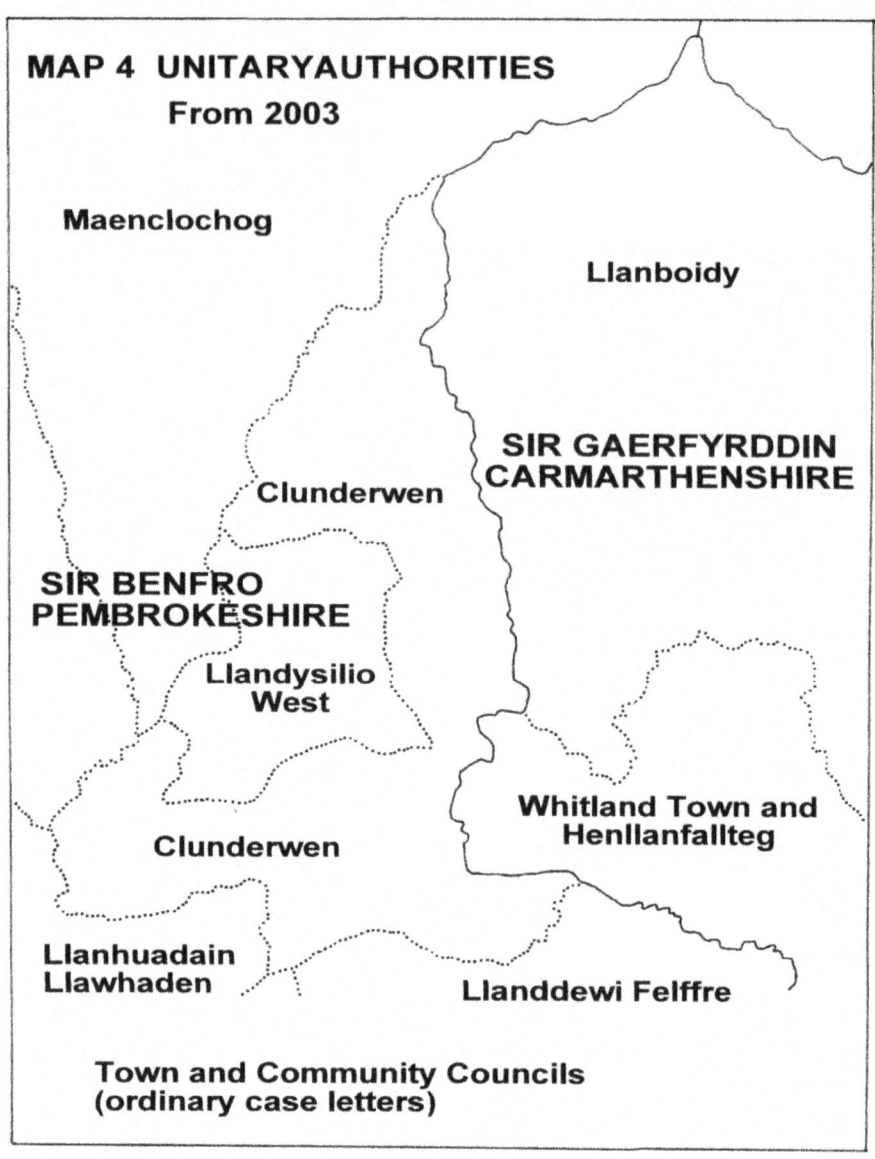

Map 4: Unitary authorites from 2003

Introduction

Editorial method

Dictionary entries comprise the following elements:

1. The lead-form in **bold** arranged according to the standard Latin-based international alphabet. Spelling generally follows that on current Ordnance Survey maps giving precedence to Welsh forms of place-names. English and anglicised names make up less than 5% of the total number of place-names discussed below. Lead-forms are followed by any other official form, eg. **Caerfyrddin, Carmarthen**. with appropriate cross-references, eg. **Carmarthen** see **Caerfyrddin**. Places *within* historic Carmarthenshire mentioned elsewhere in the text which also have non-Welsh or anglicised forms are shown thus: Gelli-aur/Golden Grove, Hendy-gwyn/Whitland, Pen-bre/Pembrey. For reasons of conciseness, similar names *outside* the county such as Swansea (Abertawe) and Brecon (Aberhonddu) are excluded from this method. Spurious forms, including those which have acquired limited currency, are discussed within lead entries and no cross-references are provided. General guidance on Welsh spellings conventions are to be found in Guidelines for Standardizing Place-names in Wales issued by the Welsh Language Commissioner (2016) (www.comisynyddygymraeg/cymru). A very small numbers of deviations from the Guidelines have been made on purely historical grounds and appropriate cross-references are given. The Ordnance Survey has issued an explanatory guide 'The Welsh origin of place names in Britain' and an explanation of its Welsh Language Scheme (www.ordnancesurvey.co.uk/resources).
2. Six-figure National Grid references for precise locations taken directly from Ordnance Survey maps and four-figure references for wider areas. No grid references are given for very large areas such as local authorities.
3. The name of the historic parish.
4. A concise interpretation of the place-name with an identification of place-name element(s) in ***bold italic*** listed in the glossary in this volume. This is based partly on the fuller glossary in *The Dictionary of the Place-names of Wales* (DPNW) supplemented by *Geiriadur Prifysgol Cymru* (GPC), the glossary in *The Place-Names of Pembrokeshire* (PNPemb), *Place-Names of Glamorgan* (PNGlam), and *English Place-Name Elements* (EPNE). These sources should be checked by anyone wishing to investigate individual elements and it needs to be emphasised that future research will add to and amend suggested meanings. All elements and words cited within individual entries are recorded in the glossary with the exception of definite articles (W *y*, *yr* and affixed '*r*, and E *the*) and most prepositions. Any qualifiers found as lead entries elsewhere in this volume appear in **bold** print, eg. Abergwili appears as *aber*, rn. **Gwili**. A more detailed discussion of this methodology is given in the glossary.
5. A sample of historic spellings in *italics* and date. No indication of the source of spellings is provided in order to reduce both the length and cost of this publication but a full record of sources is accessible online at the Centre for Advanced Welsh and Celtic Studies, Aberystwyth, and from Cymdeithas Enwau Lleoedd Cymru/Welsh Place-Name Society. It has to be stressed that the reliability of some spellings is difficult to gauge when they occur in later manuscripts since copyists sometimes amended names or miswrote them. Rather than waste space, I have ignored the date of the copy unless it is considerably later than the original form; the date in

these cases is given in brackets. Some historic forms also lack precise dates in their original sources and the usual practice, especially in edited printed sources, is to use the form 1160 x 1180 indicating the earliest and latest possible dates for a particular form. In this publication an arbitrary medial date has been adopted, ie. c.1170. The fuller covering dates appear in the online record.

6. Analysis of individual place-names. The length of this section varies according to the complexity of the name. Whenever possible, cross-references are made to similar or identical place-names found elsewhere. This section also contains relevant background information on the general history or topography of the place-name with limited referrals to secondary sources.
7. Whenever there is significant doubt over pronunciation I have added a simplified form, eg. **Caerfyrddin**, **Carmarthen** with the International Phonetic Alphabet in brackets.

The use of some technical language in a work of this nature is inevitable but the opportunity is taken to explain them where appropriate. Some specialist terms, however, occur so frequently in Welsh that repetition would waste space. The main example is lenition, more widely known as soft mutation, which belongs to a group of sound changes found in all Celtic languages. The rules are complex but consistent and should be checked in a standard Welsh grammar. It is also important to emphasise that in Welsh all nouns are masculine (***aderyn, brenin, cae***) or feminine (*banwes, cath*) though the gender may vary in different parts of Wales (***argae, cad***[1]). Gender is best identified in Geiriadur Prifysgol Cymru (http://geiriadur.ac.uk/gpc/gpc.html). The mutations commonly found in place-names may be summarised as follows:

Lenition ('soft mutation'):
Initial letters b-, c-, d-, ll-, m-, p-, and rh- are replaced by f- [v], g-, dd- [ð], l-, f- [v], b- and r-, and g- drops out altogether. In place-names these are most often caused by:

(1) Prepositions such as ***i*** 'to', ***o*** 'from', ***am*** 'around, surrounding, near to' and ***ar*** 'on, over against'
Examples: Dre-fach[1-4] (after ***i*** or ***o***), Amgoed and Emlyn (***am***) and Pontargothi (***ar***)
(2) Nouns after the definite article when it precedes a feminine singular noun and after certain prepositions. The def.art. appears as **y**, **yr** (before a vowel and -h-) and `r (after a vowel). When the definite article precedes a place-name it is now generally dropped, particularly on maps and road-signs.
Examples: Allt-y-gaer (***caer*** > *gaer*), Bancyfelin (***melin*** > *felin*), Efail-wen (***gefail*** > *efail*), Faenor (***maenor*** > *faenor*) and Felindre[1-6] (***melindref*** > *felindref, felindre*), Nant y Dresglen (***tresglen*** > *dresglen*)
(3) Qualifiers after a feminine noun
Examples: Efail-wen (***gwen*** > *wen*), Felin-foel (***moel*** > *foel*), Heol-ddu (***du*** > *ddu*) and Treberfedd (***perfedd*** > *berfedd*)
(4) After numerals ***un*** 'one' (except *ll* and *rh*) and ***dau*** feminine *dwy* 'two'
Examples: Maenor Ddwylan (***glan*** > *lan*)

Introduction

Aspirate mutation:
Most commonly occurs in place-names after *a* 'and' and numerals such as *tri* 'three'. Initial c- , p- and t- are replaced by ch-, ph- and th-
Examples: Eglwys Fair a Churig (**Curig** > *Churig*), Trichrug (***crug*** > *chrug*)

Nasal mutation:
This occurs in place-names after ***yn*** 'in' and far less commonly *a* 'and'. It is especially common in field-names and historic evidence. Initial b- , c- , d- , g- , p- and t- are replaced by m-, mh-, ngh-, n-, ng-, ph- and th-.
Examples: Maenor Rhwng Twrch a Chothi; *ynghapel llan Llyan* (Capel Llanlluan) standing for modern *yng Nghapel Llanlluan* and *Capel Cynnor ym Mhendryn* (Capel Cynnor) probably standing for *Capel Cynnor ym Mhenrhyn*.

GUIDE TO THE INTERNATIONAL PHONETIC ALPHABET

The following represents a simplified form of the IPA and should be used only as an approximate guide to pronunciation. It has to emphasised that local pronunciation in both Welsh and English often varies from standard or expected form. Differences between Welsh and English pronunciation – even when a particular place-name has a common spelling – often varies.

IPA	Welsh	English	Examples of Carmarthenshire place-names
Consonants and semi-vowels			
b	**b**ach, cw**b**l, ma**b**	**b**e	**B**erach, **B**laenynys, Nant-**b**ai
d	**d**ydd, ca**d**w, ta**d**	**d**o	**D**re-fach, Felin**d**re
dʒ	gare**j**, **di**engyd (coll.)	**j**ust	Gran**ge**, **J**ohnstown
ð	**dd**oe, be**dd**	**th**e	Llan**dd**arog, I**dd**ole, Pen**dd**eulwyn
f	**ff**enest, cor**ff**	**f**ind	**Ff**orest, **F**urnace
g	**g**lan, ce**g**in	**g**et	**G**lansefin, **G**lyn, Heol**g**aled
h	**h**aul	**h**ave	**H**ebron,
j	**i**aith, ge**i**riadur	**y**ou	**I**et-y-bwlch, Pont**y**ates, Tal**i**aris
k	**c**ig, a**c**w	**c**an	Is**c**oed, Tre**c**astell
l	**l**eicio, bo**l**	**l**ike	**L**acques, **L**augharne, Ta**ll**ey
ɬ	**ll**yfr, cy**ll**ell		**Ll**andybïe, **Ll**angynnwr
m	**m**ab, di**m**	**m**e	**M**acrels, **M**yrtle Hill
m̥	**mh**en	wor**mh**ole	Cw**mh**iraeth
n	**n**erth	**n**o	**N**antgaredig, **N**ew Mill
n̥	fy **nh**ad	u**nh**appy	Lla**ng**ynheiddon, Lla**nh**irnin
ŋ	fy **ng**wely	thi**ng**	Maenor Rhw**ng** Twrch a Chothi
ŋh	fy **ngh**ŵn	Sha**ngh**ai	Cy**ngh**ordy
p	**p**en, co**p**a	**p**lease	**P**enrhyn, **P**ontaman

Guide to the International Phonetic Alphabet

r	**r**adio, ga**rw**	**r**ight (trillled)	Ma**rr**os
r̥	**rh**estr, an**rh**eg	pe**rh**aps (rapidly)	**Rh**andir-mwyn, Porth-y-**rh**yd
s	**S**aesnes, **sws**	**s**ee	Meiro**s**, **S**andy
ʃ	**s**iarad, brw**sh**	**sh**e	La**sh**, St I**sh**mael
t	**t**atws	**t**ime	**T**alhardd
tʃ	**ts**eina, wa**ts**	**ch**ance	Ro**ch**e Castle
θ	a**th**ro, pe**th**	**th**ink	Mer**th**yr
v	**f**aint, a**f**al	**v**ery	Da**f**en, Mor**f**a, Pen-y-**f**an
w	**w**edyn, g**w**lân	**w**as	G**w**estfa
x	**ch**waer, bwl**ch**	lo**ch** (Scots)	Myna**ch**dy, Tri**ch**rug
z		bu**s**y, share**s**	Edwin**s**ford, St Clear**s**

Monopthongs

a	s**a**nt	s**a**nd	Cwm-b**a**n, Glas**a**llt, R**a**m
ɑː	m**a**b, s**â**l	f**a**ther, t**a**r	Aber-gi**â**r, Cwm-b**a**ch, Llandyf**â**n
e	p**e**th	b**e**rry	Pontyb**e**rem
eː	tr**ê**n	m**ay** (approximately)	Llwyn-t**e**g, Ty-h**e**n
ə	cym**y**dog	**a**bout	M**y**ddfai, Tal-**y**-foel
ɛ	p**e**rth, m**e**s	w**e**ll	Esgair-g**a**rn
ɛː		**air**, d**are**	Sancl**ê**r
iː	h**i**r, t**î**m	b**e**	D**i**nas
ɪ	t**i**pyn, sg**î**l	**i**t, s**i**t	Br**y**n, Capel T**y**dyst, K**i**dwelly
ɨː	ll**u**n, b**û**m	t**ea**	Cl**u**nderwen, Gland**u**ar
oː	br**o**, **ô**l	t**oe** (Scots)	Mil**o**, Rh**o**s
ɔ	br**o**n	g**o**ne	**O**gofau
ɔː		st**a**ll	M**a**rble H**a**ll
uː	c**w**ch, d**ŵ**r	p**oo**l	Llan-ll**w**ch, Pentre-c**ŵ**n
ʌ		r**u**n	B**u**rry Port
ʊ	c**w**m	p**u**t	P**w**ll
ɨː	ll**u** (north Wales)		
yː	b**û**m (north Wales)	l**u**ne (French)	

Dipthongs

aɨ	c**au**, nes**áu** t**ai**	**eye**, p**ie**	Ef**ai**l-wen, Pend**i**ne, P**u**msaint, Wernol**au**

ʌɪə		fire	
au	llaw	around	Esgairdawe, Nant-y-caws, Sawdde
aʊ		how	Mountain
aʊə		sour	'sour-whey tavern' (see Carmarthenshire Place-Names: research and analysis)
ɑːi	hael, cae		Maenor Bach-y-ffrainc, St Ishmael
eu	llew		Bwlchnewydd, Capel Dewi
ei	Seisnig		Bryn-y-beirdd, Peuliniog
eɨ	gwneud		Llanddeusant
eɪ		say	Laques
eɔ	heol		Heol-ddu
əu	bywyd		Llandeilo Abercywyn, Pont-tyweli
iu	Duw, lliw, menyw	stew	Bryn Iwan, Capel Gwynllyw
ɔi	osgoi	boy	Cloigin, Llanboidy
ɔɨ	coeden		Croesyceiliog, Goetre, Iscoed
ɔː		caw, paw	Broad Oak, Trevaughan
ʊə		poor	Ammanford
ʊi	mwy, gŵyl		Cynwyl Elfed, Gwyddgrug

ABBREVIATIONS AND BIBLIOGRAPHY

The following includes all sources cited in this publication and all abbreviations within individual entries.

AC	*Archaeologia Cambrensis*. The Journal of the Cambrian Archaeological Association (1846-) Followed by year of publication
ADEL	*Ar Drywydd Enwau Lleoedd. Ysgrifau anrhydeddu Gwynedd O. Pierce*, gol. Gareth A. Beavan and others (Talybont 2021)
ADG[1]	Gwynedd O. Pierce, Tomos Roberts a Hywel Wyn Owen, *Ar Draws Gwlad* (Llanrwst 1997)
ADG[2]	Gwynedd O. Pierce a Tomos Roberts, *Ar Draws Gwlad 2* (Llanrwst 1999)
adj.	adjective
adj.f.	feminine adjective
adv.	adverb
AFr	Anglo-French
ALaugharne	Mary Curtis, *The Antiquities of Laugharne, Pendine and their Neighbourhoods* (second edition London 1880)
AMR	Archif Melville Richards, Place-Name Research Centre, Bangor University www.e-gymraeg.co.uk/enwaulleoedd
AN	Anglo-Norman
asp.mut.	aspirate mutation
BBCS	*The Bulletin of the Board of Celtic Studies* (Board of Celtic Studies. Cardiff 1921-1992)
BEW	Thomas Rees, *The Beauties of England and Wales ... xviii. South Wales* (London 1815)
BlaenauT	*Blaenau Tywi. Enwau yn y tirwedd, Names in the landscape* (Grŵp Hanes Blaenau Tywi History Group 2014)
BowenNSW	Emanuel Bowen, *A New and accurate map of South Wales containing the counties of Pembroke, Glamorgan, Carmarthen, Brecknock, Cardigan and Radnor* (1729) NLW viewer: Wales
Bret	Breton
Brit	British
Bryn Iwan	Yvonne Francis, *Bryn Iwan 1851-2001: hanes Capel Annibynnol Bryn Iwan a'r fro ar achlysur dathlu canmlwyddiant a hanner yr achos yn y flwyddyn 2001* (Capel Annibynnol Bryn Iwan 2001)
c.	*circa*
Camden	*Britannia*, with an English translation by Philemon Holland 1610 (The Philological Museum), ed. Dana F. Sutton (University of California). Transcript at http://www.philological.bham.ac.uk/cambrit

CarlisleTD	Nicholas Carlisle, *A Topographical Dictionary of the Dominion of Wales* (London 1811)
cas.	castle
CAS	*Carmarthenshire Antiquarian Society Transactions* (Carmarthenshire Antiquarian Society 1905-39) with roman numerals and *The Carmarthenshire Antiquary* (1941 to present) with arabic numerals
CBeyond	*Carmarthenshire & Beyond: Studies in History and Archaeology in Memory of Terry James*, ed. Heather James and Patricia Moore (Carmarthenshire Antiquarian Society 2009)
CBT	Cyfres Beirdd y Tywysogion, ed. R.G. Gruffydd (six vols. Cardiff 1991-4)
CCH	Dylan Rees, *Carmarthenshire: The Concise History* (Cardiff 2005)
CDEPN	*The Cambridge Dictionary of English Place-Names*, ed. Victor Watts (Cambridge 2004)
Celtica	*Celtica: Journal of the School of Celtic Studies* (1946-)
cent(s)	century, -ies
cf.	compare
CHist	*The Carmarthenshire Historian* (Carmarthenshire Community Council, Dyfed Rural Council, Dyfed Association of Voluntary Services 1966-85)
ch(s).	church, churches
chp(s).	chapel, -s
chp(s).-of-ease	chapel(s)-of-ease
CMLlanelli	M.V. Symons, *Coal Mining in the Llanelli Area*, Vol.1 (Llanelli 1979)
cmt(s).	commote(s)
CNotes	*Carmarthenshire Notes* (1889-91), ed. Arthur Mee, vols. 1-3 (Llanelli 1889-91. Reprinted by Carmarthenshire County Council 1997)
co(s).	county, -ies
CO³	*Culhwch ac Olwen*, gol. Rachel Bromwich a D. Simon Evans (Caerdydd 1988, 1997)
Cofiadur	*Y Cofiadur, sef Cylchgrawn Cymdeithas Hanes Annibynwyr Cymru* (Undeb Annibynwyr Cymru)
Coflyfr	*Coflyfr, neu Gydymaith i'r Almanac*, gan D.R. and W. Rees (Llanymddyfri 1831, 1834)
coll.n.	collective noun
conj.	conjunction
CornPN	Oliver J. Padel, *A Popular Dictionary of Cornish Place-Names* (Penzance 1988)
Corpus	Nancy Edwards, *A Corpus of Early Medieval Inscribed Stones and Stone Sculpture in Wales. Volume II. South-West Wales* (Cardiff 2007)
cp.	civil parish
CPemb	*Calendar of Public Records relating to Pembrokeshire*, ed. Henry Owen (Cymmrodorion Record Series 7. London 1911-8)
CPNE	Oliver J. Padel, *Cornish Place-name Elements* (English Place-Name Society vol. LVI/LVII, 1985)
CRMCH	J.R. Daniel-Tyssen and A.C. Evans, *Royal Charters and Historical Documents relating to the town and county of Carmarthen* (Carmarthen 1876) Online: https://archive.org/details/royalchartersan00evangoog

Abbreviations and Bibliography

CrSG	Aneirin Talfan Davies, *Crwydro Sir Gâr* (Llandybïe 1955)
CStudies	*Carmarthenshire Studies. Essays presented to Major Francis Jones*, ed. Tudor Barnes and Nigel Yates (Carmarthen 1974)
ctf(s).	*cantref(i)*
CV	*Celtic Voices English Places. Studies of the Celtic impact on place-names in England* by Richard Coates, Andrew Breeze and David Horovitz (Stamford 2000)
Cwm Aman	Hywel Teifi Edwards (gol.), *Cyfres y Cymoedd: Cwm Aman* (Llandysul 1996)
CYB	*The Congregational Year Book 1933* (London 1933)
Cymmrodor	*Y Cymmrodor*. The Honourable Society of Cymmrodorion (London 1877)
dat.	dative
DavisSS	Paul Davis, *Sacred Springs. In search of the holy wells and spas of Wales* (Llanfoist, Abergavenny 2003)
def.art.	definite article
DEPN	Eilert Ekwall, *The Concise Oxford Dictionary of English PlaceNames* (Oxford, 4th edition, 1960)
DewiGB	welshsaints.ac.uk/edition/texts/verse/DewiGB/edited-text.eng.html
dim.	diminutive
DLlanelli	David Bowen, *Detholiad o'r Cyfansoddiadau Buddugol yn Eisteddfod Llanelli, Gorph. 1856, ynghyd a'r holl feirniadau arnynt* (John Thomas, Llanelli 1857). Translated by Ivor Griffiths, Gorseinon
DPNW	Hywel Wyn Owen and Richard Morgan, *Dictionary of the PlaceNames of Wales* (Gomer. Llandysul 2007)
DUPN	Patrick McKay, *A Dictionary of Ulster Place-Names* (Belfast 1999)
DWB	*The Dictionary of Welsh Biography down to 1940*. Honourable Society of Cymmrodorion (London 1959); online: https://biography.wales/
E	English
EANC	R.J. Thomas, *Enwau Afonydd a Nentydd Cymru* (Cardiff 1938)
ECMW	V.E. Nash-Williams, *The Early Christian Monuments of Wales* (Cardiff 1950)
Ecton 1742	John Ecton, *Liber Valorum et Decimarum* (London 1742)
Ecton 1763	John Ecton, *Thesaurus Rerum Ecclesiasticarum being an Account of the Valuations of all the ecclesiastical benefices In the Several Dioceses in England and Wales* (3rd ed. London 1763), ed. Browne Willis
Ecton 1786	John Ecton, *Liber Valorum et Decimarum* (London 1711); rev. J. Bacon, *Liber Regis* (London 1786)
ed.	edited, edition
EEW	T.H. Parry-Williams, *The English Element in Welsh* (London 1923)
MPLeg.	for example
el(s).	element(s)
Elfen	Bedwyr Lewis Jones, *Yn ei Elfen* (Capel Garmon 1992)
ELl	Ifor Williams, *Enwau Lleoedd* (Liverpool 1945, 1969)
ELlSG	J. Lloyd-Jones, *Enwau Lleoedd Sir Gaernarfon* (Caerdydd 1928)
EMESP	Neil Ludlow, *Early Medieval Ecclesiastical Sites Project. Stage 2: assessment and fieldwork Carmarthenshire. Part 2a: gazetteer of sites* (CADW 2004)

EnWales	*The Welsh Academy Encyclopaedia of Wales*, ed. John Davies, Nigel Jenkins, Menna Baines and Peredur I. Lynch (Cardiff 2008)
EPNE	A.H. Smith, *English PlaceName Elements* (English Place-Name Society, Cambridge 1956)
ERN	Eilert Ekwall, *English River-Names* (Oxford 1928)
ESeion	Parch. James Morris, Penygraig, *Efengylwyr Seion* (Dolgellau 1905)
ETG	Melville Richards, *Enwau Tir a Gwlad*, gol. Bedwyr Lewis Jones (Caernarfon 1998)
EWGT	*Early Welsh Genealogical Tracts*, ed. P.C. Bartrum (Cardiff 1966)
Fenton	Richard Fenton, *Tours in Wales (1804-13)*, ed. J. Fisher (AC supplement. London 1917)
Fenton Pemb	Richard Fenton, *A historical tour through Pembrokeshire* (London 1811) archive.org/details/b22013179
fem.	feminine
FGB	Ó Dónaill, *Foclóir Gaeilge-Béarla*, 1977. Online version of Ó Dónaill's Irish-English Dictionary
fm(s).	farm(s)
fm.n(s).	farm-name(s)
fn(s).	field-name(s)
Fr	French
GA	*Geiriadur yr Academi. The Welsh Academy English-Welsh Dictionary Online* www.geiriaduracademi.org
GazWPN	*A Gazetteer of Welsh Place-names. Rhestr o Enwau Lleoedd*, ed. Elwyn Davies (Cardiff 1967)
GBGG	J. Lloyd-Jones, *Geirfa Barddoniaeth Gynnar Gymraeg* (Caerdydd 1988)
gen.	genitive
gen.sg.	genitive singular
GHCL	*Gwaith Huw Cae Llwyd ac Eraill*, gol. Leslie Harries (Caerdydd 1953) [*floruit* 1455-1505]
GHD	*Gwaith Hywel Dafi*, gol. A. Cynfael Lake (Aberystwyth 2015. 2 vols.)
GIG	*Gwaith Iolo Goch*, gol. D.R. Johnston (Caerdydd 1988)
GLGC[1]	*Gwaith Lewis Glyn Cothi.I.*, gol. E.D. Jones (Caerdydd ac Aberystwyth 1953) [c.1447-89]
GLGC[2]	*Gwaith Lewys Glyn Cothi*, gol. Dafydd Johnston (Caerdydd 1995)
GLlF	*Gwaith Llywelyn Fardd I ac Eraill o Feirdd y Ddeuddegfed Ganrif*, gol. K.A. Bramley *et al.* (Caerdydd 1994)
GPC	*Geiriadur Prifysgol Cymru* (Caerdydd 1950) www.geiriadur.ac.uk/gpc
gr.	grange
GRO	Glamorgan Archives (formerly Glamorgan Record Office), Cardiff
GW	Chris Grooms, *The Giants of Wales. Cewri Cymru* (Lewiston 1993)
GyN	*Gwaith y Nant*, gol. Huw Meirion Edwards (Aberystwyth 2013)
Haul	*Yr Haul; neu drysorfa o wybodaeth hanesiol a gwladwriaethol* (Caerfyrddin 1835-1983)
HBryn	Enoch Rees, *Hanes Brynamman a'r cylchoedd* (Ystalyfera 1896; facsimile 1992)
HCH	Francis Jones, *Historic Carmarthenshire Homes and their Families* (Carmarthenshire Antiquarian Society and Dyfed County Council 1987)

Abbreviations and Bibliography

HChirk	C. Neville Hurdsman, *A history of the Parish of Chirk* (Wrexham 1996)
HCrm	Sir John Edward Lloyd, *A History of Carmarthenshire*, vol. I: *From Prehistoric Times to the Act of Union (1536)* (London Carmarthenshire Society 1935), vol II: *From the Act of Union (1536) to 1900* (~ Cardiff 1939)
hd(s).	hundred(s)
HEAC	Thomas Rees a John Thomas, *Hanes Eglwysi Annibynol Cymru* (Lerpwl 1871-73). 4 vols.
HEALl	Glenda Carr, *Hen enwau o Arfon, Llŷn ac Eifionydd* (Caernarfon 2011)
HEAS	D. Jones and J. James, *Hanes Eglwys Annibynol Saron, Llangeler* (Llandysul 1937)
HECI	*Hanes Eglwys Capel Iwan 1723-1923* (Caerdydd 1924)
HECS	E.T. Jones and T.R. Jones, *Hanes Eglwys Capel Sion, Llanelli* (Llanelli 1931)
HER	Dyfed Archaeological Trust Historic Environment Record, followed by Primary Reference Number for archaeological sites
HERPns	Bruce Coplestone-Crow, *Herefordshire Place-names* (Little Logaston Press 2009)
HESF	Glenda Carr, *Hen enwau o Sir Feirionnydd* (Caernarfon 2020)
HGC	*Hen Gerddi Crefyddol*, gol. Henry Lewis (Caerdydd 1931). See CBT
HMorg	David Watkin Jones (Dafydd Morganwg), *Hanes Morganwg* (Aberdâr 1874)
HMGC	*Hanes Methodistiaeth Galfinaidd Cymru*, gol. Gomer Morgan Roberts, Vol. 1 (Caernarfon 1973), Vol. 2 (Caernarfon 1978)
HMSG	Parch. James Morris, *Hanes Methodistiaeth Sir Gaerfyrddin* (Dolgellau 1911)
ho(s).	house(s)
ho.n(s).	house-name(s)
Holinshed	*The First and Second Volumes of Chronicles ... First Collected and Published by, William Harrison, and others ... 1586, by John Hooker alias Vowell Gent* [actually dates 1587] Edited by Henry Ellis and others (London 1807) in 6 vols.
HOSJ	William Rees, *a History of the Order of St. John of Jerusalem in Wales and on the Welsh Border* (Cardiff 1947)
HPLland	Gomer Morgan Roberts, *Hanes Plwyf Llandybïe* (Caerdydd 1939)
HPLlangeler	Daniel E. Jones, *Hanes plwyfi Llangeler a Phenboyr* (Llandysul 1899)
HPLlan-non	Noel Gibbard, *Hanes Plwyf Llan-non, Hen Sir Gaerfyrddin* (Llandysul 1984)
HPont	*Braslun o hanes Pontarddulais*, ed. E. Lewis Evans (Llandysul 1949)
ht(s).	hamlet(s)
HW	J.E. Lloyd, *A History of Wales* (London 1911)
HWW	Francis Jones, *The Holy Wells of Wales* (Cardiff 1954)
HYM	Glenda Carr, *Hen enwau o Ynys Môn* (Caernarfon 2015)
IMLlanelli	R.S. Craig, R. Protheroe Jones and M.V. Symons, *The Industrial and Maritime History of Llanelli and Burry Port* (Carmarthenshire County Council 2002)
Ir	Irish
JHSCW	*Journal of the Historical Society of the Church in Wales* (1949-)

John SPN	Deric John, 'Some placenames in south Wales and their etymologies': http://someplacenames.mysitecom/
LBS	*The Lives of the British Saints*, ed. S. Baring-Gould and John Fisher (CymRS. London 1907-13). 4 vols.
Leland	*The Itinerary in Wales of John Leland, in or about the years c.1538*, ed. Lucy Toulin Smith (London 1906)
len.	lenition (soft mutation)
LewisTD	Samuel Lewis, *A Topographical Dictionary of Wales* (London 1843; 3rd edn. 1845; 4th edn. 1849)
LHD	*The Law of Hywel Dda. Law texts from medieval Wales*, trans. and ed. by Dafydd Jenkins (Llandysul 1986)
LL	*Liber Landavensis. The Text of the Book of Llan Dâv*, ed. John Gwenogvryn Evans and John Rhŷs (Oxford 1893)
LlawHen	*Llawysgrif Hendregadredd*, cop. Rhiannon Morris-Jones, gol. J. Morris-Jones a T.H. Parry-Williams (Caerdydd 1933) [NLW MS. 6680C. Early 14th cent]. See CBT
LlEnwau	D. Geraint Lewis, *Y Llyfr Enwau. Enwau'r Wlad. A checklist of Welsh Placenames* (Llandysul 2007)
LMS	*London Medieval Studies* I, part 2 (1938)
lp(s).	lordship, -s
LPC	Arthur Mee, *Llanelly Parish Church: its history and records* (Llanelli 1888) www.archive.org/details/llanellyparishch00llan
LSMW	R.R. Davies, *Lordship and Society in the March of Wales, 1282-1400* (Oxford 1978)
LStD	A.W. Wade-Evans, *Life of St David* (London 1923)
LWS	G.H. Doble, *Lives of the Welsh Saints*, ed. D. Simon Evans (Cardiff 1971)
Mab²	*The Mabinogion*, trans. and ed. Gwyn Jones and Thomas Jones (London 1948; 1970)
Mab³	*The Mabinogion*, trans. and ed. Sioned Davies (Oxford World's Classics. Oxford 2007)
Malkin	Benjamin Heath Malkin, *The Scenery, Antiquities, and Biography of South Wales, I* (London 1807)
masc.	masculine
MCRM	Andrew Lucas, *Mammals of Carmarthenshire* (author 1997)
ME	Middle English (c.1150c.1500)
mess.	messuage
MethC	Rev. John Hughes, *Methodistiaeth Cymru: sef hanes blaenorol a gwedd bresenol y Methodistaid Calfinaidd yn Nghymru* (Wrexham 1851, 1854, 1856)
MMF	D. Morgan Rees, *Mines, Mills and Furnaces. An introduction to industrial archaeology in Wales* (National Museum of Wales. London 1969)
ModE	Modern English (c.1500 to present)
ModFr	Modern French
ModW	Modern Welsh
Morgannwg	*Morgannwg. The Journal of the Glamorgan Historical Society* (1957-)
MOSW	F.G. Cowley, *The Monastic Order in South Wales 1066-1349* (Cardiff 1977)

MPL	*Mélanges en l'honneur de Pierre-Yves Lambert*, ed. G. Oudäer, G. Hily and H. Leibhan (Rennes 2015), specifically Peter Schrijver, 'The meaning of Celtic **eburos*', pp. 66-76
MW	Middle Welsh
MWG	Ben Guy, *Medieval Welsh Genealogy: An Introduction and Textual Study* (Boydell Press 2020)
MSWT	Jemma Bezant, *Medieval Welsh Settlement and Territory. Archaeological evidence from a Teifi Valley landscape* (BAR British Series 487. 2009)
MyddfaiLP	David B. James, *Myddfai: Its Lands and Peoples* (Bow Street 1991)
n(s).	name(s), noun(s)
NCPN	B.G. Charles, *Non-Celtic PlaceNames in Wales* (Cardiff 1938)
NDEFN	Paul Cavill, *A New Dictionary of English Field-Names*, with introduction by Rebecca Gregory (English Place-Name Society, Nottingham 2018)
nf.	feminine noun
NH	*Northern History* (1966-)
NJacobs	Nicholas Jacobs: '*Non, Nonna, Nonnita*: confusions of gender in Brythonic hegemony': www.cymmrodorion.org/wp-content/uploads/2018/6/2
NLW	The National Library of Wales, Aberystwyth
NLWJ	*The National Library of Wales Journal (1939-)*
nm.	masculine noun
nmf.	masculine or feminine noun
n.pl.	plural noun
n.sg.	singular noun
NTC	*The Names of Towns and Cities in Britain*, ed. Margaret Gelling, W.F.H. Nicolaisen and Melville Richards (London 1970)
OBret	Old Breton
ODEPN	A.D. Mills, *Oxford Dictionary of English PlaceNames* (2nd ed. Oxford 1998)
OE	Old English
OFr	Old French
OIr	Old Irish
ON	Old Norse
ONC	*The Oxford Names Companion* (Oxford 2002)
OPemb	*The Description of Penbrokshire, by George Owen of Henllys, Lord of Kemes*, ed. Henry Owen, notes Egerton Phillimore (London 1872-1906) [1603]
OS	Ordnance Survey
OW	Old Welsh
p(s).	parish(es)
Paroch	*Parochialia*, ed. R.H. Morris (Archaeologia Cambrensis supplements 1909-11)
PBPH	Pembrey and Burry Port Heritage Group: pembreyburryportheritage.co.uk
p.ch(s).	parish church(es)
PêrG	Gomer Morgan Roberts, *Y Pêr Ganiedydd [Pantycelyn], I Trem ar ei fywyd* (Llandysul 1949)
pers.n(s).	personal name(s)
P.H., PH	Public house
pl.	plural

pn(s).	placename(s)
PNBont	Deric John, *Notes on Some Place-Names in and around the Bont* (Aberdare 1999)
PNBrec	Richard Morgan and R.F. Peter Powell, *A Study of Breconshire Place-Names* (Llanrwst 1999)
PNCrd	Iwan Wmffre, *The Place-Names of Cardiganshire* (BAR British Series 379 (I) 2004) (3 vols.)
PNDH	R.F. Peter Powell, *The Place-Names of Devynock Hundred* (Pen-pont, Brecon 1993)
PNDPH	Gwynedd O. Pierce, *The Place-Names of Dinas Powys Hundred* (Cardiff 1968)
PNEF	Hywel Wyn Owen, *The Place-Names of East Flintshire* (Cardiff 1994)
PNF	Hywel Wyn Owen and Ken Lloyd Gruffydd, *Place-Names of Flintshire* (Cardiff 2017)
PNGlam	Gwynedd O Pierce, *Place-Names in Glamorgan* (Cardiff 2002)
PNGlamorgan	Richard Morgan, *Place-Names of Glamorgan* (Welsh Academic Press, Cardiff 2018)
PNGwent	Richard Morgan, *Place-Names of Gwent* (Llanrwst 2005)
PNHer	Bruce Coplestone-Crow, *Herefordshire Place-Names* (Little Logaston 2009)
PNMont	Richard Morgan, *A Study of Montgomeryshire Place-Names* (Llanrwst 2001)
PNPemb	B.G. Charles, *The Place-Names of Pembrokeshire* (Aberystwyth 1992)
PNRad	Richard Morgan, *A Study of Radnorshire PlaceNames* (Llanrwst 1998)
PNRB	A.L.F. Rivet and Colin Smith, *The Place-Names of Roman Britain* (Cambridge 1979)
PNShr 5	Margaret Gelling in collaboration with the late H.D.G. Foxall, *The Place-Names of Shropshire*, Part Five (Nottingham 2006)
PNWales	Hywel Wyn Owen, *The Place-Names of Wales* (Cardiff 2015)
prep.	preposition
PRIS	Iwan Wmffre, 'Post-Roman Irish Settlement in Wales' in *Ireland and Wales in the Middle Ages*, eds. Karen Jankulak and Jonathan Wooding (Dublin 2007)
pron.	pronounced, pronunciation
Ptolemy	*Claudii Ptolemaei Geographia*, books I-V ed. C. Müller, (Paris 1883-1901), books VI-VIII, ed. C.F.A. Nobbe (Leipzig 1845)
PtolemyLA	*Ptolemy: Towards a linguistic atlas of the earliest Celtic place-names of Europe*, ed. David N. Parsons and Patrick Sims-Williams (CMCS Publications. Aberystwyth 2000)
q.v.	*quod vide* as a cross-referral
r(s).	river(s)
RB	Romano-British
RC	*The Religious Census of 1851. A Calendar of the Returns relating to Wales*, ed. I.G. Jones and D. Williams (Cardiff 1976). 2 vols.
RCAHM	Royal Commission on Ancient and Historical Monuments, *An inventory ... V. - County of Carmarthen* (London 1917)
ref(s).	reference(s)

RFWM	*Roman Frontiers in Wales and the Marches*, ed. Barry C. Burnham and Jeffrey L. Davies (Royal Commission on the Ancient and Historical Monuments of Wales. Aberystwyth 2010)
rn(s).	river-name(s)
SDEA	*St Davids Episcopal Acta 1085-1280*, ed. Julia Barrow (South Wales Record Society 1998)
SDL	*Survey of the Duchy of Lancaster Lordships in Wales, 1609-13*, ed. William Rees (Cardiff 1953)
sing.	singular
SFN	H.D.G. Foxall, *Shropshire Field-Names* (Shropshire Archaeological Society. Shrewsbury 1980)
SGStudies	*Sir Gâr: Studies in Carmarthenshire History. Essays in memory of W.H. Morris and M.C.S. Evans*, ed. Heather James (Carmarthen 1991)
SNW	John and Sheila Rowlands, *The Surnames of Wales. Updated & Expanded* (Llandysul 2013)
sp(s).	spelling(s)
SPLlA	Conrad Evans, *The Story of the Parish of Llanfihangel Abercywyn* (Swansea 1975)
StC	*Studia Celtica* (1993-)
StDW	*St. David of Wales: Cult, Church and Nation*, eds. J. Wyn Evans and J.M. Wooding (Woodbridge 2007)
StoryCarm	A.G. Prys-Jones, *The Story of Carmarthenshire I* (Llandybÿe 1959); *II. From the Sixteenth Century to 1832* (1972)
SWB	William Rees, *South Wales and the Border in the XIV Century* (map. OS 1933)
t(s).	township(s)
TCymm	*Transactions of the Honourable Society of Cymmrodorion* (London 1893-)
terr.n.	territorial name
TMW	Ian Soulsby, *The Towns of Medieval Wales* (Chichester 1983)
TYP	*Trioedd Ynys Prydein. The triads of the island of Britain*, ed. Rachel Bromwich (fourth edition, Cardiff 2014)
vb.	verb
VCW	Peter Lord, *The Visual Culture of Wales: Medieval Vision* (Cardiff 2003)
VO	*Voprosy onomastiki* (Russian Language Institute of the Russian Academy of Sciences)
VSB	*Vitae Sanctorum Britanniae et Genealogiae*, ed. A.W. Wade-Evans (BCS. Cardiff 1944); new edition with introduction by Scott Lloyd (Welsh Academic Press 2013) with same pagination
W	Welsh
WATU	Melville Richards, *Welsh Administrative and Territorial Units* (Cardiff 1969; 1973)
WB	Thomas M. Charles-Edwards, *Wales and the Britons 350-1064* (The History of Wales) (Oxford University Press 2013)
WCist	David H. Williams, *The Welsh Cistercians* (Leominster 2001)
WDS	*The Welsh Dialect Survey*, ed. by Alan R. Thomas (Cardiff 2000)
WLlR	*The Welsh Language before the Industrial Revolution*, ed. Geraint H Jenkins (Cardiff 1997)
WSaints	welshsaints.ac.uk

WSN	T.J. and Prys Morgan, *Welsh Surnames* (Cardiff 1985)	
WSS	Rev. Rice Rees, *An Essay on the Welsh Saints or the Primitive Christians* (London 1836)	
WTLC	T.P. Ellis, *Welsh Tribal Law and Custom in the Middle Ages* (Oxford 1926. 2 vols.)	
*	Hypothetical form	
>	Developing into	
<	Developing from	

ONLINE DATABASES AND REFERENCE RESOURCES

Archif Melville Richards, Bangor University:
e-gymraeg.co.uk/enwaulleoedd/amr/cronfa

Archives Network Wales:
www.archivesnetworkwales.info

Archives of Welsh museums, colleges and universities:
www.archiveshub.ac.uk

British History Online (miscellaneous sources):
www.british-history.ac.uk

Canolfan Bedwyr. Welsh Language Board, list of modern Welsh place-names:
www.e-gymraeg.or/enwaucymru

Cardiff University Arts and Social Studies Library
www.cardiff.ac.uk > libraries

Cardiff University Special Collections and Archives
www.cardiff.ac.uk > Special Collections and Archives

Carmarthenshire Archives
www.carmarthenshire.gov.wales/home/council-services/libraries-archives

Dyfed Archaeological Trust
www.dyfedarchaeology.org.uk

Enwau Lleoedd De Cymru (Place-names of south Wales)
sites.google.co,/site/enwaulleoedddecymru

Geiriadur Prifysgol Cymru:
www.aber.ac.uk/gpc

Glamorgan Archives, Archifau Morgannwg:
www.glamorganarchives.gov.uk, www.archifaumorgannwg.gov.uk

Historic Environments Records of Wales:
archwilio.org.uk/arch

List of Historic Place Names:
historicplacenames.rcahmw.gov.uk

Mynegai i Enwau Lleoedd ym marddoniaeth Gymraeg yr Oesoedd Canol
Ann Parry Owen (2021): geiriadura.cymru/post/mynegai-i-enwau-lleoedd-ym-marddoniaeth-yr-oesoedd-canol

National Archives, London:
www.nationalarchives.gov.uk

National Library of Scotland, Edinburgh: map images
maps.nls.uk/series

National Library of Wales, Aberystwyth:
www.llgc.org.uk

Ordnance Survey 1:50 000 scale gazetteer:
data.ordnancesurvey.co.uk/datasets/50k-gazetteer

Ordnance Survey first edition 1:63 360 maps:
www.visionofbritain.org.uk/maps

Ordnance Survey historic maps (various scales):
www.old-maps.co.uk

Ordnance Survey website:
www.ordnancesurvey.co.uk/website

Place-Name Research Centre, Archif Melville Richards:
www.e-gymraeg.co.uk/enwaulleoedd

Royal Commission on Ancient and Historical Monuments for Wales: online catalogue of archaeology, buildings, industrial and maritime heritage in Wales:
www.coflein.gov.uk

Tithe Maps of Wales, Mapiau Degwm Cymru (Cynefin):
cynefin.archiveswales.org.uk and https://places.library.wales/search

Online Databases and Reference Resources

Welsh Place-Name Society/Cymdeithas Enwau Lleoedd Cymru:
www.cymdeithasenwaulleoeddcymru/org, www.welshplace-namesociety.wales

Welsh Coal Mines:
www.welshcoalmines.co.uk

Welsh Saints:
welshsaints.ac.uk

Glossary of Place-Name Elements, Personal Names and River-Names

This glossary lists all elements found in place-names and river-names in this book and within individual entries. This is followed by separate lists of personal names and river-names. Each element is given a concise meaning, the non-Welsh language of the element and, in its gender the case of a Welsh element. This is followed by a list of personal names and a list of river-names. In an effort to reduce duplication, the glossary does not itemise place-name elements in place-names which act as qualifiers if they already possess their own entries, eg. the entry for Maenor Llansadwrn simply identifies the first element as **maenor** and cross-refers the qualifier to **Llansadwrn** where the elements are identified as *llan* and personal name **Sadwrn**. Exceptions to this rule are made for place-names such as Maenor Cefndaufynydd which lacks a separate entry for the qualifier Cefndaufynydd. These elements are specified within the body of the entry as *cefn, dau* and *mynydd*. The same method is applied to place-names such as Abercennen where the elements are identified as *aber* and **Cennen**. Ordinary case bold lettering of Cennen indicates a cross-reference to the separate entry for the river-name Cennen. In the case of names such as Aberbranddu, the elements are identified as *aber* and *Branddu*. The italic bold lettering of Branddu indicates that it does not possess a separate entry and that its make-up is analysed within the entry for Aberbranddu.

Place-name elements are followed by lists of place-names which contain these. Ordinary case lettering indicates place-names which have separate headed entries and italics indicate elements discussed within them. As an example, the place-name element ***bryn*** is followed by a list of place-names including Bryn and Brynaman which have separate entries and *Blaen-y-coed* and *Cefn-bryn-brain* which do not have separate place-name entries.

Common elements such as the definite article **y, yr** (before a vowel and h-) and the affixed form **'r** after a vowel, are omitted and examples for ***afon*** cover only those in which it is the lead element such as Afon Fawr. Lists of personal names and river-names are given at the end of the Glossary. Elements which are mentioned within individual entries are also listed in the glossary and shown under the appropriate italicised entry, eg. ***gwäell*** is cross-referenced to *Aberarad*. It must be stressed that the precise meaning of some elements is uncertain or contested and that it is sometimes difficult to square place-name evidence with definitions given, for example, in the Welsh laws; see in particular ***maenor*** and ***rhandir***. This apparent mismatch may only be explained by further investigation.

Hypothetical elements are prefixed with the asterisk * and the question mark ? is placed before elements of doubtful etymology and identity. Because this dictionary includes place-names of English, French, and Latin origin, the Welsh digraphs ch-, -dd-, -ff-, -ll- and -ng- which function as individual letters in the Welsh language are treated as separate letters according to the IPA. The same rule applies to Welsh lenited letters such as -f- for -b- and -m-. The Old English digraph æ-

Glossary of Place-Name Elements, Personal Names and River-Names

precedes -ā- and Old English ð is shown in sequence as if it were -th-. Where an element appears in its Old English form and also in its modern English form, both are listed under the Old English form, eg. **æsc**, *ash*, with a cross-reference *ash* see **æsc**. Unless stated otherwise, each element is Welsh, and Welsh elements are spelled according to the forms given in Geiriadur Prifysgol Cymru. Where there is a need to avoid ambiguity, Welsh elements are preceded with W.

Common place-name elements

a, ac before a vowel and *h-* 'and'

æsc OE, *ash* E, 'ash-tree'

Ashfield

abbey E 'abbey, monastery'

abad nm. 'abbot'

Rhandir Abad

aber, nf., earlier nm., pl. *ebyr*, 'confluence (of two rivers); mouth of a river, estuary' and 'stream, rivulet; creek'

Aberarad, Aberbranddu, Abercennen, Abergiâr, Aberglasne, Abergorlech, Abergwili, Abermarlais, Aber-nant, *Asen, Blaen-nant-llin, Blotweth, Carwe, Cennen, Crychan, Cuch, Duar, Dunant, Fanagoed, Fenni, Glan-duar, Glwydeth, Gwyddyl, Lash, Llandeilo Abercywyn, Llanfihangel Aberbythych, Llanfihangel Abercywyn, Llechach, Maenor Aberbargoed, Marlais², Nant Garenig, Nant Pibwr, Pibwrlwyd,* Pontarddulais, ?Pontyberem, *Regwm, Sannan, Talacharn², Tigen, Ydw*

āc OE, *oak* E, 'oak-tree'

Broad Oak

-ach ?Ir rn. suffix

The supposition that the suffix is Ir is likely to be reliable in most cases but the el. has been identified in areas where there is little firm evidence for Ir settlement such as Glamorgan and Gwent. The reasons for this are contentious but it is possible that the geographical distribution of the el. has shifted beyond the area of Ir settlement in historic Dyfed into adjoining areas and replaced the W cognate *-og* (< Brit *-āco-, -ācā-*). There is good evidence of bilingualism in Dyfed at least down to the sixth century. The alternative argument is that some W rns. contain a distinctive W el. *-ach* (< Common Celtic *-akkos*) used as a diminutive and later as a pejorative. See PRIS 46-61. The matter is discussed in greater depth in the main text under Cloidach

Berach, Cloidach, Clydach, Llechach, Mihartach

*ach*⁵ nf. 'lineage, descent' with extended sense 'line, scar'?

?Achddu, ?Berach, *Cloidach*

achwre, achre² nmf. 'palisade, fence'

Capel Troed-y-rhiw

-ad¹ suffix

Cwmduad

adar see *aderyn*

aderyn earlier *ederyn* nm., *adar* pl., 'bird'

Rhiw'radar

adnau¹ nmf. 'deposit, thing stored for safekeeping'

Llanarthne

aer¹ nf., pl. *aerau*, 'battle, fight'

?Nant Aerau

aeron npl. 'berries, fruits'

?Nant Aeron

afallen nf. 'apple-tree'

Llanybydder

afon nf., pl. *-ydd, -oedd*, 'river'

Afon Fach Pontgarreg, Afon Fawr, Annell, Asen, Barddu, Bele, Blotweth, Brân¹,², Camnant, Clydach¹,², Cuch, Cwm-waun-gron, Cynin, Dewi Fawr, Duar, Dulais¹⁻³, Dulas, Dunant, Fanafas, Fanagoed, *Felindre*, Fenni, Glwydeth, Gronw, Gwenlais, Gwydderig, Lash, Llechach, Lliedi, Llwchwr, Mamog, Marlais¹⁻⁴, Melinddwr, Merchon, Mihartach, Myddyfi, Mynys, Nenog, Pib, Pysgotwr, Rhydybennau, Sannan, Sawdde, Sïen, Sylgen, Taf, Talog², Teifi, Tigen, Tyweli, Tywi, Wysg / Usk, Ydw

xliii

ager Latin nm. 'land, open country'
Hernin

-ai suffix
?*Aberglasne*, Sawdde

aidd nmf. 'ardour, zeal'
Nant Eiddig

aigre OFr 'sharp'
Egrmwnt / Egremont

aith n.pl. 'furze, gorse'
Cwmhiraeth, ?Hiraeth

*alabon 'hill, crest'
Llanfair-ar-y-bryn

albus, -a, -um Latin adj. 'white'
Hendy-gwyn

āld, *eald* OE 'old, former'

allt, *gallt* nf., *alltau*, *elltydd*, *eillt* pl., 'hill(side); wooded slope'
Aber-giâr, Allt, Alltwalis, Alltyferin, Allt-y-gaer, *Caerfyrddin*, *Capel Cefnberach*, *Capel Penarw*, *Capel Teilo*², *Cilymaenllwyd*, *Cwmysgyfarnog*, Glasallt, Hendy, Llwynyrebol, *Maestreuddyn*, Penallt

am¹ prep. 'around, surrounding; beyond; near, at'
Amgoed, Emlyn, Llanymyddyfri

amanw see also ***banw²***
Aman

-an dim. suffix
Cathan, *Cil-y-gell*, Crychan, *Glanbrydan*, *Maenor Cilnawen*, *Maenor Garllegan*, *Pontyberem*, *Sannan*, ?*Waun Baglam*, *Talacharn²* (Corran), *Ydw*

ar prep. 'near, at, by, on'
Capel Gwynllyw¹, *Dafen*, *Eglwys Fair Glan-taf*, *Hendy-gwyn*, *Llanfair-ar-y-bryn*, *Llanfihangel-ar-arth*, *Llanglydwen*, *Nant Pedol*, ?*Pantarfon*, *Pontarddulais*, *Pontargothi*, *Pontarllechau*, *Pont-ar-sais*, *Talog²*

-ar(a) Brit suffix
Llwchwr

aradr nmf, 'plough'
Aberarad

araith nmf. 'speech or address, speech'
Pontbrenaraeth

ardd¹ nf., adj., 'hill, highland'; cf. I *ard*, 'height, hill'
Talhardd

argae nmf., pl. *argaeau*, 'dam, weir, floodgate'
Rhydargaeau

argoed (*ar* + *coed*) coll.n., pl. *-ydd*, 'trees, forest, edge or border of forest'
Cathargoed

arian, *ariant* nm. 'silver, silvery; money'
Annell

arllwys² vb. 'to pour out; pour (down)'
Cwm-marles

arydd nm. 'ploughman'
?*Nant Cwm-merydd*

asen¹ nmf., *asyn* nm. 'donkey, ass'
Asen

asen² nf. 'rib; breast'
?Asen

ash see ***æsc*** OE

aur nm. 'gold; gold (colour)'
Gelli-aur

awen¹ nf. 'muse, inclination, desire
Maenor Cilnawen

Awst 'August', the eighth month
Miawst

bæc OE, ***back*** E, 'back, ridge'
Bacau

Babel 'Babel, Babylon', sometimes derogatory
Babel

babi nm. 'baby', also fig.
Mamog

bac nm., pl. *bacau*, dialect *bace*, 'back, ridge', borrowed from E *back*, ME *bakke*
Bacau

bach¹ adj. 'small, little; lesser'

Glossary of Place-Name Elements, Personal Names and River-Names

Afon Fach Pontgarreg, *Afon Fawr, Blaenau*[3,4], *Blaen-waun, Brechfa, Bryn, Brynhafod,* Capel Bach, Capel Bach y Drindod, *Capel Brynach, Capel Erbach, Capel Ifan*[1], *Capel Isaac, Capel y Croesfeini, Capel y Drindod, Capel yr Ywen, Carnwyllion, Cathilas*[2], *Cefn Llwydlo, Cilcarw, Cilgryman, Crachdy, Cross Inn,* Cwarter-bach, Cwm-bach[1,2], *Cwm-ban, Cwm-ffrwd,* Cynnull-bach, *Dewi Fawr, Dinas,* Dre-fach[1-4], *Dryslwyn, Esgairdawe,* Ffair-fach, *Garnant, Gelli-ddu, Glan-tren, Glasallt, Glyn*[3], *Glyn-tai, Goetre, Groesffordd, Hengil, Hengoed, Llandeilo Abercywyn, Llanmilo, Machynys, Maenor Cilau, Maerdy, Maes-gwyn, Meinciau,* Morfa-bach, *Nant Rhydw, Nant Thames,* Penddeulwyn, Pentre-bach, *Pentywyn, Piodau, Porth Tywyn, Pysgotwr, Rhos, Rhos-goch, Talhardd, Tor-y-coed, Travellers Rest, Trefechan*[1], *Ysgubor-fawr*

bach[2] nmf. 'hook, nook, angle, bend'

Bachsylw, Bach-y-ffrainc, *Machynys*

back E see **bæc** OE

bae[1] nm. 'bay'

Bae Caerfyrddin

bagl nmf. 'crozier, crook or staff of a bishop, etc' and fig.

?*Waun Baglam*

bai[1] nm. 'fault, failing; transgression'

?Nant-bai

bainc see **mainc**

ban[1] nf., *bannau, baniau, *benni* pl., 'top, summit, peak; horn; corner, angle; arm, branch'

Bynea, Cwm-ban, Mynydd Du, Pen-y-fan, *Trichrug*

ban[5] adj. 'loud, noisy'

Cwm-ban

banc[1] nm. 'rising ground, hillock, hill, upland pasture, bank (of a river)'

Bancffosfelen, Bancycapel, Bancyfelin, Banc-y-ffordd, Broadway[2], Bryn-y-beirdd, Glyn[5], Pen-y-banc[1,2], *Tymbl*

banw[2] nm. '(young) pig, piglet'

Aman, Fanafas, Fanagoed

banwes variant *manwes* nf. 'young sow, gilt'

Fanafas

bar[2] nm., *bariau, barrau* pl., 'head, top, summit'

?Berach, Crug-y-bar

bardd nm., pl. *beirdd,* 'poet, bard'

Bryn-y-beirdd

bargod[1] nm. 'eaves'

Nant Bargod

barton ME, E see OE **beretūn**

baw nm. 'dirt, mud; excrement'

Llanymddyfri

bay E 'broad inlet of the sea where the land curves in'

Carmarthen Bay

bearach, Ir *biorach*[1] adj. 'pointed'

Berach

bechan see **bychan**

bedwen nf., *bedw* pl. 'birch, birch-grove', *bedwen* nf. 'birch-tree'

Cathilas[1], Esgair-gam

beili nm. 'bailey; enclosure, yard; farmyard' and extended to mean 'farm, farm with a yard'

Esgair-gam

beirdd see **bardd**

belau[1], **bele**[2] nmf. 'wolf, predatory beast; marten, pine marten, polecat'

Bele

ben[1] nf., pl. *benni, bennau,* 'cart, wagon'

Rhydybennau

beorg OE 'hill, tumulus', E *burrow*, in pl. *borowes, burrows* 'sand-dunes'

Porth Tywyn

ber see **byr**[1]

bêr nmf. 'spear, lance'

?Berach

***bêr** ?el. in *beraf*[1], *beru* vb. 'to flow, drip'

Berach, Cellifor

berem, *berm* 'yeast'
Pontyberem

beretūn OE 'barley farm or settlement' later 'outlying grange, demesne farm'
Bertwn

bere-wīc OE, ***berwick*** ME, 'barley farm' later 'outlying farm'
Berwig

*****berran*** (***ber*** + ***rhan*** 'part, share') nf. 'short share-land'

berwick see OE ***bere-wīc***

Bethlehem the biblical city
Bethlehem

betws[1] (< OE *bed-hūs*) nm., 'house of prayer, chapel of ease'
Betws, Betws Ystumgwili, Capel y Betws, Llangynheiddon, Tre-lech a'r Betws

beudy nm.'cowshed, cow-house'
Brynaman, Llanboidy

bicws 'type of food consisting of oatbread crumbled into buttermilk'
Trichrug

bilain, *bilaen* nm. 'villein, serf; labourer, peasant'
Drefelin, Felindre[1, 3]

black E see ***blæc*** OE

blæc OE 'black, dark-coloured, overgrown'
Black Mountain

blaen nm., *blaenau* pl., and adj., 'end, point, tip, apex, summit; source or upper reaches of river or stream; extremity, uplands', etc'
Aber-giâr, Abermarlais, Achddu, Afon Fach Pontgarreg, Bele, Berach, Blaenau[1-4], Blaen-nantllin, Blaensawdde, Blaen-waun, Blaen-y-coed, Blaenynys, *Blotweth, Capel Maesnonni, Carwe, Cathan, Cennen, Cil-y-gell, Cuch, Cwm Cynnen, Cwmisfael, Dafen, Dewi Fawr, Esgair-garn, Fforest*[3]*, Fforest Bedol, Glan-tren, Glyn*[3]*, Gorlech, Gwestfa Blaen-twrch, Gwyddyl, Hiraeth, Llechach, Lliedi, Llwyn-swch, Maenor Blaen-cuch, Maenor y Merydd, Marlais*[4]*, Nant Bargod, Nant Cynnen, Nant Dâr, Nant Gwythwch, Nant Melyn, Nant Rhydw, Pen-y-garn, Pib, Pontbrenaraeth, Sannan, Sïen, Sylgen, Waunclunda, Wysg, Ydw*

blaidd nm., pl. *bleiddiaid, bleiddiau,* 'wolf'
Cefn-blaidd

blanche Fr 'white'
Hendy-gwyn

blawd[2] npl. 'flowers, blooms'
Blotweth

blewog adj. and sometimes nf., 'hairy, furry, bushy'
Cwmliog

blodau see ***blodeuyn***

blodeuyn nm., *blodeuen* nf., ***blodau*** pl., 'a flower'
Blotweth

blotwaith, *blodwaith* nm. 'meal-dust'
Blotweth

bocs[1], *bocys*[1] pl., *bocsen, bocysen* nm., 'box-trees'
Bocs

bol[1], *bola* nm., pl. *byly*
Twyn

bôn[1] nm. 'bottom, base; tree-trunk, stump'
?Pantarfon

box E 'box-tree'
Bocs

brād OE, ***brode*** ME, 'broad'
Broadlay, Broad Oak, Broadway[1, 2], Derwenfawr

brain see ***brân***

braith see ***brith***

brân nf., *brain* pl., 'crow, rook, raven'
Aberbranddu, Brân[1, 2]*, Cefn-bryn-brain,* Clog-yfrân*, Croesyceiliog,* ?Cwmdwyfran*, Cynheidre, Miawst, Taliaris, Tŷ'r-frân*

*****brau*** adj. 'brittle, weak; free'
Groesffordd

bre[1] nmf. 'hill, hillock, mountain'; high'
Groesffordd*, Llanybri,* Moelfre[1-3]*,* Pen-bre

Glossary of Place-Name Elements, Personal Names and River-Names

brech see **brych**

brefan see **breuan**²

brenin nm. 'king, sovereign'
Ffaldybrenin

breuan¹ nf., *breuanau* pl., 'hand-mill, quern; mortar (for grinding)'
Llyn Brianne

breuan² nf., *breuain* pl., 'carrion crown, raven'
Trefreuan

breuant nmf., *breuannau* pl., 'windpipe, throat'
?*Llyn Brianne*

breuer, **breufer** ?adj. and n. 'loud, roaring'
Groesffordd

briallu¹ n.pl. 'primrose'
Crosshands

bridge see **brycg** OE

brith¹ adj., **braith** adj.f., 'marked with different colours, mottled, speckled'
Llech-fraith

bro nf., *broydd* pl., 'vale, lowland' and 'region, country, land'
Fro

broad see **brād** OE

brōc OE, *brook* E, 'brook, stream'
Brook, *Llanymddyfri*

brode ME see **brād** OE

bron¹ nf., *bronnydd* pl., 'breast (of a hill), hillside, slope'
Bronysgawen, *Capel Teilo*², *Castellcosan*

bronwydd (**bron**¹, **gwŷdd**) n.pl. 'hillside trees; wooded hillside'
Bronwydd

brook see **brōc** OE

brwd adj. 'hot, warm'
?*Glanbrydan*

brycg OE, 'a bridge'; *brigge*, *brugge* ME

brych adj., **brech** adj.f., 'brindled, spotted, speckled'

Brechfa, Nant Brechfa

bryn nm., *-iau, -nau* pl., 'hill, mount, rise, bank; heap' etc
Blaen-y-coed, Bryn, Brynaman, Brynaman-uchaf, Bryn-du, Bryngwynne, Brynhafod, Brynhyfryd, Bryn Iwan, Bryn Nicol, Bryn-y-beirdd, Cae'r-bryn, *Caerfyrddin*, *Cathilas*¹, Cefn-bryn-brain, Crachdy, Cwmcarnhywel, Dol-y-bryn, Efail-wen, *Esgair*¹, Fanagoed, Henllan², Heol-ddu, Hirfryn, Llanfair-y-bryn, Llyn Brianne, Nant Aerau, Nant Aeron, Nant-bai, Pentrefelin, Pentre-wyn, Rhydybennau, Tachlouan, Telych, Travellers Rest

buarth nm., *buarthau* pl., 'fold or enclosure to milk cows, etc; penfold; court-yard, farmyard'
?Mihartach

buches² nf. 'milking-fold, cow-house; herd (esp. of cattle)'
Taliaris

buddair, **byd(d)air** ?nf. 'bird of prey' and 'bittern'
?Llanybydder

burh OE, dat. *byrig* 'fort'
Porth Tywyn

bwla, *bwly* nm. 'bull, castrated bull'
Sarnau

bwlch nm., *bylchau* pl., 'breach, gap; pass'
Bwlchnewydd, Iet-y-bwlch

bwr adj. 'fat, strong, big'
Porth Tywyn

bychan adj., **bechan** adj.f., *bychain, bychan* pl., 'little, small, lesser'
Cantref Bychan, *Cefn Llwydlo*, Llandeilo Abercywyn, Nant Thames, *Trefechan*¹,²

byddar pl. *byddair*
?Llanybydder

byr¹ adj., **ber** adj.f., 'short, small'
Cellifor, Cwm-byr, *Pontyberem*

byrig see **burh**

byth adv. and nm. 'ever, always; eternity'

xlvii

PLACE-NAMES OF CARMARTHENSHIRE

Llanfihangel Aberbythych

cad[1] nf. 'battle, conflict, war'

Cathan, *Llangadog*[1], *Llangathen, Nant Aerau, Nant Cynnen*

cadair[1] nf. 'chair, seat; chair-shaped hill, elevated place'

Llan-gan, Pencader

cadno nm. 'fox'

Cwrtycadno

cae nm., *caeau* pl., 'hedge, hedge-row, fence' but more widely 'field, enclosure' in fns.

Cae'r-bryn, *Capel Cynfab, Capel Dewi*[2], *Capel Gwilym Foethus, Capel Gwynllyw*[1], *Capel Sant Silin, Capel Troed-y-rhiw*, Cefncaeau, *Cellifor, Cilgwyn,* ?Cryngae, *Llanybydder, Maenor Castell Madog, Miawst, Pengwern, Pont Sbwdwr, Rhydargaeau*

caer nf., *caerau* pl., 'a fortified place, a fortification, a fort' and 'city'

Allt-y-gaer, *Betws Ystumgwili, Bronysgawen,* Caerfyrddin/Carmarthen, *Dolaucothi,* Fforest Gaerdydd,/Cardiff Forest, *Maenor Grongar, Ogofau*

cagl, *cagal* nm. 'dung, esp. of sheep or goats'

Pentrecagal

cain[1] adj. 'fine, fair, bright'

Cilfargen, Coed-gain

calch nm. 'lime, chalk'

Meidrum

caled adj. 'hard; rough'

Heol-galed, Salem

call adj. 'sharp, wily'

Cil-y-gell

cam[2] adj. 'crooked, bent'

Camnant, *Cwmgwili*[2], *Esgair-gam*

can[1] adj. 'white, shining'

Llan-gan

cannwr nm 'bleacher'

Cilgannwr

canol, *cenol* nm, -au, -ydd pl., and adj. 'middle, centre'

Bryngwynne, Cefncaeau, Cil-y-gell, Crachdy, Ganol, *Halfway*[3], *Maenor Cilau, Penddeulwyn, Pistyll*, Rhandir Ganol, *Tre-garn*

cantref nm. 'cantref, hundred', a large administrative division thought to have developed during the 9th and 10th centuries as a unit for collection of tribute within a system of overkingship

Cantref Bychan, Cantref Eginog, Cantref Gwarthaf, Cantref Mawr

canwr nm. 'singer; songbird'

Cilgannwr

canwr y coed 'wood warbler'

Cilgannwr

canwr yr ardd 'garden warbler'

Cilgannwr

capel nm., pl. -au, -i, -ydd, -oedd, 'chapel', esp. 'chapel of ease, Nonconformist meeting-house or chapel'

Abergorlech, Babel, Bancycapel, *Bryn Iwan,* Capel, Capel Bach, Capel Bach y Drindod, Capel Begewdin, Capel Brynach, Capel Cadfan, Capel Cefnberach, Capel Ceinwyry, Capel Coker, Capel Crist, Capel Cynfab, Capel Cynnor, Capel Dewi[1-3], Capel Dyddgen, Capel Dyddgu, Capel Erbach, Capel Gwenlais, Capel Gwilym Foethus, Capel Gwyn, Capel Gwynfe, Capel Gwynllyw[1, 2], Capel Hendre, Capel Iago, Capel Ifan[1, 2], Capel Isaac, Capel Iwan, Capel Llanddu, Capel Llanlluan, Capel Maesnonni, Capel Mair[1-3], Capel Mihangel, Capel Newydd, Capel Penarw, Capel Peulin, Capel Sant Silin, Capel Seion, Capel Teilo[1, 2], Capel Troed-y-rhiw, Capel Tydyst, Capel y Betws, Capel y Croesfeini, Capel y Drindod[1, 2], Capel yr Ywen, Cwm Capel, *Eglwys Fair a Churig, Graig, Llandyry, Llanfihangel, Llangynheiddon, Llanwrda, Llidiardnenog,* Maenor Capel, *Maenor Tregelyn,* Peniel, *Rhydcymerau,* Soar, *Sylen, Wern*[2]

caput Latin, *capitis* gen. 'head, top'

Pencader

caredig, *ceredig* adj. and nm. 'kind, gentle'

Glossary of Place-Name Elements, Personal Names and River-Names

Nantgaredig

carfan nmf. 'weaver's beam, rail; ridge, boundary'

Cwm-waun-gron

Carmel Biblical mountain near the Mediterranean Sea

Carmel

carn nmf., *-au* pl., and adj., 'cairn, barrow, tumulus, mound, rock; heap, pile'; nm. 'hoof, foot'

Cennen, Cwmcarnhywel, Esgair-garn, *Fferemfawr*, Garn, *Maesllanwrthwl*, *Penrhiw-goch*, *Pen-y-garn*, ?*Talacharn¹*, Tre-garn

carreg nf., *cerrig* pl., 'stone, rock'

Afon Fach Pontgarreg, Carreg Cennen, Cefncerrig, Crug-glas, *Dolaucothi*, Gorsaf Llwyfan Cerrig, *Llandyfân*, *Llan-y-crwys*, Mynyddygarreg, *Nant Brechfa*, *Ogofau*, Pencarreg, *Pumsaint*, Tre'rcerrig

carth nm. 'hemp; hards, oakum' and 'offscourings, sweepings, offal'

Cynin

carw nm., *ceirw* pl., 'deer, hart, stag'

Carwe, Cilcarw

castel(l) ME 'castle'

Castell Pigyn, Castell Toch, Castle Ely, Green Castle, Newcastle Emlyn, Roche Castle

castell nm. 'castle, stronghold; castellated mansion; fortified town or city'

Capel Bach y Drindod, Capel Maesnonni, Castellcosan, Castelldwyran, Castellheli, Castell Llwyd, Castell Moel, Castell Newydd Emlyn, Castell Pigyn, Castell Toch, Castellyrhingyll, *Dinefwr*, Foelgastell, *Maenor Betws*, *Maenor Castell Draenog*, *Maenor Castell Madog*, Rhos, Trecastell¹⁻³

castellum Latin 'castle, fortress'

Castellnewydd Emlyn

castle see ***castel(l)*** ME

cath nf. 'cat'

Cathan, Cathargoed, *Cathilas¹*, Llangathen

cath Ir

Llangathen

cathedra Latin nf., *-ae* genitive, 'chair'

Pencader

cau¹ vb. 'to close, enclose, hedge around'

Parc-y-rhos

caw nm. 'band, bandage, knot'

?*Cwmcawlwyd*

cawl nm. 'soup, pottage; cabbage, colewort'; *cawl y môr* 'sea-cabbage, sea-kale'

?*Cwmcawlwyd*

cawr nm. 'giant'

Llanedi

caws nm. 'cheese, curds'

Nant-y-caws, Pant-y-caws

ceann Ir 'head'

Cenarth

cefn, dial. *cefen* nm., *-i*, *-au* pl., 'back, fig. support; ridge, butt of ploughed land'

Aberbranddu, *Bacau*, Capel Cefnberach, Capel Cynfab, Cefn-blaidd, Cefn-bryn-brain, Cefncaeau, Cefncerrig, Cefn-coed, Cefneithin, Cefn Llwydlo, Cefn Padrig, Cefn Sidan, Cefn-y-pant, Cenarth, Crosshands, Egrmwnt, *Fforest²*, Henllan¹, Hernin, Hirfryn, Llanfihangel, *Maenor Cefndaufynydd*, *Meiros*, *Pentre-wyn*, Tachlouan, Telych, Traean March, Trefynydd

ceiliog, *ceilog* nm., pl. *ceil(i)ogod*, *-au*, 'cock'

Aber-giâr, Croesyceiliog, ?Penceiliogi

ceiliogi vb. 'to copulate, to impregnate'

Penceiliogi

ceiniog, *ceinog* nf. 'penny'

Cwrtycadno

ceisiad² nm., pl. *ceisiaid*, 'sergeant of the peace, catchpole; tax gatherer; treasurer'

Rhydyceisiaid

cell¹ nf. 'cell, bower'

?Cil-y-gell

celli nf. 'grove, copse, woodland'

PLACE-NAMES OF CARMARTHENSHIRE

Cellifor, *Cil-march, Crachdy, Cynheidre, Fforest², Gelli-aur, Gelli-ddu, Gellidigen, Gelligati, Gelliogof, Gelli-wen, Tanglwst*

celynnen nf., *celyn¹* n.pl. or coll.n. 'holly', celynnen nf. 'holly-tree'

Clynennos, *Maenor Tregelyn*

cen¹ nm. 'skin, scale; lichen'

Cenarth, ?Cennen

cenaw 'cub, puppy'

Maenor Cilcenawedd

cenol see **canol**

cen y cerrig nm. 'lichen'

Cennen

ceunant nm. 'ravine, deep dingle, gorge'

Llanboidy

chastel Fr 'castle, fortress'

Castellnewydd Emlyn

church see **cirice** OE

chwerw adj. 'bitter, acrid'

Capel Troed-y-rhiw

chwith adj. 'left, left-handed; strange, unusual, sinister'

Dyffryn

ci nm., pl. **cŵn**, 'dog, hound'

Pentre-cŵn

cib nmf. 'vessel, bowl'

Tre-gib

cil¹ nm., sometimes nf., *ciliau, cilion* pl., 'retreat, corner, nook' and, with a rn., 'source'

Cilcarw, Cilellyn, Cilfargen, Cilgannwr, Cilgryman, Cilgwyn, Cilhernin, Ciliau, Cil-march, Cilmaren, Cilnawen, Cilrhedyn, Cilsân, Cilsant, ?Cil-wr, Cil-y-cwm, Cil-y-gell, Cilygernant, Cilymaenllwyd, Cilyrychen, *Cwm-pen-graig, Hengil, Maenor Cilcenawedd, Maenor Cilgynfyn, Maenor Cilhengroes, Maenor Cilhernin,* Maenor Cilnawen, Waungilwen

cilfach nf., *-au, -oedd, -fechydd* pl., 'nook, recess, corner, sheltered or secluded spot; retreat' etc

Bryn Nicol, Capel y Betws, Cathan

ciliwr, *cilwr* nm. 'pursuer, one who puts to flight; recluse; who who goes or falls back'

?*Cil-wr*

cirice, *cyrice* OE 'church'

Eglwys Gymyn, Llan-saint, Llan-y-bri, Llan-y-crwys, Newchurch

cladach Ir 'shore; rocky foreshore'

Cloidach

cláidigh, *clóidigh, cláideach* Ir

?*Cloidach*. See Cloidach¹ in main text

clas nm. 'monastic community'

?*Tre-clas*

clau 'swift, ready'

Cloidach

clawdd nm., *cloddiau,* 'trench, earthwork; ditch'

Capel Troed-y-rhiw

cleien nf. 'clayey soil; soft stone'

Capel Gwyn

clog² nf. 'rock, cliff, precipice'

Clog-y-frân, Tre-glog

cloigyn dim. of *cloig, clöig* nmf. 'hasp, clasp, spindle'

?*Cloigyn*

clun² nm., 'meadow, moor; brake'

Aber-giâr, Alltwalis, Clunderwen, ?*Glyn-tai, Glynyrhenllan, Maenor Clun-tŷ, Waunclunda,* Waun-y-clun

clwt nm. 'piece, patch of land'

Cwrtycadno

clwyd nf., *clwydau* 'movable hurdle; gate, door'

Glwydeth

clwydedd 'door, gate'

Glwydeth

clywed 'to listen'

Cloidach, Clydach¹,²

cnwc² nm., pl. *cnyciau,* 'hillock, knoll; swelling, lump'

Maenor Pen-y-cnwc, Nant Pen-y-cnwc

Glossary of Place-Name Elements, Personal Names and River-Names

coal see **col** OE

coch adj. 'red, ruddy; brown' etc.

Castell Toch, Gors-goch, Maenor Tre-goch, Penrhiw-goch, Rhos-goch

cochyn nm., *cochen* nf., 'red-haired person; hare'; ?'red-coloured thing'

?Nant Gochen

coed nm. and coll.n. 'forest, wood, trees', pl. of *coeden* nf., double pl. *coedydd*, 'tree'

Amgoed, *Blaenau³*, Blaen-y-coed, Cefn-coed, Coed-gain, Coedmor, Dugoedydd, ?Fanagoed, Hengoed, Iscoed[1-4], *Maenor Cadwgan*, *Malláen*, *Nant Eiddig*, *Nant Melyn*, Tor-y-coed, Traws-mawr, Uwchcoed Morris

coedcae, *coetgae* nm., 'quickset, a hedge of trees; land enclosed with a hedge, field, close' etc.

Coedcae

coetref nm. 'woodland homestead or dwelling'

Goetre, *Gwydre*

coll² coll.n., *collen* nf., **cyll** pl., 'hazel; sapling'

?*Gwestfa Pont Rhyd-coll*, ?*Hengil*

common E adj. 'ordinary, shared by, familiar'; n. 'piece of public land' etc

Cynnull-bach, Eglwys Gymyn

cor nm. 'dwarf, pigmy'

?Nant Corrwg, *Talacharn²*

corach Ir 'marsh'

Fforest²

corn nm., *cyrn* pl., 'horn; point'

Allt, Llanfihangel Rhos-y-corn

cornant nmf. 'little brook, swift stream'

Nant Corrwg

cornel nmf. 'corner, nook'

Bryn

cornor¹ nm., pl. *cornorion*, 'hornblower, bugler, trumpeter; chief, leader of a host'

Cornorion

corrach nm. 'dwarf, pygmy'

Nant Corrwg

cors nf. and coll.n. 'swamp, bog, marsh'

Gors-goch, Gors-las, *Maenor Tre-goch*, *Marlais*[4]

cors, -e E 'marsh, moor, bog' (from **cors**)

Honeycorse

coth-

Cothi

court OFr, ME, 'a space enclosed by walls or houses, a yard' and 'a large house, a manor' (from late 13th cent); sometimes confused with *cort(e)* OE, ?'a short plot of ground, a piece of land cut off'. Locally 'monastic grange'; cf. **cwrt¹** nm. 'enclosure, yard, farm-yard; grange, court'

Faenor

cow see **cū** OE

cowyn, *cywyn* nm. 'plague, pestilence'

?Cywyn

crach pl. 'scabs; scabby eruption'; adj. 'scabby'

Crachdy

cragen nf., *cregyn* pl., 'shell; shell-fish'

Dol-gran

craig nf., *creigiau*, *creigydd* pl., 'a rock, cliff'

Crachdy, Cwm-pen-graig, Graig, *Nant y Rhaeadr*

creigiau see **craig**

Crist nm. 'Christ, the Lord's Anointed'

Capel Crist

croes nf., *crwys*, later *croesau*, pl.; 'a cross, crossroads'

Capel Bach, Croesyceiliog, Llan-y-crwys, *Maenor Cilhengroes*, *Maenor Tre-goch*, Pen-y-groes, *Rhydaman*, Tŷ-croes

croesfaen nmf., pl. *croesfeini*, 'stone or wooden cross used to mark a boundary; stone cross'

Capel y Croesfeini

croesffordd nf. 'cross-road'

Groesffordd

cron see **crwn**

cros OE, **cross** E 'a cross, the Cross; crossroads'

Crosshands, Cross Hands, Cross Inn, Glandy Cross, *Rhydaman*

li

crug nm., *-(i)au* pl., 'hillock, knoll; cairn, tumulus; heap, mass'

Bryn Iwan, Capel y Croesfeini, Cruclas, *Crugglas*, Crug-y-bar, Gwyddgrug, Maesycrugiau, Trichrug

crwb¹ nm. 'hump, lump' in GPC

Crwbin

crwbyn dim. of *crwb¹* nm. 'hump, lump'

?Crwbin

crwn adj., *cron* adj.f., 'round, circular'

Cryngae, *Cwm-waun-gron*, Llech-gron, *Maenor Grongar*

crwys see *croes*

crych adj. and nm., pl. *crychiau*, 'wrinkled, crumpled; rough; rippling, bubbling'

Crychan, Nant Crychiau

cryd nm. 'a shivering, trembling; ague, fever'

Maenor Tregelyn

cryman nm. 'reaping-hook, sickle, bill-hook'

Cilgryman

cuwch¹, *cuch* nmf. 'frown, scowl; grimace' also fig.

Cuch, Cwm-cuch

cwarter nm. 'quarter; region lying about a point of the compass; district'

Cwarter-bach, Cwarter-mawr, *Maenor Cadwgan*, *Maenor Cwarter Trysgyrch*

cwm nm., *cymoedd, -au, -ydd* pl., 'a deep, narrow valley, coomb, glen, dale; hollow, bowl-shaped depression'

Aber-giâr, *Achddu*, *Berach*, *Blaen-y-coed*, *Bynea*, *Cathan*, Cil-y-cwm, Cwmaman, Cwm-ann, Cwm-bach¹, ², Cwm-ban, Cwmbyr, Cwm Capel, Cwmcarnhywel, Cwm Cathan, Cwmcawlwyd. Cwmcothi, Cwm-cuch, Cwm Cynnen, Cwm-du, Cwmduad, Cwmdŵr, Cwmdwyfran, Cwmfelin-boeth, Cwmfelinmynach, Cwm-ffrwd, *Cwmgwili¹*, Cwmgwili², Cwmhiraeth, Cwm Hwplyn, Cwmhywel, Cwmifor, Cwmisfael, Cwmliog, Cwmllethryd, Cwm-marles, Cwm-mawr, Cwm-miles, Cwm-morgan, Cwmoernant, Cwm-pen-graig, Cwm-twrch, Cwm-waun-gron, Cwm-y-glo, Cwmysgyfarnog, Dol-gwm, *Dunant*, Felin-gwm Isaf, Felin-gwm Uchaf, *Fforest Bedol*, *Garnant*, *Glyn y Swisdir*, *Gwenffrwd*, ?*Gwestfa Maesllangelynyn*, *Maesllanwrthwl*, *Mynys*, Nant Aerau, Nant Cwm-merydd, *Nant Cynnen*, Nant Garenig, Nant Tawe, Pen-cwm, Pentre-wyn, Pontyberem, *Tre-gib*, *Ydw*

cwmwd 'commote', a division of a cantref. A cantref might contain two or more commotes

Caeo¹, *Carnwyllion*, *Maenordeilo*, *Perfedd*

cŵn see *ci*

cwrt¹ nm., 'court, mansion; courtyard, enclosure, farmyard' and 'monastic grange'

Bryn-y-beirdd, Cwrt, Cwrt Henri, Cwrtycadno, *Derllys*, *Gransh*, *Nant-bai*, Pentre-cwrt

cwter nf. 'gutter, channel; rain-trough'

Brynaman

cyd¹ adj. 'a joining; union; common'

Cynnull-bach, Dolhywel

cydblwyf variants *cydblwydd, cytblwyf* nm. 'district belonging to or forming part of two or more parishes'

Cydblwyf

cyff nm., *cyffiau, cyffion*, 'trunk (of a tree), stump, log'

Cyffig

cyfor¹ nm. 'flow, flood; a surging'

Cwm-ffrwd

cyll see *coll²*

cymer¹ nm., *cymerau* pl., 'confluence of two or more rivers or streams, meeting of waters'

Rhydcymerau

cymyn¹ 'endowment, gift by will'

Eglwys Gymyn

cynfab nm. 'eldest son, first-born'

Capel Cynfab

cyngor nm. 'advice, counsel; council'

Cynghordy

Glossary of Place-Name Elements, Personal Names and River-Names

cynhaeaf, *cynhaef* nm. 'harvest, autumn'
Cynheidre

cynhordy (*cynnor, tŷ*) nm. 'dog-kennel, dog-house; gate-house, gateway, porch' (GPC). Thought to be composed of *cynnor* 'door-post, side-post of door' and *tŷ* which suggests that 'dog-kennel' is an extension of meaning. This has been complicated by a suggestion that some medieval instances of **cynhordy** are miscopyings of earlier forms of ModW by *cyrfdy, cwrfdy, cwrwdy* meaning 'a beer-house'. The matter has been investigated by Dafydd Jenkins in LHD, p.238. Clearly, **cynhordy** had already acquired the sense of 'dog-kennel' by the 13th cent.
Cynghordy

cyning OE, **king** E
Kingsland

cynnen nf. 'contention, strife, battle'
?*Cwm Cynnen*, Nant Cynnen

cynnull² nm. 'collection, a gathering (of) harvest, etc.'
Cynnull-bach, Cynnull-mawr, *Gwestfa Trefgynnull*

cyntaid variants *cynhaid, cynnaid* nf. 'first swarm of bees'
?Cynheidre

cyrch¹ adj. 'direct, aggressive' and nm. 'attack, assault'
Maenor Cwarter Trysgyrch

cyw nm., *cywion* pl., 'young bird, chick; young animal'
Aber-giâr

da adj. 'good, beneficial'
Waunclunda

daf- see **dof-**

dafad nf., **defaid** pl., 'sheep, ewe'
Dafen

dafaty (*dafad, tŷ*) nm. 'sheepfold, sheepcote'
Dyfatty

dail coll. 'leaves; foliage'
Tir-y-dail

dan see **tan**

dâr nf., **deri** pl., 'oak-tree'
Derllys, *Nant Dâr*

dau variant **deu**, number and adj., **dwy** fem., 'two; pair'
Castelldwyran, ?Cwmdwyfran, Llanddeusant, Maenor Cefndaufynydd, Maenor Ddwylan Isaf, Maenor Ddwylan Uchaf, Penddeulwyn, *Penrhyn²*

de Fr 'of'
Talacharn²

deau, *de³*, *deheu* nmf. and adj. 'the south'
Llanddeusant

defaid see **dafad**

Deheubarth 'south Wales', lit. 'south part'
Llanddeusant

delli, *-ni* nm. 'blindness'
Aberglasne

***Demet-** tribal name
Dyfed

deri see **dâr**

derwen nf., **derw** pl., 'oak-tree'
Clunderwen, Derwen-fawr, *Miawst, Nant Dâr*

derwydd² npl. 'oak-trees'
Clunderwen, Derwydd

deu see **dau**

diflas adj. and nf. 'tasteless; disagreeable, disgusting'
Tafarn Ddiflas (in Introduction. Carmarthenshire Place-Names: research and analysis)

din (< Br **dūno-*) nm., *-au* pl., 'city, fort, fortress, fastness, stronghold (eg. defensive hill)'
Caerfyrddin/Carmarthen, Dinefwr, ?Pen-dein

dinas nm., later nf., *-oedd* pl., 'city, large town, town (fortified or unfortified); fortress; refuge'
Dinas, *Dinefwr*

liii

PLACE-NAMES OF CARMARTHENSHIRE

do- Brit intensive prefix
Dyfed

dof-, *daf-* el, with sense 'tame, domesticated'
?*Dafen*

dof[1] adj. 'tame, domesticated'
Dafen

dog 'to take, snatch'
Llan-dawg

dôl[1] nf., *dolau* pl., 'meadow, water-meadow'
Cwrt, *Dolaucothi*, *Dol-gran*, *Dol-gwm*, *Dolhywel*, *Dol-wyrdd*, *Dol-y-bryn*, *Iddole*

domus Latin 'a' house, home'
Hendy-gwyn

down see OE *dūn*

draen[1], *drain* nmf., *drain* pl., 'thorn(s), thornbush'
Ffynnon-ddrain

draenog nm. 'hedgehog, urchin' and adj. 'thorny, prickly'
Maenor Castell Draenog

drain see *draen*[1], *drain*

drain E (OE *drē(a)hnian*) 'channel, conduit; outflow'
Swan Pool Drain

drum, *drumau* see *trum*, *drum*

dryslwyn, *dyryslwyn* nm. 'tangled bush, thicket, bramble-brake; place full of brambles'
Dryslwyn

dryw[1] nmf. 'wren'
Maenor Castell Draenog

du, *duf* adj., *duon* pl., 'black, sable, dark; overgrown; bitter'
Aberbranddu, *Achddu*, *Bryn-du*, *Capel Llanddu*, *Clydach*[1], *Cwm-du*, *Cwmduad*, ?*Dewi Fawr*, *Duar*, *Dugoedydd*, *Dulais*[1-3], *Dulas*, *Dunant*, *Esgairifan*, *Gelli-ddu*, *Heol-ddu*, *Mynydd Du*

duad[1] nm. 'a blackening, a darkening'
Cwmduad

dūn OE 'a hill', *doun* ME, 'a hill, an expanse of open hill-country'
Whitehill Down

*$d\bar{u}no$-, *$d\bar{u}non$ see *din*

duon see *du*, *duf*

dŵr, *dwfr* nm., *dyfroedd* pl., 'water'
Crychan, *Cwm-dŵr*, *Felin-gwm Isaf*, *Llanddowror*, *Llanymddyfri*, *Melinddwr*, *Nant Tridwr*, *Penrhyn*[2], *Pysgotwr*, *Sardis*[1]

dwy see *dau*

dwyran (*dwy*, *rhan*) compound and adj. 'two parts'
Castelldwyran

dyar, *dear* adj. 'loud, noisy'
Duar

Dydd Gŵyl 'feast-day of a saint; holiday'
Cenarth, *Cynwyl Elfed*

Dyfed
Dyfed

dyffryn nm. 'valley, vale, bottom' etc
Cilellyn, *Cynin*, *Dyffryn*, *Dyffryn Ceidrych*, *Gronw*, *Gwyddyl*, *Hengil*, *Nant Cynnen*, *Teifi*

dyn nm. 'man, person'
Gelli-ddu

dywal adj. 'fierce, cruel, furious'
Alltwalis, ?*Tyweli*

ēast, *ēasterne* OE adj., adv., 'eastern, east'. The adj. is applicable where the place named lies eastwards of some older place or faces east; as an adv. elliptically '(place) east of (something)'
East Marsh

ebol nm. 'colt, foal'
Llwynyrebol

ebyr see *aber*

-ed[3] suffix
?*Carwe*, *Lliedi*

efwr, *ewr*[1] (< Celt *eburo*-) nm. and coll.n. 'cow-parsnip, hogweed'

Glossary of Place-Name Elements, Personal Names and River-Names

Dinefwr

egin npl. 'shoots, sprouts; descendants'

Cantref Eginog

Eginog

Cantref Eginog

eglwys nf. 'church'

Capel Ifan¹, Capel Troed-y-rhiw, Castelldwyran, Cyffig, Eglwys Fair a Churig, Eglwys Fair Glan-taf, Eglwys Gymyn, Eglwys Trefwenyn, Henllan², Llan-gain, Llannewydd, Pentre-wyn

eiddig adj. 'voracious, greedy; ardent'

Nant Eiddig

***eiskā** Celt ?'to move' or to move swiftly'.

Wysg

eisteddfa nf. 'seat, chair, bench' and topographically 'place resembling a chair'

Llanegwad, Llangynnwr, Maenor Gynnwr

eithaf adj. 'extreme, farthest, most distant'

Gwlad

eithin n.pl. or coll.n.

Cefneithin, *Hiraeth*

elestr, *gelestr*, dial. *geletsh*, coll.n. 'sword-flag, fleur-de-lis, iris; lily'

Blaen-nant-llin

Elfed

Elfed

-ell dim. suffix

Annell

ellyn nmf. 'razor'

Cilellyn

-en suffix

?Cennen, Dafen, ?*Mynydd Sylen*, ?Sïen

ende OE 'end, the end of something'

Pen-sarn

erbarch nm. 'great honour, respect; adoration; object of worship'

Capel Erbach

esgair nf. 'ridge, mountain spur; leg, shank'

Capel Sant Silin, Esgair¹,², Esgairdawe, Esgair Ferchon, Esgair-gam, Esgair-garn, Esgairifan, Esgeirnant, *Llan-y-crwys, Maenor Castell Madog, Nant Dâr, Nant Tawe, Pentre-wyn, Tanglwst*

esgob nm. 'bishop'

Tiresgob

estyn¹, *ystyn¹* nm. 'extension, a stretching, prolongation'

Maenor Glynystyn

-fa, see **ma(n)**, **-fa**

farmer pl. *-s* 'farmer'

Ffarmers

feld OE, **field** E, 'open land, land for pasture or cultivation', later 'enclosed land, field'

Ashfield, *Capel Mihangel*

ferry E

Ferryside, *Penrhyn²*

ffair¹ nf., *ffeiriau* pl. 'fair, market'

Abercennen, Bynea, Ffair-fach

ffald nf. 'fold, pen, pound'

Ffaldybrenin

fferi nf. 'ferry; ferry-boat'

Glanyfferi

fferm, *ffarm, fferem* (ME *ferm(e)*) 'farm, holding; rent, tax'

Fferem-fawr, *Garn*

ffin¹ nf. 'boundary, border'

Nant-y-ffin, Ystrad-ffin

ffordd nf., *ffyrdd* pl., 'road, way, street'

Banc-y-ffordd, *Rhos*, Tyn-y-ffordd

fforest¹ nf. 'forest, park for hunting'

Dyffryn Ceidrych, Fforest¹⁻³, Fforest Bedol, Fforest Crychan, Fforest Gaerdydd

ffos nf. 'ditch, dike; trench, furrow'

Bancffosfelen

ffranc¹ nm.. *ffrainc* pl., 'foreign mercenary, enemy; Frenchman'

PLACE-NAMES OF CARMARTHENSHIRE

Bach-y-ffrainc, *Maenor Bach-y-ffrainc*

ffrwd nmf. 'swift stream, torrent'

Cwm-ffrwd, *Felindre⁴*, Gwenffrwd, *Trichrug*

ffwlbart, *ffwlbard* nm. 'polecat, foulmart'

Esgair¹

ffwrnais, *ffwrnes* nf. 'furnace'

Ffwrnes

ffynnon nf., *ffynhonnau* pl., 'spring, fountain, well; source, origin'

Banc-y-ffordd, Capel Cynfab, Capel Dewi¹, Capel Erbach, Capel Gwenlais, Capel Iwan, Capel Maesnonni, Capel Mair¹, Capel Mihangel, Capel Penarw, Capel Sant Silin, Cenarth, Cilymaenllwyd, Esgair¹, Ffynnon-ddrain, Ffynnonhenri, Ffynnon-oer, *Llandeilo'r-ynys, Llandybïe, Llandyfân, Llanelli, Llan-gan, Llangathen, Llangeler, Llangynheiddon, Llan-y-bri,* Maesyffynnon, *Pant-y-caws,* Pantyffynnon, *Pentregwenlais, Pumsaint, Sannan, Sïen*

field see OE **feld**

five E 'five'

Five Roads

ford OE, E 'shallow place in a river by which a crossing can be made; way, road'

Ammanford, *Edwinsford*

forest E 'forest, area reserved for hunting'

Cardiff Forest

foul E see **fūl** OE

four E number and n.

Four Roads

fūl OE, E **foul**

Llanymddyfri

furnace E 'furnace, structure for heating metals, etc' with extended sense 'ironworks, metalworks'

Furnace¹,², *Ffrwnais*

furze E see **fyrs** OE

fyrs OE 'furze'

Halfpenny Furze

gaing nf. 'chisel; wedge'

Coed-gain

gallt see **allt**

-gar (= *(i)âr*)

?Gwydderig

garan¹ nmf., pl. *garanod, garnau* 'crane, heron'

Nant Garenig

gardd nf. 'garden; enclosure'

Tanerdy

garlleg¹ npl. 'garlic'

Maenor Garllegan

garth¹ nmf. 'mountain ridge, promontory'

Capel Teilo², Cenarth, Cynin, Fforest², Gwestfa Penarth, Llanfihangel-ar-arth

garth² nm. 'field, enclosure'

Cynin, Fforest², Gwestfa Penarth

garw adj. 'rough, rugged; coarse'

Garnant

gate ME, E see **gatu** OE

gatu OE, **gate** ME, E, *gates* pl., 'opening, gap; gate'

Caerfyrddin, ?Pinged

gefail¹ nf. 'smithy, forge'

Bryn Iwan, Efail-wen, *Foelgastell, New Inn¹*

gelau¹, *gele* nmf. 'leech

?Llethrgele

gelau² nm. 'blade, weapon, spear'

Llethrgele

geletsh see **elestr**

ger prep. 'near, close to'

?Cilygernant

gïach nmf. 'snipe'

Croesyceiliog

glain¹ nm., *gleiniau, gleinion* pl. 'gem, precious stone'

Bynea

glais variant of **glas²** nm. 'stream'

Glossary of Place-Name Elements, Personal Names and River-Names

Alltwalis, Dulais[1-3], *Dulas, Gwenlais, Lash, Marlais*[1-4], *Morlais*

glan nf., *-nau, glennydd* pl., 'river-bank, edge; slope'

Annell, Betws Ystumgwili, Capel Gwynllyw[1], *Capel Llanddu, Capel Mair*[2], *Cathilas*[2], *Crychan, Cynin, Dafen, Dunant, Duar,* Eglwys Fair Glan-taf, *Fanafas, Felin-gwm Isaf, Garnant, Glanaman, Glanbrân, Glanbrydan, Glan-duar, Glangwili, Glanmôr, Glansefin, Glan-tren, Glanyfferi, Glanyrannell, Groesffordd, Lash, Llangennech, Llanmilo,* Maenor Ddwylan Isaf, Maenor Ddwylan Uchaf, *Maenor Tre-goch, Maesllanwrthwl,* Mountain, *Myddyfi,* Nant Aeron, Nant Thames, Nant Treuddyn, *Pentregwenlais, Rhydybennau, Talacharn*[2] (Glancorran), *Telych*

glân adj. 'clean; clear of sin; holy; fair'

?*Glandy Cross*

glas[1] adj., *gleision* pl., 'blue, greenish blue; green, verdant'

?*Aberglasne,* Cruclas, Crug-glas, Glasallt, Gorslas, Rhiw-las, *Rhos*

glas[2] see **glais**

glo nm., coll.n., 'coal; charcoal'

Cwm-y-glo

glyn nm. 'narrow valley, glen, dingle'

Alltwalis, Brynaman, Bynea, *Cwm-cuch,* Emlyn, Glyn[1-5], Glynaman, Glyncothi, Glyn-hir, ?Glyn-tai, Glynyrhenllan, Glyn y Swisdir, *Maenor Clun-tŷ*, Maenor Glynystyn, Sylen

go[1], **go-** prep., adv., prefix 'under; rather, somewhat'

Gothylon

***gobann-** Brit 'smith'

?*Fenni*

godre, *godref*[2] nm. 'skirt, border, edge; foor or bottom (of mountain, hill, Etc)'

Cwm Cynnen

golau, *goleu* adj. 'light, source of light, bright'

Gronw, Maenor Gwernolau, Wernolau

golden E adj. 'made or consisting of gold; coloured or shining like gold'

Golden Grove

goleuni nm. 'light, illumination, brightness'

Wernolau

gor-, gwor-, gwar- 'over, very, exceedingly'

Fforest[2], ?*Gorlech,* Llanfihangel-ar-arth

gordd see **ordd**

gorsaf nf. 'station, standing-place' etc

Gorsaf Llwyfan Cerrig

gorthir nm. 'uplands, highland'

Talacharn[1]

graean npl. and coll.n. 'gravel, coarse sand'

Dol-gran

grāf(a), *græfe* OE 'grove, copse, thicket'

Golden Grove

gransh nf (GPC), nm. 'grange'

Gransh

grange ME 'granary, barn, grange, farm belonging to a religious house or feudal lord'

Grange

grēne OE, E *green* 'green, verdant'

Green Castle

gro coll.n. 'coarse mix of pebbles and sand in river-bed, gravel; gravelly shore'

Dyffryn

grove E see **grāf(a)** OE

grug coll.n. 'heather, ling'

Clynennos

gwäell nmf. 'knitting-needle, skewer'

Aberarad

gwair[1] nm. 'grass (grown for harvesting); hay'

Hebron[1]

gwaith[1] nm. 'work; a working place, works, factory'

Blotweth, Bynea, Dinefwr

gwarthaf[2] nm. 'uppermost part'

Cantref Gwarthaf

gwastad variant *gwastod* adj. 'flat, level' and nm.pl. *gwastadoedd, gwastadau*

Gwastade

gwaun nf., ***gweunydd*** pl., 'high and wet level ground, moorland, heath'

Bacau, Blaen-waun, *Blaen-y-coed, Cwm-waun-gron, Cynnull-bach,* Foelgastell, Llangadog², Miawst, *Morfa¹, Myhathan, Nant Rhydw, Penrhiw-goch,* Waun Baglam, Waunclunda, Waungilwen, Waun-y-clun

gwedd¹ nmf. 'sight, appearance; face'

Maenor Cilcenawedd

gweirglodd nmf. '(lowland) hay-field, meadow'

Cornorion

gwely nm. 'bed; group of persons who, as descendants of a common ancestor, were joint occupiers of land, tract of tribal land held in joint ownership and called by the name of the stock-father of a particular progeny'

Cilcarw

gwen see ***gwyn***

gwenith nmf., pl. 'wheat'

Pontyfenni

gwennol nf. 'swallow, martin'

Nant Gwennol

gwenyn npl. 'bees, wild bees'

Maenor Castell Draenog, Pentre-wyn

gwern nmf. 'alder-tree(s); alder-grove, alder-marsh, swamp'

Maenor Gwernolau, Pengwern, *Pontyfenni,* Wern¹,², Wernolau

gwernog adj. 'swampy, fenny' and nf., pl. *gwernogau*

Gwernogle

gwestfa 'entertainment, hospitality, food-render': LHD, p.351 (possibly to be equated with *twnc*) or an area levy paid by groups of *trefi* WTLC I, pp. 213-4. Glanville R. Jones, StC 28 (1994), p.89 (and refs. cited), states that the *gwestfa* was a 'lesser administrative subdivision ... literally the place, *ma*, whose noble proprietors had originally supplied the [Welsh] king and his entourage when on circuit with hospitality, namely *gwest*, which included the provision of a meal and also sleeping quarters'. The *gwestfa* was later converted into a food-render which was to be delivered to the king's court (***llys¹***) and later commuted to a cash payment. This term was also applied to the district responsible for the payment for which see, for example, MWST 76-83. Some *gwestfâu* cannot be located geographically particularly when *gwestfa* is combined with a pers.n., apparently indicating a *gwestfa* payment made by what we may suppose is a kindred group. It is open to question whether this particular category had fixed geographical boundaries or definitions.

Faenor Isaf, Gwestfa Blaen-twrch, Gwestfa Bleddyn, Gwestfa Cadwgan ap Cynon, Gwestfa Cadwgan ap Tegwared, Gwestfa Cenarth, Gwestfa Cilfargen, Gwestfa Cilsân, Gwestfa Cwmblewog, Gwestfa Cwmcothi, Gwestfa Cwm-twrch, Gwestfa Cynwyl Elfed, Gwestfa Dinefwr, Gwestfa Gruffudd ab Elidir, Gwestfa Gwion Sais a Maredudd ap Heilyn, Gwestfa Hengoed, Gwestfa Llangathen, Gwestfa Llanhirnin, Gwestfa Llannewydd, Gwestfa Llanwrda, Gwestfa Llywelyn Gwynnau, Gwestfa Llywelyn ap Heilyn, Gwestfa Maesllangelynyn, Gwestfa Maestreuddyn, Gwestfa Merthyr ac Aber-nant, Gwestfa Moreiddig, Gwestfa Owain ap Rhydderch, Gwestfa Penarth, Gwestfa Perth-lwyd, Gwestfa Pont Rhyd-coll, Gwestfa Rhingylliaid, Gwestfa Tre Cynwyl Gaeo, Gwesta Trefgynnull, Gwestfa Tre-lech a'r Betws, Gwestfa Wyrion Idnerth, Gwestfa Wyrion Ieuan ac Wyrion Seisyll, Gwestfa y Faenor Isaf, Gwestfa Ysgolheigion, Gwestfa Ystradfynys, Westfa

gwesty nm. 'lodging, guest-house, inn'

Rhydaman

gweunydd see ***gwaun***

gwinau adj. 'bay, reddish brown'

?Bryngwynne

gwinllan

Maesllanwrthwl

gwiwer nf. 'squirrel'

Maenor Castell Draenog

Glossary of Place-Name Elements, Personal Names and River-Names

gwlad nf. 'country, land'
Gwlad, *Ystrad Tywi*

gŵr nm, pl. *gwŷr*, double pl. *gwyrion*, 'man'
Gwestfa Wyrion Idnerth, Gwestfa Wyrion Ieuan ac Wyrion Seisyll, Llanddowror, *Llanwrda*

gwrach nf. 'ugly old woman, crone, witch'
Maenor Tregelyn

gwrda nm. 'nobleman, lord; good man'
Llanwrda

gwŷdd coll.n. and pl. 'trees; forest, woods'

gŵydd[1], nm. 'presence; sight, face' and as a qualifier 'openly, prominent'
Gwydderig, Gwyddgrug, ?Gwydre

gŵydd[2] nf. 'goose'
Penceiliogi

gŵydd[3] adj. 'wild, untamed, savage'
Gwyderig

gŵydd[4] nm. 'grave, burial mound'
Gwyddgrug

gwyddwal, *gwyddel* nmf. 'thicket, bush, brambles'
?Gwyddyl

gŵyl[1] nf. 'holiday, holy-day, feast of patron saint; watch, guard'
?*Gwili*[1], Llanedi

gŵyl[2] adj. 'modest, gentle'
?*Gwili*[1]

gwyllt, *gwyll* adj. 'wild, as opposed to tame; untamed'
Pengwern

gwymp adj., *gwemp* adj.f., 'excellent, splendid, comely and fair'
?Gwempa

gwyn adj., adj.f. *gwen*, pl. *gwynion*
Alltwalis, *Bryngwynne*, Capel Gwyn, Capel Gwynfe, Cilgwyn, *Clydach*[1], Efail-wen, Felin-wen, Gelli-wen, *Goetre*, Gwenffrwd, Gwenlais, Hendy-gwyn, *Llanwrda*, Maes-gwyn, Maesllanwrthwl, *Nant Pibwr*, Pant-gwyn, Pentre-tŷ-gwyn, *Pentregwenlais*, *Pentrewyn*, Pibwrlwyd, Rhyd-wen, *Trechgwynnon*, Waungilwen. ?Wen, ?Ysgwyn

gwyndwn, *gwndwn* nm. 'unploughed land'
Llechdwni

gwynion see **gwyn**

gwynlliw[1] adj. 'white-coloured', nm. and nf. *gwenlliw* 'whiteness, brilliance'
Capel Gwynllyw[1]

gwynnau[1], *gwyniau* adj. 'spirited, lively'
Gwestfa Llywelyn Gwynnau

gwynnon, *gwnnon* coll.n. 'fog, long white straw; dry twigs, straw'
Trechgwynnon

gwyrdd adj., *gwerdd* adj.f., 'green, verdant'
Dol-wyrdd

gwyry, *gwyryf* nm. 'virgin'
Capel Ceinwyry

gwŷs[1] nmf. 'sow; pig'
Twrch[2]

gwythol as a variant of *gwythog*[1] adj. 'fierce, angry'
Gothylon

gwythwch nm. 'wild pig'
Nant Gwythwch

haearn nm. 'iron; object made of iron; fig. harness, strength'

haen nf. 'stratum, layer', dim. *haenen*, ?adj. *haenenog*
Nenog

hafod nf. 'summer residence, farmstead occupied in summer months'. A location of additional pasture typically in upland areas in Carmarthenshire
Brynhafod, *Hendy*, Pentrefelin, *Talog*[1]

halfway ME, E '(place) half way (to another place)', often applying to an inn
Halfway[1-3]

halfpenny ME, E num. 'halfpenny, halfpenny coin'

Halfpenny Furze

hālig OE, pl. *hālge*, 'holy'

Llan-saint

hall OE, E 'hall, large dwelling, large building'

Marble Hall, Middleton Hall

halog adj. 'dirty, soiled'

Login, Talog[1]

hamlet E

Egwad

hand E, pl. *hands*

Crosshands, Cross Hands

hār OE, E *hoar*

Capel Mihangel

hardd adj. 'beautiful, fair, fine'

Harddfan, *Talhardd*

Hebron the biblical city in Canaan and modern Palestine

Hebron[1,2]

heli nm. 'brine, salt water'

Castellheli

helyg[1] coll.n. and adj., sing. *helygen* 'willow-tree, sallow-tree'

Brisgen, Capel Gwyn, Clynennos

hen adj. 'old, aged, former'

Capel Iago, Capel Troed-y-rhiw, Castell Moel, Glynyrhenllan, *Goetre*, Hen Briordy, Hendy, Hendy-gwyn, Hengil, Hengoed, Henllan[1,2], Henllan Amgoed, *Llanarthne, Llwynhendy, Maenor Cilhengroes, Pentre-wyn*, Trostre, Tŷ-hen

hendref nf., 'winter dwelling located in the valley to which the family and its stock returned after transhumance, permanent residence; home farm'

Capel Hendre, *Hendy, Maes-y-bont*

heol nf., *heolydd* pl., 'way, road, street'; locally *hewl* and 'path, track'

Heol-ddu, *Heolgaled*, Pedair-heol, Pump-heol, Salem, Sardis

herber[1] (< ME *erber* < OFr *erber*) nmf. 'arbour, herb- or flower-garden'. ME *erber* means 'grassy piece of land, garden'

?Penrherber

herber[2] (< OE *here-berg*) ?nm. 'shelter, lodging'. This is treated in GPC as a single element with **herber**[1] but it may derive from ME *herber* (OE *here-beorg*) 'land by a shelter, or affording shelter' or ME *herberge* (Fr *auberge*)

Penrherber

Hermon Biblical city located on a hill

Hermon[1,2]

hill see **hyll** OE

hir adj., pl. *hirion*, 'long, tall, extensive'

Cwmhiraeth, Glyn-hir, Hiraeth, Hirfryn, ?Penboyr, *Tir Rhoser*

hiraeth nm. 'grief or sadness, longing'

Cwmhiraeth

hoar see OE **hār**

hogl nmf. 'ill-designed, ramshackle; shed, shelter for cattle, etc'

Gelliogof

honnye see **hūnig** OE

Horeb biblical mountain

Horeb[1,2]

hūnig OE 'honey; sweet (like honey)' or 'sticky' as in 'sticky, muddy land'

Honeycorse

hist see **ust**

hwch nmf. 'sow'

Aman

hwīt OE, 'white; infertile, dry'; in fns. sometimes 'dry, open pasture'

White Mill, Whitland, Whitehill Down

hwrdd nm. 'a ram'

Allt

hyfryd adj. 'pleasant, agreeable'

Brynhyfryd

hyll, *hull* OE, **hill** 'hill, natural eminence or elevated piece of land'

Glossary of Place-Name Elements, Personal Names and River-Names

Caerfyrddin, Llandre, Myrtle Hill[1, 2], Pentywyn[2], Whitehill Down

i prep. 'to'

-i suffix

Cothi, Cydweli, Dewi Fawr, ?*Fenni*, Gwili, Llanymddyfri, Lliedi, *Nant Cynnen*, *Burry Port*, Tawe, Teifi, Tyweli

iâr[1], *giâr* nf., pl. *ieir*, 'hen, hen-bird'

Aber-giâr, Gwydderig

iares (*iâr*[1], *-es*[2]) nf. 'flock of chickens'

Taliaris

iet[1] nf. 'gate'

Iet-y-bwlch, *Pontyates*

-ig (Brit *-īko*, *-īkos*) suffix found in pers.ns., adjs. and pns. (EANC 180)

Gwydderig, *Llanfihangel Aberbythych*, Nant Garenig

in E prep. 'in, within'

Castellnewydd Emlyn

-in[1] adj. suffix

inn E 'inn, public house, tavern'

Cross Inn, New Inn[1, 2], *Rhydaman*

-iog[1] terr. suffix. '(land) belonging to, territory of'

Catheiniog, *Elfed*, Peuliniog

-ion, -on territorial suffix

Carnwyllion, *Elfed*

is 'below'

Is Cennen, Iscoed[1-4], Is-morlais, *Malláen*

isaf adj. 'lower, lowest'

Aber-giâr, *Berach*, *Brechfa*, *Brisgen*, *Bryngwynne*, *Capel Dewi*[3], *Capel Gwyn*, *Capel Gwynllyw*[1], *Capel Hendre*, *Capel yr Ywen*, *Cathargoed*, *Cefncaeau*, *Cil-y-gell*, *Crachdy*, *Cruclas*, *Cwm Cathan*, *Cwm-waun-gron*, *Cynnull-bach*, *Dol-gwm*, *Dryslwyn*, *Emlyn*, *Faenor Isaf*, *Felin-gwm Isaf*, *Felin-wen*, *Glan-tren*, *Glanyrannell*, *Goetre*, *Gronw*, *Gwestfa Blaen-twrch*, *Gwestfa Ystradmynys*, *Halfway*[3], *Hengil*, *Llanybydder*, *Maenor Cilau*, *Maenor Cilhengroes*, *Maenor Ddwylan Isaf*, *Maenor Lleision*, *Maenor Tegfynydd*, *Maenor Tre-goch*, *Maes-gwyn*, *Moelfre*[2], *Pengwern*, *Pentywyn*, *Pistyll*, *Rhandir Isaf*, *Rhos*, *Trostre*, *Twynmynydd*, *Tŷ-isaf*, *Wen*

king see OE ***cyning***

la, le Fr def.art.

Glanyfferi

lacu OE, ***lake***[2] ME, drainage channel, side-channel of a river; water-course'

Lacques, *Macrels*

lake[1] ME, E 'pool, lake' (< Fr *lac*)

lake[2] see ***lacu*** OE

land, lond OE, ***land(e)*** ME, 'land; a tract of land' and 'a strip of land in a common field'

Kingsland, *Maenor Betws*, Whitland

landa latinicisation of ***land***

Hendy-gwyn

lang OE 'long'

leacach Ir nm. 'area of flat rocks', adj. 'strewn with flat stones'

Llechach

lēah OE,'a wood'; esp. 'clearing in a wood'; later 'piece of open land, a meadow, a lea'

Broadlay

*leuco- Brit 'bright, shining'

Llwchwr

ley, lea E see ***lēah*** OE

little see ***lȳtel*** OE

llachar adj. 'bright, shining'

?*Talacharn*[1]

llafar adj. 'loud, clear' and n. 'capacity to speek, speech'

Cloidach

llaid nm. 'mud, dirt, clay; ?swampy'

?Lliedi

llain nmf. 'blade, spear' usually in sense 'strip of land'

Cwm-pen-graig, *Parc-y-rhos*

llan nf., *-nau, -noedd, llennydd* pl., originally 'clearing, open space' and piece of 'enclosed land' surviving in the last sense in compounds (*coedlan, perllan, gwinllan*) and later 'a piece of consecrated ground, churchyard, church' etc.

Bachsylw, Capel Crist, Capel Cynfab, Capel Dyddgen, Capel Llanddu, Capel Llanlluan, Capel Tydyst, Capel y Croesfeini, Cil-y-cwm, Cilygernant, Cyffig, Egwad, Glansefin, Glynyrhenllan, *Gwestfa Maesllangelynyn,* Henllan[1, 2], Henllan Amgoed, Llan[1-3], Llanarthne, Llanbedr, *Llanboidy,* Llandawg, Llanddarog, Llanddeusant, Llanddowror, Llandeilo, Llandeilo Abercywyn, Llandeilo'r-ynys, Llandeulyddog, Llandingad, Llandybïe, Llandyfaelog, Llandyfân, Llandyfeisant, Llandyry, Llandysilio, Llanedi, Llanegwad, Llanelli, Llanfair-ar-y-bryn, Llanfallteg, Llanfihangel, Llanfihangel Aberbythych, Llanfihangel Abercywyn, Llanfihangel-ar-arth, Llanfihangel Cilfargen, Llanfihangel Rhos-y-corn, Llanfihangel-uwch-Gwili, Llanfynydd, Llangadog[1, 2], Llan-gain, Llan-gan, Llangathen, Llangeler, Llangennech, Llanglydwen, Llangyndeyrn, Llangynheiddon, Llangynin, Llangynnwr, Llangynog, Llanhernin, Llanismel, Llanllawddog, Llan-llwch, Llanllwni, *Llanmilo,* Llannewydd, Llannon, Llanpumsaint, Llansadwrn, Llansadyrnin, Llan-saint, Llansawel, Llansteffan, Llanwinio, Llanwrda, Llan-y-bri, Llanybydder, Llan-y-crwys, Llanymyddyfri, Maenor y Llan, ?*Maesllanwrthwl, Pentywyn*[2]

*****llanan** (***llan, -an***) ?'small enclosure'

Twynllanan

llandref (***llan, tref***) nf. 'church township', typically the township possessing the parish church

Egrmwnt, Llandre

llannerch nmf., *llanerchau, -i,* etc. pl., 'a clearing, glade, oasis, pasture' etc.

Llannerch, *Maenor Berwig*

llath nf. 'rod, staff; sail-yard, spar'

Capel Hendre

lle[1] nm. 'locality; a specific place; residence; position'

Gwernogle, Maenor Betws

llech[1] nf., pl. *llechau,* 'slate; slab of stone; rock'

Gorlech, Llanfihangel-uwch-Gwili, Llechach, Llechdwni, Llech-fraith, Llech-gron, *Maesllanwrthwl,* Pontarllechau, Talyllychau, Tre-lech, Tre-lech a'r Betws

llefrith nm. 'milk'

Nant-y-caws

llethr[1] nmf. 'slope, hillsdie, steep ascent'

Betws Ystumgwili, Capel Cadfan, ?*Cwmllethryd,* Llethrgele

llety nm. 'lodging, home'

Myddfai[1],*Tanglwst*

lliant nm. 'flood, flow, sea'

Lliedi

llidiart*, llidiard* nmf. 'gate'

Llidiardnenog

llif[2] nm. 'stream, flow'

Berach, Cellifor

llifeiriaint (?*llif*[2], ***bêr**) nm and pl. 'a flowing, flood, deluge'

Berach, Cellifor

llin[1] nmf. 'lineage, pedigree; line, streak, groove'

?*Blaen-nant-llin*

llin[2] nm. and adj. 'flax; thread or cloth made of flax'

?*Blaen-nant-llin*

llipryn nm. and adj. 'limp, soft, drooping'

Tor-y-coed

llodre nm. ?'place, location, building'

Coedmor

llwch[2], variant *llych,* pl. *llychau,* 'lake, pool, stagnant water'

?*Gorlech,* Llan-llwch, *Llwchwr, Sawdde,* Talyllychau

llwyd adj., *llwydion* pl., 'grey; pale; muddy (of water); holy, blessed'

Berth-lwyd, *Capel Mihangel,* Castell Llwyd, Cilymaenllwyd, ?*Cwmcawlwyd, Gwestfa Perth-lwyd,* Pibwrlwyd, *Pontyfenni*

Llwydlo transferred pn., Ludlow

Cefn Llwydlo

llwyfan² nmf. 'stage, raised floor, platform'

Gorsaf Llwyfan Cerrig

llwyfen nf., pl. *llwyfeni, llwyfenni*, 'elm(-tree); elm-bark'

Gwestfa Ystradmynys

llwyn nmf., *-au, -i, -ydd* pl., 'bush, plant', or coll. sense, 'grove, copse'. OBret *loin, loen*, Co **lon*, 'grove, thicket'

Aber-giâr, Capel Ceinwyry, Capel Dewi², Capel Llanddu, Cynheidre, Llanmilo, Llwynhendy, Llwyn-swch, Llwyn-teg, Llwynyrebol, *Moelfre²*, *Mynachdy²*, *Pant-y-llyn*, Penddeulwyn, *Piodau*

llygad nmf., pl. *llygaid*, 'eye; source of river'

Capel Gwenlais, Llwchwr, Wysg

llygoden nf., pl. *llygod*, 'mouse, rat'

Maenor Castell Draenog

llyn nmf., *-noedd, -nau, -iau, -ydd* pl., 'lake, pool, pond, puddle' etc.

?Cilellyn, Llyn Brianne, ?Pant-y-llyn, *Teifi*

llys¹ nmf., *-oedd, -au* pl., 'court, palace, manor-house, hall, imposing building, habitation of king, prince, nobleman, etc.; manorial seat of administration'

Derllys, *Dinefwr, Maenor Lleision*, Maenor Llys

long see ***lang*** OE

Ludlow

Cefn Llwydlo

lȳtel OE 'little, small'

ma-, -ma, -fa nmf., *-au, mai, mei* pl., 'a plain, field; place, spot; ?small, flat piece of land'

Brechfa, *Capel Gwynfe*, Cynheidre, ?Gwempa, ?Machynys, ?*Maenor Meddyfnych*, Malláen, Myddfai¹,², Myhathan, Nant Brechfa

mab nm., *meib(i)on* pl., 'boy, son, infant'

Gwestfa Cadwgan ab Einion, Gwestfa Cadwgan ap Tegwared, Gwestfa Gruffudd ab Elidir, Gwestfa Gwion Sais a Maredudd ap Heilyn, Gwestfa Llywelyn ap Heilyn, Gwestfa Owain ap Rhydderch, Mabelfyw, Mabudrud, Maenor Meibion Seisyll

machwy nm. 'bay (part of sea)'

Bae Caerfyrddin

mael² nm. 'prince, chieftain, lord'

Llangadog¹

maen nm., dial. *mân, main, mein, meini* pl., and adj., 'stone, esp. one having some speciality or a particular use; large stone as used, e.g. in building, rock'

Capel Mihangel, Cilymaenllwyd, *Maesllanwrthwl*, Rhos-maen

maenor nf., 'administrative unit', a subdivision of a commote, later associated with E *manor* which has a distinct origin. The south Wales recension of the Welsh laws suggest that a ***maenor*** might contain anything between seven and twenty-four *trefi* (***tref***) but this does not tally with place-name evidence. This shows that in the greater part of the county the ***maenor*** was a smaller unit often coterminous with a township or vill (sometimes described as a hamlet or *amlwd*), eg. Maenor Bach-y-ffrainc, Maenor Cilhengroes and Maenor Cilnawen. In the lordship of Llandovery, for example, the *maenor* had just fifty free tenants and paid a toll or *potura* (for two serjeants) known in 1396 as *bwyd teulu* (HCrm I, p.235) originating as a food-rent. The location of some *maenorau* is uncertain because of the sparsity of evidence and several examples in historical sources have been omitted such as Maenor Gelynnos (in Llanegwad) probably located near Ynys-rhyd (SN 533275). The latter is recorded in company with Tir y Marchog (unlocated) in 1622 located in *Maynor glininnos* in 1613.

Faenor, Faenor Isaf, *Faerdref, Hernin, Iddole,* Llanddarog, Llandeilo'r-ynys, *Llan-gain*, Llechgron, *Llethrgele*, Maenor Aberbargoed, Maenor Bachsylw, Maenor Bach-y-ffrainc, Maenor Berwig, Maenor Betws, Maenor Blaen-cuch, Maenor Brwnws, Maenor Cadwgan, Maenor Capel, Maenor Castell Draenog, Maenor Castelldwyran, Maenor Castell Madog, Maenor

Cefndaufynydd, Maenor Cenarth, Maenor Cilau, Maenor Cilcenawedd, Maenor Cilellyn, Maenor Cilgynfyn, Maenor Cilhengroes, Maenor Cilhernin, Maenor Cilnawen, Maenor Clun-tŷ, Maenor Cwarter Trysgyrch, Maenor Ddwylan Isaf, Maenor Ddwylan Uchaf, Maenordeilo, Maenor Egrmwnt, Maenor Egwad, Maenor Fabon, Maenor Forion, Maenor Fouwen, Maenor Gain, Maenor Garllegan, Maenor Glynystyn, Maenor Grongar, Maenor Gwempa, Maenor Gwernolau, Maenor Gwynfe, Maenor Gynnwr, Maenor Hengoed, Maenor Henllan Amgoed, Maenor Iscoed, Maenor Llanddeusant, Maenor Llanedi, Maenor Llanfihangel, Maenor Llanglydwen, Maenor Llangoedmor, Maenor Llangynin, Maenor Llan-non, Maenor Llansadwrn, Maenor Llansawel, Maenor Lleision, Maenor Llys, Maenor Meddyfnych, Maenor Meibion Seisyll, Maenor Myddfai, Maenor Pencarreg, Maenor Penrhyn, Maenor Pen-y-cnwc, Maenor Rhiwtornor, Maenor Rhwng Twrch a Chothi, Maenor Tallwn, Maenor Tal-y-fan, Maenor Tegfynydd, Maenor Tre-dai, Maenor Tregelyn, Maenor Tre-goch, Maenor y Llan, Maenor y Maes-gwyn, Maenor y Merydd, Maenor y Mynachdy, Maenor Ysgwyn

maerdref nf. 'land adjacent to the court worked by unfree tenants supervised by the *maer biswail* (dung bailiff) to provide food, etc., for the court, demesne, home farm; hamlet attached to chief's court'

Faerdref

maerdy nm. 'farm supervised by *maer* or steward; farm, dairy farm'

Maerdy, *Taliaris*

maes nm., *meysydd* pl., and adv., 'open country as opposed to woodland, expanse of open land, level land, plain, open field', later 'field'

Abercennen, Bryngwynne, Capel Maesnonni, Dafen, Esgairifan, ?Fanafas, Ffair-fach, Gronw, Gwestfa Maesllangelynyn, Maes-gwyn, Maesllanwrthwl, Maestreuddyn, Maes-y-bont, Maesycrugiau, Maesyffynnon

magwyr nf. 'wall; ruin'

Alltwalis

maharen, *myharen* nm. 'ram'

Cilmaren

mai see **ma-**

maidd nm. 'whey, curds and whey'

Tafarn y Maidd Sur: See Introduction. Carmarthenshire Place-Names: research and analysis

mainc nf., pl. *meinciau*, 'bench, long seat' also fig.

Meinciau

mam nf. 'mother; dam; queen bee'

?Mamog

mamog nf. 'dam, esp. in-lamb or breeding ewe, brood-mare'

?*Mamog*

man^1, **-fa** nmf. 'particular place'

Harddfan, *Meinciau*

mangoed, *mân goed* npl. 'small trees, young wood, undergrowth'

Fanagoed

manor E 'manor, lordship'

Maenordeilo

marble E 'marble, resembling marble'

Llan-y-bri, Marble Hall

march nm., *meirch* pl., 'horse, stallion'; 'great, large' with extended sense 'strong, vigorous'

Cil-march, Marros, *Traean March*

March title

Traean March

marchog^1 nm. 'horseman, rider, mounted warrior; nobleman'

Llansadwrn

mare Latin 'sea'

Caerfyrddin

marsh E see **mersc** OE

mawr, *mor* adj. 'big, great, high'

Afon Fawr, *Blaenau$^{3, 4}$, Blaen-waun, Bryn, Brynaman, Brynhafod,* Cantref Mawr, *Capel Ifan^1, Capel Mihangel, Carnwyllion, Cilcarw, Cilgryman,*

Glossary of Place-Name Elements, Personal Names and River-Names

Coedmor, *Crachdy*, *Cross Inn*, Cwarter-mawr, *Cwm-ffrwd*, Cwm-mawr, Cynnull-mawr, Derwen-fawr, Dewi Fawr, *Dryslwyn*, Fferem-fawr, *Gelli-ddu*, *Glan-tren*, *Glasallt*, *Glyn-tai*, *Groesffordd*, Hengoed, Lacques, Llandeilo, Llanegwad, *Llangadog¹*, Maenor Cilau, Marlais¹⁻⁴, Meinciau, *Morfa²*, Morlais, *Pontarddulais*, Pysgotwr, *Rhos-goch*, Tor-y-coed, Traws-mawr, *Trefechan¹*, Tŷ-mawr, Ysgubor-fawr

medaf vb. 'I reap'

Dyfed, Elfed

mei-, meidd- 'half, mid, middle'

Meidrum, Meiros, Miawst

meidr², **meidir²** nf. '(narrow country) lane, track leading to a farm'

Llanboidy

melen see **melyn**

melin nf., -au pl., 'mill, water-mill'

Bancffosfelen, Bancyfelin, *Bynea*, *Capel Llanddu*, *Capel Llanlluan*, *Cloigyn*, Cwmfelin-boeth, Cwmfelinmynach, *Cwm-mawr*, *Cwrt*, *Dol-y-bryn*, Drefelin, *Felindre³,⁴*, Felin-foel, Felingwm-isaf, Felingwm-uchaf, Felin-wen, *Gwyddyl*, Maenor Cwarter Trysgyrch, Melinddwr, *Nant Melyn*, *Nant Pibwr*, Pentrefelin, *Trap*, *Traws-mawr*, Trecastell³

melindref (**melin**, **tref**) nf. 'mill settlement, settlement owing suit to a mill'

Felindre¹⁻⁶, Afon Felindre

melyn adj., **melen** adj.f., 'yellow; brown'

Bancffosfelen, Heol-ddu, ?Nant Melyn

melys, melus adj. and nm. 'sweet, pleasant-tasting'

Nant Felys

*men- Brit ?'flow, current'

?Fenai

menyn see **ymenyn**, **menyn**

merchan (**merch**, dim. **-an**) nf. 'little girl; little daughter'

?Merchon

mersc OE, ME mersh, E marsh

East Marsh

merthyr¹ nm. 'martyr'

Llangeler

merthyr² nm. 'graveyard or shrine consecrated with the bones of a saint; martyr'

Capel Tydyst, *Derllys*, *Llangeler*, Merthyr, Merthyr Cynog

merydd¹ adj. 'slow, sluggish' and 'stagnant'

Maenor y Merydd, Nant Cwm-merydd

merydd² nm. 'fat beast, animals'

Nant Cwm-merydd

met- see **medaf**

mign nf., pl. mignau, etc, 'marsh, bog; bog moss'

Mynydd Fign

milain 'villein, peasant'

Felindre¹

mill see **myln**, mylen OE

Milo part of biblical Jerusalem

Milo

moel¹ adj. 'bare', nf. 'bare mountain or hill', pl. moelydd

Castell Moel, *Dinas*, Felin-foel, Foelgastell, Moelfre¹⁻³, *Nant Pedol*, Tal-y-foel

moethus adj. 'fond of luxury or a life of ease; pampered; refined'

Capel Gwilym Foethus

mont Fr 'mount, hill'

Egrmwnt / Egremont

moor see **mōr** OE

mōr OE, 'a moor', earlier 'barren waste-land'

Llangadog², Morfa¹

mor see **mawr**

môr¹, mor OW (< mori-) nm. 'sea, ocean'

Caerfyrddin/Carmarthen, Glan-mor, *Llanybri*

morfa nmf. '(sea) marsh, salt-marsh, land (or marsh) by the sea-shore' later 'wet, marshy land', sometimes applied to inland locations

Dafen, Morfa¹,², Morfa-bach, *Pen-sarn*, *Pentowin*

mori- Br see **môr**

mount E 'a hill, a rise'

Mount Pleasant

mountain ME, E 'mountain, lofty hill'

Black Mountain, Mountain

mud[1] adj. 'dumb, mute; silent'

?Ydw

mud- stem in *mudaf*[1], *mudo*[1] vb. (a) 'to move away (b) to move, to convey, to bear away'

?Ydw

mwyalch npl. 'blackbirds'

Croesyceiliog

mwswm, *mwswn* nm. 'moss, lichen'

Fforest[1]

mwyn[2] nm. 'mineral, ore; mine'

Rhandir-mwyn

mydd, *midd* nm. 'vessel, dish, vessel'

Myddfai[1, 2]

myln, *mylen* OE 'a mill'

Felindre[4], *Llangadog*[2], *Maenor Tegfynydd*, *Nant Pibwr*, New Mill, White Mill

mynach[1], *manach* nm., -od, -iaid, mynaich, etc pl., 'monk, sometimes also a friar'

Cwmfelinmynach, *Merthyr*

mynachdy nm. lit. 'monk's house' (**mynach**, *tŷ*) with extended use of 'house or farm owned by a monastery, a grange'

Mynachdy[1, 2]

mynwent nf. 'graveyard, cemetery'

Capel Crist, Capel Iago, Capel Mair[2], *Capel Sant Silin, Llanfihangel*

mynydd nm., -oedd, -au pl., 'mountain, (large) hill' with extended sense 'exposed area, esp. a heath'

Alltwalis, Capel Isaac, Carnwyllion, Cwm-ffrwd, Hiraeth, Llanfihangel, Llanfynydd, *Llwyn-teg, Maenor Cefndaufynydd, Maenor Tegfynydd,* Mynydd Du, Mynydd Figyn, Mynydd Llanllwni, Mynydd Llanybydder, Mynydd Malláen, Mynydd Myddfai, Mynydd Pen-bre, Mynydd Pencarreg, Mynydd Sylen, Mynyddygarreg, Penymynydd, *Traws-mawr,* Trefynydd, *Trostre,* Twynmynydd

myrtle E evergreen shrub of genus *Myrtus* with aromatic foliage and white flowers

Myrtle Hill[1, 2]

-nai suffix

?*Aberglasne*

nant nf., earlier nm., *nentydd*, poet. *naint* pl., earlier 'valley, ravine, glen'

Aber-giâr, Aber-nant, *Afon Fawr, Berach, Blaen-nant-llin,* Camnant, *Capel Gwynllyw*[1], *Cefn-bryn-brain,* Cilygernant, *Clydach*[1], *Crychan, Cwmcawlwyd, Cwm-ffrwd, Cwmoernant, Dol-gran,* Dunant, *Esgairdawe, Esgair-garn,* Esgeirnant, *Fenni,* Garnant, *Glan-duar, Gwenlais, Gwyderig, Gwyddyl, Hebron*[1], Llanboidy, *Llanmilo, Llanybydder, Maenor Cwarter Trysgyrch, Maenor y Merydd, Miawst,* Nant Aerau, Nant Aeron, Nant-bai, Nant Bargod, Nant Brechfa, Nant Corrwg, Nant Crychiau, Nant Cwm-merydd, Nant Cynnen, Nant Dâr, Nant Eiddig, Nant Felys, Nantgaredig, Nant Garenig, Nant Gochen, Nant Gwennol, Nant Gwythwch, Nant Hust, Nant Melyn, Nant Pedol, Nant Pen-y-cnwc, Nant Pibwr, Nant Rhydw, Nant Tawe, Nant Thames, Nant Treuddyn, Nant Tridwr, Nant-y-caws, Nant y Dresglen, Nant-y-ffin, Nant y Rhaeadr, Pennant, *Pentregwenlais, Pen-y-garn, Pontarllechau, Trostre, Twrch*[2], *Ydw, Ystrad-ffin*

naw- el. in *dineuo* 'to pour, flow; stream out' and *nawes* 'running, pouring'

Maenor Cilnawen

nen nmf. 'roof; top, summit'

?Nenog

neuadd

Neuadd Middleton

new see OE *nīwe*

newydd adj. 'new'

Bwlchnewydd, Capel Newydd, Castellnewydd Emlyn, Drenewydd, *Felin-foel,* Llannewydd, *New Inn*[1], *Rhos, Traws-mawr, Tre-lech, Trostre*

nīwe OE 'new'

Newcastle Emlyn, Newchurch, New Inn[1, 2], New Mill, Newton[1, 2]

Glossary of Place-Name Elements, Personal Names and River-Names

noef Fr 'new'
Castellnewydd Emlyn

noethni, *noethi*² vb. 'to bare, to expose', *-ni* nm. 'nakedness'
Aberglasne

norð OE 'north, northern'

novel Fr 'new'
Castellnewydd Emlyn

novum Latin 'new'
Castellnewydd Emlyn

o prep. 'from'
Glansefin, Trap

oak see **āc** OE

odyn nf. 'kiln; (lime-)kiln
Rhydodyn

oen nmf., pl. *ŵyn*, 'lamb'
Pentre-wyn

oer adj. 'cold, cool'
Cwmoernant, Ffynnon-oer

-og adj. suffix
Cantref Eginog, Cloidach, Mamog, ?Nenog

ogof, archaic *gogof* nf., pl. *gogofau*, 'cave, cavern; cavity'
Dolaucothi, ?Gelliogof, Llanedi

oistre AFr 'oyster'
Ysterlwyf

old see **āld**, *eald* OE

on E prep.

-on¹, **-ion¹** noun suffix
?*Gothylon*

onn coll.n., *onnen* nf., *ynn* pl., 'ash (tree(s), wood'
Llan-non, Maenor Lleision

ordd, *gordd* < Brit *ordo-*, nf. 'hammer, mallet'
Dyfed

-os dim. pl. suffix
Clynennos

ostium Latin 'mouth of a river'
Abergwili

pabell see **pebyll**

pandy nm. 'fulling-mill'
Meidrum

pant nm. 'hollow, depression, valley'
Capel Gwyn, Cefn-y-pant, Fanagoed, Gwestfa Ystradmynys, Pantarfon, Pant-gwyn, Pant-y-caws, Pantyffynnon, Pant-y-llyn, Pentowin, Rhos, Rhydargaeau, Soar

parc nm., *-(i)au* pl., 'enclosed land, field, paddocks', later 'recreation area' and 'housing estate'
Alltwalis, Berach, Betws Ystumgwili, Capel Iwan, Capel Mihangel, Capel Troed-y-rhiw, Castellcosan, Cwm-pen-graig, Eglwys Gymyn, Fanagoed, Llanarthne, Llan-y-bri, Merthyr Cynog, Mynyddygarreg, Parc Howard, Parc-y-rhos, *Pontyfenni*

parc AFr see **park** ME, E

parddu nm. 'soot, smut, blackness'
?Barddu

park ME, E, **parc** AFr 'an enclosed tract of land for beasts of the chase' though the usual sense in west Wales and much of co. Carmarthen is 'field, enclosure' later 'recreation area' and 'housing estate'
Glanbrydan

parth nmf. 'area, region; part'
Llanddeusant

pau nmf. 'country, land, district; habitation'
?*Pen-boyr*

parsel nmf. 'parcel, division'
Cynnull-bach

pedol nf. 'horseshoe'
Fforest Bedol, ?Nant Pedol

pedwar adj., *pedair* adj.f., 'four'
Pedair-heol

lxvii

pen**¹** nm., *pennau* pl., 'a head, height, hill; chief, supreme' and 'end of'. It is not always easy to distinguish these senses

Aber-giâr, Annell, Bacau, Betws Ystumgwili, Broadway², Bryn Iwan, Bryn-y-beirdd, Capel Penarw, Cathilas², Cwm-pen-graig, Esgair¹, Esgair-garn, Felin-gwm Isaf, Foelgastell, Gwestfa Penarth, Hendy, Maenor Betws, Maenor Pen-y-cnwc, Morfa¹, Nant Pen-y-cnwc, Penallt, Pen-bre, Pen-boyr, Pencader, Pencarreg, Penceiliogi, Pen-cwm, Penddeulwyn, Pen-dein, Pengwern, Pennant, Penrherber, Penrhiw-goch, Pen-rhos, Pen-sarn, *Pentowin,* Pen-twyn, Pentywyn, Pen-y-banc¹,², Pen-y-fan, Pen-y-garn, Pen-y-groes, Penymynydd, *Rhos, Telych, Trichrug, Twynllanan*

Peniel, *Penuel* Biblical city on the east side of the river Jordan

Peniel

penrhyn (*pen¹, rhyn¹*) nm. 'cape, promontory, headland'

Penrhyn¹,²

pentref, *pentre* nm. 'chief farm, chief settlement' later 'village'

Drefelin, Maenor Cilau, Maenor Tre-goch, Pentre, Pentre-bach, Pentrecagal, Pentre-cŵn, Pentre-cwrt, Pentrefelin, Pentregwenlais, Pentre Morgan, Pentre-poeth, Pentre-ty-gwyn, Pentre-wyn, *Tre-garn*

perfedd nm. and adj. 'middle or centre, heartland'

Perfedd, Treberfedd

perllan nf. 'orchard'

Maesllanwrthwl

perth nf., *perthi* pl., 'wood' later 'bush, hedge'

Berth-lwyd, *Gwestfa Perth-lwyd*

pi² nmf., pl. *piod, *piodau*

?Piodau

pib nf. 'musical wind instrument; (water-) pipe, tube'

Pib

pibwr 'one who has diarrhoea; piper'

Nant Pibwr, Pibwrlwyd

pica¹ adj. 'pointed, spiked'

Garn

picyn, ***pigyn¹*** nm. 'vessel for eating or drinking'

Castell Pigyn

piggin E 'small pail or vessel'

Castell Pigyn

pigwn² a variant of *pigwrn¹* nm. 'cone, spire, point'

Pigwn

pigyn¹ nm. 'pointed or tapering end or object, (sharp) point; spire, peak'

Bigyn, Castell Pigyn

pingot 'small enclosure'

Pinged

pioden nf., pl. *piod* 'magpie'

Piodau

pistyll nm., *-au* pl., 'spout, well'; occasionally 'rill, stream'

Capel Dewi¹, Capel Teilo¹,², Cilgwyn, Llangynnwr, Llansawel, Pistyll

plas nm. 'mansion, large house, hall; place'

Alltyferin, Egrmwnt, Gronw, Llangeler, Maenor Tal-y-fan, Pontantwn, Pontyfenni, Trimsaran

pleasant E 'agreeable, pleasant'

Mount Pleasant

poeth adj. 'hot, burning; burnt'

Cwmfelin-boeth, Pentre-poeth

pōl OE, 'pool'

Pwll, Swan Pool Drain

pont nf. 'bridge'

Aber-giâr, Afon Fach Pontgarreg, Annell, Capel y Drindod, Gwestfa Pont Rhyd-coll, Hendy, Llandeilo'r-ynys, Maes-y-bont, *Miawst, Myddyfi, Nant Pibwr,* Pentrefelin, Pont Abram, Pontaman, Pontantwn, Pontarddulais, Pontargothi, Pontarllechau, Pont-ar-sais, Pont-henri, Pont-iets, Pont Sbwdwr, Pont-tyweli, Pontyberem, Pontyfenni, *Rhydowen, Talog²*

pontbren, *pompren* nmf., *pontbrenni, -ni* pl., '(wooden) foot-bridge'

Glossary of Place-Name Elements, Personal Names and River-Names

Felin-gwm Isaf, Pontbrenaraeth

pool E see **pōl** OE

port ME, ModE 'town, often a market town; port' Burry Port
Glanyfferi

porth² nm. 'gate; door, porch' and 'gateway'
Caerfyrddin, Porth-y-rhyd¹,²

porth³ nmf. 'harbour, haven'
Porth Tywyn

porthladd nmf. 'harbour, port; sea-port'
Porth Tywyn

pound see **pynd** OE

priordy nm. 'priory; monastery'
Hen Briordy

prysg, *prys* nm. 'copse, grove'
Llanybydder

prysgen nf. 'shrub, bush'
?Brisgen

pump, *pum* number and nm. 'five'
Cilymaenllwyd, Llanpumsaint, Pump-heol, Pumsaint

pund ME see **pynd** OE

pwll nm., *pyllau* pl., 'deep hole or shaft, a pit; pool, pond, a pool in a river' and occasionally 'stream, ditch'
Brynaman, Pwll, Pwll-trap

pynd OE, *pund* ME 'a pound, enclosure'
?Pinged

pysgod n.pl., sing. *pysgodyn* nf.
Pysgotwr

'r see **y**, **yr**

ram E 'ram, male sheep or goat'
Ram

rēad OE, *rede* ME, 'red-coloured'
Red Roses

red E see **rēad** OE

rest 'resting-place'
Travellers Rest

rex Latin gen.pl. *regum*
Regwm

rhaeadr nmf. 'waterfall, cataract, torrent'
Nant y Rhaeadr

rhag prep. 'before, in front of, in the face of'
Machynys

rhan nf. 'part, share, portion, division'
Castelldwyran

rhandir nmf. 'part of a country, area, territory; territorial unit'. In the Welsh laws, the *rhandir* was a very small unit, two or three making up a *tref*, but this sense does not seem to appear in place-name evidence for co. Carmarthen Rhandir Abad, Rhandir Ganol, Rhandir Isaf, Rhandir-mwyn, Rhandir Rhydodyn, Rhandir Uchaf

rhedyn n.pl. 'ferns, bracken'
Cilrhedyn

rhingyll nm. 'court official; ringild; bailiff'
Castellyrhingyll, Gwestfa Rhingylliaid

rhiw nf., -(i)au pl., 'slope, ascent, hill'
Alltwalis, *Bryn Iwan*, *Capel Troed-y-rhiw*, *Llanfallteg*, *Maenor Rhiwtornor*, Penrhiw-goch, Rhiw-las, Rhiw'radar

rhos¹ n.pl. 'roses'
Red Roses

rhos² '(upland) moor, heath(land)'
Cwmliog, Llanfihangel, *Llanfihangel Rhos-y-corn*, *Llangeler*, *Maenor Cilhengroes*, Marros, Meiros, Parc-y-rhos, Pen-rhos, Rhos, Rhosaman, Rhos-goch, Rhos-maen, *Tir Rhoser*, *Trefynydd*

rhwd¹ nmf. 'rust; filth, mud'
Nant Rhydw

rhwmp¹ nm. '(large) auger'
Aberarad, Berach

rhwng prep. 'between'
Maenor Rhwng Twrch a Cothi

rhyd nf., -(i)au pl., 'ford'

lxix

PLACE-NAMES OF CARMARTHENSHIRE

?*Cwmllethryd, Dol-y-bryn, Esgair-garn, Glyn⁵, Gwestfa Pont Rhyd-coll, Llanybydder, Maenor y Merydd, Meidrum, Pen-boyr, Porth-y-rhyd*[1,2], Rhydaman, Rhydargaeau, Rhydcymerau, Rhydodyn, Rhydowen, Rhydsarnau, Rhydwen, Rhydwilym, *Rhydybennau*, Rhydyceisiaid, Talog[1]

rhyd- see **rhwd**

rhyfel nmf. 'war(fare), conflict'

Capel Troed-y-rhiw

rhyg nm. 'rye (grain)'

Maenor Castell Madog

river E 'river'

road E 'road, way'

Five Roads, Four Roads

roche Fr 'a rock'

Roche Castle

rood E 'cross, crucifix'

Llan-y-crwys

rōse OE, **rose** pl. *roses*, 'rose (the flower); rose-bush'

Red Roses

rupa Latin 'a rock'

Roche Castle

saen[1] nf. 'cart, wagon'

Cilsân

saer nm. 'carpenter, joiner; craftsman, builder', pl. *seiri*

Rhydargaeau

Saeson see **Sais**

saileach Ir 'salty'

Dyffryn

saint see **sant**

saint E, Fr

Pentywyn[2], St Clears, St Ishmael

Sais, pl. **Saeson**, 'Englishman'

Gwestfa Gwion Sais a Maredudd ap Heilyn, Pont-ar-sais

Salem Biblical short form of Jerusalem

Salem

san see **sant**

sân, *saen*[2] nm. 'seine(-net)'

Cilsân

sandig OE, **sandy** E 'sandy'

Sandy

sandy see **sandig** OE

sant, san, etc. nm.; **saint** pl., 'saint, holy person'

Capel Sant Silin, ?Cilsant, *Cilymaenllwyd, Cynwyl Elfed*, Llanddeusant, Llandyfeisant, Llanpumsaint, Llan-saint, *Pentywyn*, Pumsaint, Sanclêr, Sannan

Sardis A city in western Turkey mentioned in Classical writings and the Bible and a common Nonconformist chapel-name

Sardis[1,2]

sarn nmf., p. *sarnau* 'causeway, stepping-stones, path, road'

Pen-sarn, *Rhos*, Rhydsarnau, Sarnau, Talsarn

Saron Biblical Saron was an area of beauty and fertility stretching along the Mediterranean. A common Nonconformist chapel-name

Saron[1,2]

sawdd[1] nmf. 'a sinking, drowning, torrent'

Sawdde

sbeit, spite E, a derogatory n. often applied to inns and interpreted as 'a place built in spite', generally explained as one built 'in spite' of a neighbouring inn often generating explanatory tales which can rarely be proven. It may simply be used to describe a building – inn, chapel or house – built in an unfavourable location

Tafarn-sbeit/Tavernspite (in Introduction. Carmarthenshire Place-Names: research)

sea E 'sea' (OE *sǣ*)

Seaside

Seion The biblical Mount Sion or Zion

Capel Seion

seren nf. 'a star'

Trimsaran

lxx

Glossary of Place-Name Elements, Personal Names and River-Names

serfel[1], *sierfel*[1] nm. 'garden chervil'
Cwmcarnhywel

seri[1] ?nm., 'paved path, causeway; horse, steed'
Rhydargaeau

sgothi see **ysgothi**

si[1] nmf. 'rumour, whisper; hiss, mumble
Sïen

sidan[1] nm., 'silk', adj. 'silken, silky, smooth'
Cefn Sidan

side E (OE *sīd*) 'side, the long part; land alongside, beside'
Ferryside, Seaside

sigl[1] nm. 'a swinging, a swing'
Capel Hendre

Siloh Biblical Siloh or Shiloh
Siloh

Soar Biblical Zoar, one of the five cities of the plain in the Jordan valley
Soar

spite E see **sbeit**

stān OE, E stone
Capel Mihangel

station E 'stopping place on a railway, esp. with a platform'
Llwyfan Cerrig Station

sticil, *sticill* nf. 'stile'
Hengil

stone see **stān**

street E
Pentrecagal

strife E, ME < Fr *estrif* 'conflict, struggle'
Maenor Betws

sūð OE 'south'

sūðer OE 'southern'

sur adj. 'sour, bitter'
Tafarn y Maidd Sur: See Introduction. Carmarthenshire Place-Names: research and analysis

swan E
Swan Pool Drain

swch[1] nmf., 'ploughshare, snout, (pointed) end or tip'
Llwyn-swch

swfr nm. 'noise, din, murmur, a rustling'

Swistir nm. Switzerland
Glyn y Swisdir

Swiss 'pertaining to Switzerland' (fanciful)
Swiss Valley

syddyn, *eisyddyn* nm. 'tenement, land; dwelling-place'
Capel Hendre, Goetre, Hendy, Pantyffynnon, Tŷ-isaf

*****syl-** as in *sylfaen* 'foundation(-stone), base' and and *sylwedd* 'substance; solidity'?
?Mynydd Sylen, ?Sylen

-taf
Llanwrda

tafarn nmf. 'pub, tavern, inn'
Halfway[2], Piodau, Tafarn-sbeit/Tavernspite, Tafarn Ddiflas and Tafarn y Maidd Sur (in in Introduction. Carmarthenshire Place-Names: research and analysis), Tymbl

tai see **tŷ**

tair see **tri**

tâl[2] nm. 'forehead, front, end' and in topographical sense 'end, brow of a hill'
Cynheidre, Dol-y-bryn, Maenor Tal-y-fan, ?Talacharn[1, 2], Talhardd, Taliaris, Talsarn, Tal-y-foel, Talyllychau

talog adj. 'jaunty, cheerful'
Talog[1], Talog[2]

*****tam-** Br ?'flowing'
?Nant Tawe, Taf, Teifi

tan[1], *dan* prep. 'under, below, underneath'
Capel Ceinwyry, Capel Mair[1], Tal-y-foel

tanerdy nm. 'tannery, tanhouse'
Tanerdy

lxxi

PLACE-NAMES OF CARMARTHENSHIRE

teg adj. 'fair, fine, beautiful'

Llwyn-teg, *Maenor Tegfynydd, Miawst, Nant Pedol*

Tellwn ?pn.

Maenor Tellwn

****telych*** ?'hill, slope'

See Telych for a full discussion

?Tachlouan, ?Telych

telynor, *-ior* nm., pl. *telynorion*, 'harpist'

Cornorion

Temple Bar transferred n. from Temple Bar, London

Temple Bar

terra Latin 'land, territory'

Regwm

-teu see ***-tou***

****teu̯ā, teu̯-, tu-*** IE 'strong'

Tywi

the E def.art.

tir nm. 'land, ground, territory'

Alltwalis, Camnant, Crachdy, Cynheidre, Dolaucothi, Dolhywel, Esgairifan, Fanagoed, Fenni, Fforest Bedol, Garnant, Glyn², Gwestfa Blaen-twrch, Gwestfa Maesllangelynyn, Gwyddyl, Halfway², Hernin, Llanybydder, Llan-y-crwys, Maenor Tregelyn, Nant Dâr, Nant Melyn, Pantyffynnon, Pentre-wyn, Pontantwn, Tiresgob, *Tir Rhoser, Tir-y-dail, Tŷ'r-frân*

tōh, OE, ME *togh* 'steadfast, tough'

Castell Toch

ton² nm., *tonnau* pl., 'lay-land' and nf., 'skin, surface'; variant *twn²* compounded in *gw(y)ndwn*, 'unploughed land'

?*Llechdwni*

-ton see ***tūn*** OE

****tonni***

?*Llechdwni*

tor¹ nmf. 'a breaking; gap, breach, break'

Tor-y-coed

tor² nf. 'a swelling, bulge, protuberance; breast, slope, flank, or side (of mountain, hill, etc.), (river) bank'

Tor-y-coed

-tou or ***-teu*** suffix

town see ***tūn***

tra² adv. 'very, extremely'

Capel Troed-y-rhiw

traean nmf. 'a third, third part'

Amgoed, Gransh, Traean Clinton, Traean March, Traean Morgan

trafn¹ nmf. 'leader, lord' or 'home. dwelling'

Abermarlais

trallwng, *trallwn* nm. 'dirty pool, boggy spot (on road, etc)'

Bancffosfelen, Cwmcarnhywel, Traethnelgan

transh, trensh, etc nm. 'trench; dyke, bulwark'

Sylgen

trap¹ nm. 'trap; trapdoor' and fig. for 'place which attracts and detains' such as an inn

Pwll-trap, Trap

travellers 'travellers, passers-by'

Travellers Rest

traws adj. 'across, opposite; strong, powerful' sometimes as a variant of ***tros*** adj. 'over, across, above' as in Trostre. The meaning is sometimes ambiguous and pns. which contain this el. should be treated individually and topography may prove critical. B.G. Charles (PNPemb 46) suggests that it may mean 'a feature, such as a piece of land, ridge lying athwart' and see DPNW. That makes good sense with Traws-mawr since the ho. bearing this n. lies on the slopes of a promontory between streams. In other pns. ***traws*** is used adjectivally: in the case of Trawsfynydd, co. Merioneth, where it appears to describe a mountain route (***mynydd***) revising the suggestion in DPNW, p.463, that it is 'across the mountain'. In Trawsgoed, cos. Brecon, Cardigan and Montgomery, it applies more likely to an area adjoining or extending over woodland (***coed***).

Glossary of Place-Name Elements, Personal Names and River-Names

Traws-mawr, Trostre

trech[1] adj. 'mightier, stronger'

Trechgwynnon

tref nf., *-i, -ydd, -oedd* pl., earlier 'house, dwelling place, homestead' but most often 'hamlet, township' and later 'town'. This element is generally translated as 'township' where the name refers to an administrative unit or less specifically as 'settlement' and in a few instances as 'town'

Carwe, Cynheidre, Dre-fach[1-4], Drefelin, Drenewydd, *Gelligati,* Gwestfa Tre Cynwyl Gaeo, *Gwestfa Trefgynnull,* ?Gwydre, *Machynys, Maenor Cwarter Trysgyrch, Maenor Tre-dai, Maenor Tregelyn, Maenor Tre-goch, Myhathan, Pentrecagal, Pentrecŵn, Pentre-wyn,* Treberfedd, Trecastell[1-3], Tre-clas, Trefechan[1], Trefechan[2], Treforis, Trefreuan, Trefynydd, Tre-garn, Tre-gib, Tre-glog, Tregynin, Tregynnwr, Treherbert, Trehopcyn, Tre-lech, Tre-lech a'r Betws, Tremoilet, Tre'rcerrig, Trostre, *Twyn,* ?Wen

trefddyn (**tref**, *dynn*) nm. 'protected homestead'

Maestreuddyn, Nant Treuddyn

tren adj. 'strong, powerful'

Glan-tren

trensh see **transh**

tres[2] nm. 'battle, raid'

Maenor Cwarter Trysgyrch

tresglen nf. 'thrush, esp. mistle thrush'

Nant y Dresglen

tri, variant **try-**, fem. **tair**, 'three'

Nant Tridwr, Trichrug

trindod nf. 'Trinity; Trinity Sunday'

Capel Bach y Drindod, Capel y Drindod

troed nmf. 'foot, base'

Alltwalis, Capel Troed-y-rhiw

tros see **traws**

trum, drum nmf., *drumau* pl., '(crest of) mountain or hill, peak, ridge'

Meidrum, Trimsaran

try- intensifying prefix

?Trichrug

tulach[1] Ir 'low hill, hillock'; see **tyle**

Tachlouan, Telych

tumble E 'to fall', fig. 'a place to fall, perilous hill or slope'

Tymbl / Tumble

tūn OE, ***-ton*** ME, *town* 'an enclosure, a farmstead, an estate' later 'village, town, settlement'

Johnstown, Newton[1,2], Hopkinstown

twll[1] nm., *tyllau* pl., 'hole, aperture, hollow'

Cwm-pen-graig

twrch gen.sg. and pl. *tyrch*, nm. 'hog, (wild) boar'

Aman, Twrch[1,2]

twyn nm., *twyni* pl., 'hill, hillock, tump, knoll, rising; (sand-)dune"

Carwe, Pentowin, Pen-twyn, Twyn, Twynllanan, Twynmynydd

tŷ nm., **tai** pl., 'house, building'

Castelldwyran, Crachdy, Cynghordy, *Felindre*[6], Felin-foel, Fforest Bedol, *Gelligati,* Glyn-tai, Groesffordd, Halfway[2], Hendy, Hendy-gwyn, Llwynhendy, *Maenor Clun-tŷ, Maenor Tre-dai,* Mynachdy[1,2], Pentre-tŷ-gwyn, *Rhos, Tanglwst, Tir Rhoser, Tir-y-dail, Trap,* Tre-garn, Tŷ-croes, Tŷ-hen, Tŷ-isaf, Tŷ-mawr, Tŷ'r-frân, Tre-lech

ty- honorific prefix to pers.ns.

Llan-dawg, Llandyfaelog, Llansteffan

tyddyn, tyn nm., 'smallholding, small farm', earlier ? 'building'

Aberarad, Annell, Bocs, Cuch, Lacques, Llanwrda, *Maenor Tregelyn, Moefre*[3], Pontyfenni, Tyn-y-ffordd

tyle [< ?Ir *tulach,* 'hill, knoll'] nm., 'slope, ascent, hill'

Tachlouan, Telych

tyn[1] adj. 'tight, stretched; dense (of growth etc.)'

?Gelli-ddu

tyn[3] see **tyddyn**

PLACE-NAMES OF CARMARTHENSHIRE

tywod, variant *tyfod* nm., '(stretch of) sand; sand'

Afon Fach Pontgarreg

tywyn² nm. 'beach, seashore, sand-dune'

?Pentowin, Pentywyn, Porth Tywyn

uchaf adj., 'highest; higher (of two)'

Aber-giâr, Berach, Brechfa, Brisgen, Brynaman-uchaf, *Bryngwynne,* Capel Dewi³, Capel Gwynllyw¹, Castellcosan, Cathargoed, Cefncaeau, Cil-y-gell, Cilymaenllwyd, Crachdy, Cwm Cathan, Cwm-ffrwd, Cwm-waun-gron, Cynnull-bach, Dol-gwm, Dryslwyn, Felin-gwm Uchaf, *Glyn¹, Gronw, Gwestfa Blaen-twrch, Gwestfa Ystradmynys, Halfway³, Hengil, Iscoed²,* Llanybydder, Maenor Cilau, Maenor Cilhengroes, **Maenor Ddwylan Uchaf,** Maenor Lleision, Maenor Tegfynydd, Maenor Tre-goch, Miawst, Moelfre², Penddeulwyn, Pengwern, Pentywyn², Pistyll, Rhandir Uchaf, Trostre, Wen

uchel adj. and nm. 'high, tall'

Aber-giâr

udd nm. lord, chief'

Iddole

uisc OW

Wysg / Usk

ulmētum Br

Elfed

up E 'up, upper, higher'

upon prep. 'on, upon'

upper ME 'upper, higher'

Upper Brynaman

urbs gen. *urbis*, Latin 'town, city'

Llanddowror

ust, *hist* nmf. 'hush; a hush, silence'

?Nant Hust

uwch, *uch¹* prep. 'above, on top of, beyond; above' and adj. 'higher'

Emlyn, Llanfihangel-uwch-Gwili, *Malláen,* Uwchcoed Morris, Uwch Sawdde

vacca Latin nf. 'a cow'

Iddole

vale E 'valley, vale'

valley ME (*valey* AFr) 'valley'

Swiss Valley

***walis** W borrowing from ME **weall(e)s**

Alltwalis

wall pl. *walls* see OE **weall**

way E see **weg** OE

weall OE, **wealles**, *walles* pl., 'wall'

Alltwalis, Macrels

weg OE, *way* E, 'a way, a road'

Broadway¹,², *Glanyfferi*

well E see **wella** OE

wella, *wælle, wella, wielle* OE 'a well, a spring, a stream'

Macrels, Merthyr Cynog

west OE 'west, western'

East Marsh

-wg suffix

Maenor Meibion Seisyll, Nant Corrwg

white E see **hwīt** OE

-wy¹ r.n. suffix 'bending, turning'

Nant Rhydw, ?Nant Tawe, ?Ydw

ŵyr nm., pl. 'grandson, grandchild, descendant'

Gwestfa Wyrion Idnerth, Gwestfa Wyrion Ieuan ac Wyrion Seisyll

y, **yr** before vowel and h- '**r** after a vowel, def. art. 'the'

ych¹ nm. *ychen* pl., 'ox'

Cilyrychen

-ydd¹ territorial suffix, '(land) belonging to, territory of'

-ydd³ noun suffix

ymenyn, **menyn** nm. 'butter'

Pant-y-caws

ymryson¹ nmf. 'contention, strife, dispute'

Glossary of Place-Name Elements, Personal Names and River-Names

Halfway²

yn, yng, `n (after a vowel) prep. 'in'
Castellnewydd Emlyn, Llanymyddyfri, Tir-y-dail

-yn suffix
Login

ynys nf., *-oedd* pl., 'island; river-meadow'
Alltyferin, Blaenynys, Llandeilo'r-ynys, Machynys, Maenor Brwnws, Morfa², ?Ysgwyn

ysbydwr nm. '(Knight) Hospitaller, hospitaller', pers.n. or epithet
Pont Sbwdwr

ysbyty nmf. 'hospital, hospice; lodging-house for pilgrims (esp. one of the establishments of the Knights Hospitallers)'
Ysbyty

ysgawen nf., pl. *ysgaw*, 'alder-tree'
Bronysgawen

ysgolhaig, *sgoláig* nm., *ysgolheigion* pl., 'scholar, learned person; cleric; clerk'

Gwestfa Ysgolheigion

ysgothi, *sgothi* vb. 'to defecate, scour (of animals); eject, squirt'
Cothi

ysgubor nf., *ysgubor(i)au* pl., 'barn, granary'
Tor-y-coed, Ysgubor-fawr

ysgyfarnog nf. 'hare'
Cwmysgyfarnog

ystlys nmf. 'side, flank, edge, bank (of river)'
Gwyddyl

ystrad nm., 'vale, valley bottom, river-valley; wider part of a valley'
Gwestfa Ystradmynys, Llangennech, Strade, Ystrad¹,², Ystrad-ffin, Ystrad Tywi

ystum nmf. 'curve, meander (of a r, etc)'
Ystumgwili

ywen nf., pl. *ywenni*, coll.n. *yw²*, 'yew-tree'
Capel yr Ywen

PLACE-NAMES OF CARMARTHENSHIRE

List of personal names and surnames

Comprehensive lists of Welsh personal names are lacking and the following list is compiled from the dictionary entries. Many of the individual names can be identifed in other historical records, particularly pedigrees and poetry, and other place-name studies. Conjectural names are prefixed *. Brief passing references to personal names, eg. Brychan, are omitted.

Abram, Abraham	Pont Abram
Aeron	goddess of battles: ?Nant Aeron
Ann	?Cwm-ann
Antwn, Anton	Pontantwn
?*Arddneu* or ?*Arthneu*	Llanarthneu
Arthur	Trecastell[3]
*Ballteg	?Llanfallteg
?*Begewdin*	?Capel Begewdin
Bifan, Bevan	Twyn
Bleddyn	Gwestfa Bleddyn
Blodeuedd, Blodeuwedd	Maenor Cilcenawedd
*Borion	Maenor Forion
*Breuan	?Trefreuan
*Brwnws	Llandeilo'r-ynys, Maenor Brwnws
Brynach	Capel Brynach
Cadfan	Capel Cadfan
Cadog	Llangadog[1, 2]
Cadwal	Cydweli
Cadwgan	Blaenau[1] Gwestfa Cadwgan ap Cynon, Gwestfa Cadwgan ap Tegwared, Maenor Cadwgan
Caeo	Caeo[1, 2]
Cain	Capel Ceinwyry, Llan-gain, Maenor Gain
Callan	Cil-y-gell
Can, Cann	Llan-gan
Carnwall	Carnwyllion
Cathan	Llangathen, Myhathan
Cathen	Catheiniog, Llangathen
Cati	Gelligati
Caw	Cwmcawlwyd
Ceidrych	Dyffryn Ceidrych
Ceinwyry	Capel Ceinwyry
Ceitho	Llanpumsaint
Celer	Llangeler
Celynnin	Llanpumsaint
Cenawedd	Maenor Cilcenawedd
Cennech	Llangennech
Clear (< Clair), *Clêr*	Sanclêr, St Clears
Clinton	Traean Clinton
Cloufan	?Tachlouan
Clydwen	Llanglydwen

Glossary of Place-Name Elements, Personal Names and River-Names

Clydwen	Llanglydwen
Co(i)nin Ir	Llangynin
Coker	?Capel Coker
*Cosan, -m	Castellcosan
Curig	Eglwys Fair a Churig
Cyffig	Cyffig
Cynan	Nant Cynnen
Cyndeyrn	Llangyndeyrn
Cynein, Cynen	Nant Cynnen
Cynfab	Capel Cynfab, *Capel Newydd*
Cynfyn	*Maenor Cilgynfyn*
Cyngar	*Maenor Grongar*
Cyngen	*Nant Cynnen*
Cynheiddon	Llangynheiddon
Cynin ?< OW *Cunein*	Cynin, ?Eglwys Gymyn, Llangynin, Tregynin
Cynnen	Nant Cynnen
Cynnwr < Cynfor, -wr	Capel Cynnor, Llangynnwr, *Maenor Gynnwr*, Tregynnwr
Cynog	Llangynog, *Maenor Tregelyn*, Merthyr Cynog
Cynon	*Gwestfa Cadwgan ap Cynon*
Cynwyl	Caeo (Cynwyl Gaeo), Cynwyl Elfed
*Cynys	?Machynys
Darog	Llanddarog
Dewi	Capel Dewi[1-3], *Dewi Fawr*
Dingad	Llandingad
Distinnic	Cywyn
Durant AFr, Fr	Castelldwyran
?Dyddgen	*Capel Dyddgen*
Dyddgu	Capel Dyddgu
*Dyfnych	?*Maenor Meddyfnych*
Edi	Llanedi
Edwin	Edwinsford
Efwr	Dinefwr
*Egin	?*Cantref Eginog*
Egwad	Egwad, Llanegwad, Llanfynydd
Einion, Einon	Gwestfa Cadwgan ab Einion, *Maenor Tre-goch*
Elfed	Cynwyl Elfed, ?Elfed
*Elfyw	Mabelfyw
Elgan	Traethnelgan
Elidir	Gwestfa Gruffudd ab Elidir
Elli	Llanelli
Euddogwy, latinised Oudoceus	Llan-dawg
Goronwy, Gronw	Gronw
Gruffudd	Gwestfa Gruffudd ab Elidir
*Gryngar	?*Maenor Grongar*
Gwgan	*Blaenau*[3], *Maenor Cadwgan*
*Gwidigada	?Gwidigada
Gwilym	Capel Gwilym Foethus, Rhydwilym

PLACE-NAMES OF CARMARTHENSHIRE

Gwineu	Gwestfa Llywelyn Gwynnau
Gwinio, Gwynio	Llanwinio
Gwion	Gwestfa Gwion Sais a Maredudd ap Heilyn
Gwrdaf	Llanwrda
Gwrfyl	*Capel Cynfab*
Gwrthefyr	*Castelldwyran*
Gwrthwl	*Maesllanwrthwl*. For the conjectural pers.n. v. TYP 383
Gŵydd(i)ar (< OW *Guidcar*)	Gwydderig. For instances see EANC 190 and TYP 390-1
Gwyn	*Llanpumsaint, Llanwinio, Llanwrda, Pumsaint,* ?*Ysgwyn*
Gwyndaf	Llanwrda
Gwynfyw	Gwestfa Llywelyn Gwynfyw
Gwynllyw	Capel Gwynllyw[1,2], *Llangadog*[1]
Gwynnau	Gwestfa Llywelyn Gwynnau
Gwynno	Llanpumsaint
Gwynno	Llanpumsaint
Gwynoro	Llanpumsaint
Haearn	Llanhernin
Harvey	Cilcarw
Heilyn	*Cilellyn*, Gwestfa Gwion Sais a Maredudd ap Heilyn, Gwestfa Llywelyn ap Heilyn
Heli	Castellheli, Castle Ely
Henri, Henry	Cwrt Henri, Ffynnonhenri, Pont-henri
Herbert	Treherbert
Hernin, Hiernin (? < *Heyernin*)	?*Cilhernin*, Hernin, Llanhernin
Hoedlyw	*Crachdy*
Hopcyn, Hopkin	Hopkinstown, Trehopcyn
Howard	Parc Howard
Hwplyn	Cwm Hwplyn
Hywel, Howel	*Capel Llanddu*, Cwmcarnhywel, Cwmhywel, Dolhywel
Iago	Capel Iago, *Cilymaenllwyd*
Iddole	Iddole
Idnerth	Gwestfa Wyrion Idnerth, *Iddole*
Ieuan	Capel Ifan[2], Gwestfa Wyrion Ieuan ac Wyrion Seisyll
Ifan, Iwan	Bryn Iwan, Capel Ifan[1,2], Capel Iwan, Esgairifan
Ifor	Cwmifor
Ioan	Tre Ioan
Iorath, Ieroth, Iorwerth	*Llanfihangel-ar-arth*
Isaac	Capel Isaac
Ismael, Ishmael	Llanismel, *Llandyfeisant*, St Ishmael
Iwan see *Ifan*	
Jenkin	see Siencyn
John	Capel Ifan[1,2], Capel Iwan, Johnstown
Laundry	Llandre
Llaen	Malláen
Llawddog, Llawyddog	Cenarth, Llanllawddog
Llawen	Llanllwni
Lleision	?Maenor Lleision
Llewenni	?Llanllwni

Glossary of Place-Name Elements, Personal Names and River-Names

Llonio	Llanllwni
Lluan	Capel Llanlluan
Llwydog	Llanllawddog
*Llyddgen	Capel Dyddgen
Llywelyn	Capel Gwilym Foethus, *Dyffryn Ceidrych*, Gwestfa Llywelyn Gwynfyw, Gwestfa Llywelyn ap Heilyn
*Llywenni	?Llanllwni
Mabon	Maenor Fabon
Madog	*Maenor Castell Madog*
Maelog	see *Tyfaelog*
Mair	Capel Mair[1-3], Eglwys Fair a Churig, Eglwys Fair Glan-taf, Llanfair-ar-y-bryn, *Llan-y-bri*
*Mallteg (or Ballteg)	?Llanfallteg
Maquerel Fr	?Macrels
Maredudd	Gwestfa Gwion Sais a Maredudd ap Heilyn
*Maren	?Cilmaren
*Margain	?Cilfargen
Maurice	*Abergwili*
*Meddyfnych	*?Maenor Meddyfnych*
Meilir	*Llanfihangel-uwch-Gwili*
Merin	*Alltyferin*
Middleton	Neuadd Middleton, Middleton Hall
Mihangel	Capel Mihangel, *Cil-y-cwm*, Llanfihangel, Llanfihangel Aberbythych, Llanfihangel Abercywyn, Llanfihangel-ar-arth, Llanfihangel Cilfargen, Llanfihangel Rhos-y-corn, Llanfihangel-uwch-Gwili
Miles	Cwm-miles
*Moilet	?Tremoilet
Môr	*Coedmor*
*Morbri	?Llan-y-bri
Moreiddig	Gwestfa Moreiddig
Morgan	Cwm-morgan, Pentre Morgan, Traean Morgan
Moris, Morris	Iscoed[4] (Morris), Treforis, Uwchcoed Morris
*Mouwen	?Maenor Fouwen
Mynys	?Mynys, ?Ystrad
Myrddin	*Caerfyrddin, Carmarthen*
Natali	*Llanddeusant*
Nicol	Bryn Nicol
Non	Llan-non
Nonni or None	Capel Maesnonni
Nonnita	Llan-non
Nyniaw	Llanwinio
Oudoceus see *Euddogwy*	
Owain, Owen	Gwestfa Owain ap Rhydderch, *Maenor Cilnawen*, Rhydowen
Padrig, Patrick	Cefn Padrig
Paulinus	*Capel Peulin, Peuliniog*
Pemberton	Pemberton

lxxix

PLACE-NAMES OF CARMARTHENSHIRE

Petheloc	*Llanddeusant*
Peulin	Capel Peulin, Peuliniog
Rhoser	Tir Rhoser
Rhydderch	Gwestfa Owain ap Rhydderch
Rhys	*Pontantwn, Pont Sbwdwr*
Roche, de la	Roche Castle
Sadwrn	*Llanddowror*, Llansadwrn
Sadyrnin	Llansadyrnin
*Saeran, Saran	?Trimsaran
Samuel	*Llansawel*
*Sân	?Cilsân
Sannan	Sannan
*Sawyl, Sawel < *Safwyl < Sawl*	Llansawel
Sbydwr	Pont Sbwdwr. See **ysbydwr** in Glossary
Seisyll	Gwestfa Wyrion Ieuan ac Wyrion Seisyll, Maenor Meibion Seisyll
*Sïen	Sïen. Possible variant of *Sïan*
Siencyn	Felindre[3]
Silin	Capel Sant Silin
Steffan, Ystyffan	Llansteffan
Stephen, Steffan	*Llansteffan*
Sulfyw	Bachsylw
Sylen	?Mynydd Sylen, Sylen
Tanglwyst	Tanglwst
Tegwared	Gwestfa Cadwgan ap Tegwared
Teilo	*Brechfa*, Capel Teilo[1, 2], *Llanddowror*, Llandeilo, Llandeilo Abercywyn, Llandeilo'r-ynys, *Llandeulyddog*, Maenordeilo
Teulyddog	Llandeulyddog
Thomas	*Myddfai*[1]
*Trechgwyn	?Trechgwynnon
Tudwystl	Capel Tydyst
Tybïe, Tybïau	Llandybïe
Tyddawg	?Llan-dawg
*Tydyri	?Llandyry
Tyfaelog	Llandyfaelog
Tyfái, Tyfeisant	Llandyfeisant
Tyfân	Llandyfân
Tysilio	Llandysilio
*Uderydd	Mabudrud
Watcyn, Watkin	*Trap*
Yates	Pontyates
Ysfael	Cwmisfael, Llanismel
*Ystlwyf (?< *Ystrlwyf)*	?*Ysterlwyf
Ystrwyth	Ysterlwyf

Glossary of Place-Name Elements, Personal Names and River-Names

List of river-names

Some stream-names are likely to be purely descriptive, eg. Nant Melyn (***melyn***) and Nant Pen-y-cnwc are clearly qualified by an adjective and by a place-name, but this is not always the case. It is sometimes difficult to separate descriptive elements from names which qualify ***afon*** or ***nant***. Nant Eiddig, for example, may possess Eiddig as the river-name or ***eiddig*** as a qualifying adjective. For the purposes of this list, qualifying elements such as Melyn, Pen-y-cnwc and Eiddig are treated as *names*. Where the rn. contains an element which is clearly a qualifier such as ***bach*** and ***mawr*** then the rn. is found in the general glossary above.

Aerau	Nant Aerau
Aeron	Nant Aeron
Aman	Aman, Ammanford, Brynaman, Brynaman-uchaf, Cwmaman, Glanaman, Glynaman, Pontaman, Rhosaman, Rhydaman
Annell	Annell, *Glanyrannell*
Arad	Aberarad
Araeth	Pontbrenaraeth
Arwy, Arrow	*Capel Penarw*
Asen	Asen
?Bai	Nant-bai
?Ban	?Cwm-ban
Barddu	Barddu
Bargod	Maenor Aberbargoed, Nant Bargod
Bawddwr	Llanymddyfri, Llandovery
Bedol	Fforest Bedol
Bele	Bele
Berach	Berach, *Capel Cefnberach*
Berem	Pontyberem
*Beyr	?Pen-boyr
Blotweth	Blotweth
Brân	Brân[1, 2], *Cefn-bryn-brain*, ?Cwmdwyfran, Glanbrân
Branddu	*Aberbranddu*
Bre	*Groesffordd*
Brechfa	Nant Brechfa
Breuan	*Groesffordd*
Brianne	Llyn Brianne
Brydan	Glanbrydan
Burry	see Byrri
*Bydderi	Llanybydder
Byrri, Burry	Burry Port
Bytherig	Llanybydder
Bythych	Llanfihangel Aberbythych
Cadnant	Nant Aerau
Camnant	Camnant
Carfan	Cwm-waun-gron
Carwed	Carwe/Carway

lxxxi

PLACE-NAMES OF CARMARTHENSHIRE

Cathan	Cathan, Cwm Cathan
Cawlwyd	Cwmcawlwyd
Caws	Nant-y-caws
Cedi	Cynnen
Ceidrych	Dyffryn Ceidrych
Ceiliog	*Aber-giâr*
Cennen	Abercennen, Carreg Cennen, Cennen, *Cwm Cynnen*, Is Cennen
Cib	Tre-gib
Cloidach	Cloidach
Clydach	Clydach[1, 2]
?*Cochen*	?Nant Gochen
Corran	Talacharn[2]
Corrwg	Nant Corrwg
Cothi	*Brechfa*, Cothi, Cwmcothi, Dolaucothi, Glyncothi, Maenor Rhwng Twrch a Chothi, Pontargothi
Crychan	Crychan
Crychddwr	*Crychan*
Crychiau	Nant Crychiau
Crychnant	*Crychan*
Cuch	Cuch, Cwm-cuch, *Emlyn, Maenor Blaen-cuch*
?Cwm-merydd	Nant Cwm-merydd
Cwm-waun-gron	Cwm-waun-gron (earlier Carfan)
Cyngar	?Maenor Grongar
Cynin	Cynin, *Eglwys Gymyn*
Cynnen	Cwm Cynnen, Nant Cynnen
Cyw	*Aber-giâr*
Cywyn	Cywyn *Llandeilo Abercywyn, Llanfihangel Abercywyn*
Dafen	Dafen
Dâr	Nant Dâr
Dewi	Dewi Fawr
Dresglen	Nant y Dresglen
Duad	Cwmduad
Duar	Duar, Glan-duar
Dulais¹	Dulais¹
Dulais²	Dulais²
Dulais³	Dulais³, Pontarddulais
Dulas	Dulas
Dunant	Dunant
?*Dwyfran*	?Cwmdwyfran
Dyfod	*Fach Pontgarreg*
Dyfri	*Llanymddyfri*
Eiddig	Nant Eiddig
Eregwm	see Yregwm
Fach Pontgarreg	Fach Pontgarreg
Fanafas	Fanafas
Fanagoed	Fanagoed
Felys	Nant Felys

Glossary of Place-Name Elements, Personal Names and River-Names

Fenni	Fenni, Pontyfenni
Fferws	*Marlais²*
Ffin	*Nant-y-ffin, Ystrad-ffin*
Garenig	Nant Garenig
*Garllegan	?Maenor Garllegan
Garnant	Garnant
Giâr	Aber-giâr
*Glasne	?Aberglasne
Glwydeth	Glwydeth
Gochen	*Afon Fawr*
Gorlech	Abergorlech, Gorlech
Grân	Dol-gran
Gronw	Gronw
?Gwair	*Castellheli*
Gwenffrwd	Gwenffrwd
Gwenlais	Capel Gwenlais, Gwenlais, *Pentregwenlais*
*Gwenlliw, *Gwynlliw	Capel Gwynllyw¹
Gwennol	Nant Gwennol
Gwili	Abergwili, *Betws Ystumgwili*, Cwmgwili¹,², Glangwili, Gwili¹, Llanfihangel-uwch-Gwili, Ystumgwili
*Gwthwl, *Gwrthwl	?*Maesllanwrthwl*
Gwyddyl	Gwyddyl
Gwydderig	Gwydderig
Gwythwch	Nant Gwythwch
Hernin, Hiernin	?Hernin
Huerdic	*Marlais²*
Hust	Nant Hust
Iâr	*Aber-giâr*
Lash	Lash
Llethryd	Cwmllethryd
Lliedi	Lliedi
Llin	*Blaen-nant-llin*
Llwchwr, Loughor	*Llangennech*, Llwchwr
Marlais¹	Marlais¹, Cwm-marles
Marlais²	Marlais²
Marlais³	Marlais³
Marlais⁴	Marlais⁴, Abermarlais
Marlais⁵	See *Gronw*
Melyn	Nant Melyn
Melys	Nant Felys
Merchon	Merchon
Merin	*Alltyferin*
Merydd	Maenor y Merydd, ?Nant Cwm-merydd
Mihartach	Mihartach
Morlais	Is-morlais, Morlais
Mudan	*Ydw*
Myddyfi	Myddyfi
Mynys	*Gwestfa Ystradmynys*, Mynys

PLACE-NAMES OF CARMARTHENSHIRE

Nawen	*Maenor Cilnawen*
Nenog	Llidiardnenog, Nenog
?*Oernant*	?Cwmoernant
Pedol	Fforest Bedol, Nant Pedol
Pen-y-cnwc	Nant Pen-y-cnwc
Pib	Pib
Pibwr	Nant Pibwr, Pibwrlwyd
Pysgotwr	Pysgotwr
Regwm	?Regwm
Rhaeadr	Nant y Rhaeadr
Rhydw	Nant Rhydw
Sannan	Sannan
Sawdde	Blaensawdde, Felindre[1] (Felindre Sawdde), Sawdde, Uwch Sawdde
Sefin	Glansefin
Selach	*Dyffryn*
Sïen	Sïen
Sylgen	Sylgen
Taf	Eglwys Fair Glan-taf, *Fforest Gaerdydd / Cardiff Forest, Hendy-gwyn, Llanddowror*, Taf
Talog	Talog[2]
Tawe	Esgairdawe, Nant Tawe
Teifi	Teifi
Thames	Nant Thames
Tigen	Tigen
Tren	Glan-tren
?*Treuddyn*	?Nant Treuddyn
Tridwr	Nant Tridwr
Trysgyrch	*Felindre*[4], Maenor Cwarter Trysgyrch
Twrch	Cwm-twrch, Gwestfa Blaen-twrch, Maenor Rhwng Twrch a Chothi, Twrch[1, 2]
Tyweli	Pont-tyweli, Tyweli
Tywi	Glantywi, Tywi, Ystrad Tywi
Wysg, Usk	Usk, Wysg
Ydw	Ydw
**Yregwm, Eregwm*	?Regwm

Aberarad to Ashfield

Abergorlech. The former Wheaten Sheaf public house c.1935.

Aberarad   SN 316402
'mouth of (r.) Arad': *aber*, rn. *Arad*
Aberarad 1563-1762, Abarorarde, Abererard 1558
The rn. may contain the el. *ar-* found in *aradr* 'plough' or *aradr* with colloquial loss of *-r* describing a r. which ploughs through the landscape (EANC 100) comparable with Gwachell (*gwäell* nmf. 'knitting-needle, skewer') and Rhymni, cos. Glamorgan and Monmouth (*rhwmp*[1] 'auger'). There are some doubts, however, with this explanation because forms recorded in 1558 and 1574 have *arard* rather than *arad*. Possibly the third *-r-* has intruded as an echo of the first two. The name is recorded as that of a ho. *Tethin Aber Arad* (Cenarth) 1573 and *Tythin Aberarad* 1629 (*tyddyn*). The village developed near Castell Newydd Emlyn/Newcastle Emlyn workhouse.

Aberbranddu   SN 704456
Caeo
'mouth of (stream called) Branddu': *aber*, rn. *Branddu*
Aber Branddy 1603, Aberbrandddy 1559, Aber Brandû 1819, Aber-bran-ddu 1834
A hill west of the stream is *Keven Branddy* 1605, *Cefn Bran ddû* 1819, *Cefn-bran-ddu* 1834 (*cefn*). Branddu, literally 'black crow' (*brân*, *du*) perhaps one which is 'dark and dips like a

crow', must refer to the small unnamed stream which enters Afon Cothi here. A ch. or chp. is recorded here in 1836 in an unreliable source (WSS 329) and no supporting evidence has been found.

Abercennen ?SN 629215
Llandeilo Fawr
'mouth of (r. called) Cennen': *aber*, rn. **Cennen**
(ch. at) *Aberkennen* 13th cent (1332), (fair) *ffair Maes Aberkennen* 1609
The fair (***ffair***) was held on 11 November evidently on an adjoining field (***maes***). The same source (SDL) in 1609 records another fair in the same ctf. Is Cennen at Llandybïe. Abercennen is no longer found on OS maps and appears to have been replaced by Ffair-fach (q.v.).

Aber-giâr SN 501409
Llanllwni
'mouth of (stream called) Iâr: *aber*, rn. ***Giâr***
(mess. etc) *Aber gyar a Chilog* 1632, *Aber-cwm* 1831, *Aber-Iâr* 1888, *Aber-Giâr* 1905
Nant Iâr rises (SN 513388) on the hill Pen Llwyn-uchel (***pen***[1], ***llwyn***, ***uchel***) less than a mile from Nant Ceiliog and they meet (SN 500411) north of the modern ht. It is possible that the first historical mention specifically applies to the confluence of Nant Ceiliog with Teifi near the ho. Aberceiliog (SN 499423) recorded as *Abercylog* 1811, *Aber-ceiliog* 1831. If so, then Iâr a Cheiliog meaning 'hen and cock' (***iâr***[1], ***ceiliog***, *ceilog*) referred to the stream below the confluence of Iâr and Ceiliog. The form *giâr* has a prosthetic *g-* based on the common supposition that it has been lost by len.; cf. *gallt* for *allt*. The modern ht. developed where the Llanybydder-Caerfyrddin/Carmarthen road crosses the valley (***cwm***) of Nant Iâr extending eastwards to Pont Ceiliog (***pont***) over the stream Ceiliog. Iâr is otherwise recalled in the ns. of hos. *Clyn-y-iar-isaf, ~ -uchaf* 1831 and *Blaencwmyar* 1810, *Blaen-cwm-iar* 1831 (***blaen***, ***clun***[2], ***isaf***, ***uchaf***). Iâr and Ceiliog have identical rns. in co. Brecon (PNDH 13.83, 87) but their union there creates Nant Cyw, 'stream of the chick' (***nant***, ***cyw***).

Aberglasne SN 581221
Llangathen
'mouth of (stream called) ?Glasne': *aber*, lost rn. ?***Glasne***
Aberglasne 1619, *Aberglasney* 1635, 1720, *Aber Glaseney* 1668, *Aber Glasney* 1811
There seems to be no record of any stream called Glasne or Glasnai here unless it is a lost n. of the small stream which rises in a pool near Llangathen church meeting Tywi (SN 581214) less than 1km south of the mansion. R.J. Thomas (EANC 25) compared Glasne, apparently a former n. of Cletwr Fach, in Llandysul, co. Cardigan, and suggested ***glas*** and a suffix ***-nai*** as a variant of ***-ai*** comparable with ***dellni***, *delli* and ***noethni***, *noethi*.

Abergorlech SN 586337
chp. Llanybydder
'mouth of (r.) Gorlech': *aber*, rn. **Gorlech**
Abergorluch 1411, *Abergorlewch* 1544, *Abergorlech* 1567, 1740, *Abergorlech Chap.* 1742
Located where Gorlech (q.v.) enters Cothi. Notable for a chp. (***capel***) recorded in 1552.

Abergwili SN 437210
'mouth of the (r.) Gwili: *aber*, rn. **Gwili**
Aberwili, Aberguili 1222, *Abergwely, Abergwily* 1270, *Abergwely* 1281, (ch. of) *St. Maurice, Abergwylly* 1331, *aber gwyli* c.1566
Located where Gwili (q.v.) meets the r. Tywi. The home of a former collegiate ch., palace and lp. of the bishop of St Davids, and a former borough with a market and fair. Abergwili was one of the principal chs. dedicated to Dewi/David in Wales and the poet Gwynfardd Brycheiniog refers to *aber gwyli* in his Canu y Dewi, thought to date to the 1170s (CBT II, 26.94n). Abergwili also appears in Latin form as *in ostio guili* late 13th cent (***ostium*** 'mouth of a river').

Abermarlais SN 691294
Llansadwrn
'mouth of (r.) Marlais': *aber*, rn. **Marlais**[4]
(*Trafn*) *Abermarlais* c.1400, *Abbermerles, Abber Marles* 1532, *Abermarlesse* 1553, *Abermarles*

1615, 1736, *Abermarleys* 1669, *Abermarlais Park* 1852-3

Also a manor but best known for its mansion (HCH 3-4) described by the antiquarian John Leland c.1537) as a 'well favourid stone place motid, new mendid and augmented' by Sir Rhys ap Thomas (1449-1525) and possessing a park (mentioned in 1644). The r. rises at Blaenmarlais (SN 674346) (*blaen*) and meets Tywi at (SN 69582889). The first ref. has *trafn¹* nmf. 'leader, lord' or 'home. dwelling'. There are at least five rs. Marlais in co. Carmarthen including **Marlais**¹⁻⁴ (q.v.).

Aber-nant SN 339231
'mouth of the streams': *ebyr* pl. later *aber* sg., *nant*
Hebernat 1264, *Obernaund* 1266, *Ebernant* 1290-c.1700 (frequent), *Abernant* 1535, 1671, *ebyr nant* c.1566
The n. may have the less specific sense of 'area or valley where streams meet'. *ebyr* has been displaced by *aber* probably under the influence of pns. such as Abergwili and Abermarlais. Several small streams meet the brook Henllan just above the village.

Achddu SN 447016
Pen-bre
'black ?scar': ?*ach⁵*, *du*
Hachthy 1549, *hachthey* 1551, *Achthey* 1567, *Achthû* 1696, *Achddy* 1742, *Achddu* 1745, (hos.) *Ach-ddu-isaf, -uchaf* 1833
The area is now part of Porth Tywyn/Burry Port. The suggested meaning is tentative since *ach⁵*, pl. *achau*, does not have this sense in literary sources (see DPNW 451 for Talachddu) but it could describe a r. cut deeply into the landscape and cf. Blaenachddu (SN 299371) at Cilrhedyn recorded as *Blaenachddu* 1811 (*blaen*) where it appears to refer to a small unnamed tributary of Afon Mamog. The same el. seems to occur in the rn. Achau recorded in Cwmachau (SO 010380), co. Brecon, as *Cwm Achey* 1812, *Cwm-achau* 1832 (*cwm*).

Afon Fach Pontgarreg SN 3225
rn. SN 315280 to Afon Cywyn SN 324249
'small river at Pontgarreg': *afon, bach¹* len. *fach*, ?pn. *Pontgarreg*
Afon Fach Pont-garreg 1889
Pontgarreg is 'stone bridge', *pont, carreg* but there is now no br. over the r. apart from the footbridge at Dinas. The r. rises above Blaendyfod (SN 314275) recorded as *Blaendyfod* 1831 (*blaen*) which suggests the r. is properly Dyfod probably meaning 'sandy river': *tywod, tyfod* '(stretch of) sand; sand'.

Afon Fawr SN 3330
rn. SN 330330 to Nant Gochen SN 351303
'big, greater river': *afon, mawr* len. *fawr*
Afon Fawr 1889
Earlier forms have not been identified. The comparison is with smaller side-streams or with the upper reaches of Nant Gochen (q.v.). Afon Fawr may be properly Afon Gochen since *afon* typically refers to larger rs. and *nant* to smaller rs. and streams. Nant Gochen now applies to the entire r. from SN 343329 down to Afon Duad but it may once have applied only to that part of the stream down to its junction with Afon Fawr.

Allt SN 555025
Llangennech
'wooded-slope': *allt*
(coalworks at place) *Allt Llangennech* 1609, (commons) *the Allt* 1665, *Allt* 1892
Applied in 1879 and 1907 to a line of hos. on the south slopes of a small hill and along the road leading from Salem Baptist chp. to the hos. Corn-hwrdd (SN 549026) (*corn, hwrdd*). The current OS Landranger extends it to the area around the chp. and Nant-y-gro and to hos. along Troserch Road.

Alltwalis SN 445317
Llanfihangel-ar-arth, Llanllawddog
'wooded-slope at the walls': *allt*, *walis
Tire Allt dwalais 1624, *Tyr Allt y walysh* 1637, *Altywallis* 1719, *Allt y wallis* 1765, *Allt Wallis* 1831, *Troed-Allt-Walis* 1873

The first historic form has *tir* 'land'. The pn. is now applied on OS maps to the locality recorded as *Troed-y-rhiw* 1831 (***troed, y, rhiw***) and is taken from a fm.n. Alltwalis (SN 45053212) near the road leading towards Llanybydder. Melville Richards (AMR) suggested the second el. is a pers.n. Wallis but ***walis** occurs elsewhere in pns. such as Wallis (SN 0126), co. Pembroke, at Ambleston (PNPemb 397), (field) *Park y Wallis* (at Sanclêr/St Clears) 1753 (***parc***), *Magwr y Walis* 1653 (***magwyr***) (at Fishguard: PNPemb 56), Clun-Wallis (SN 537334, at Llanfihangel Rhos-y-corn) recorded as *Cleen Wallis* 1584, *Klyn Wallys* 1626 and *Clyn-Wallis* 1887-9 (***clun²***), and Glynwalis (SN 593068, at Llanedi), recorded as *Glynywalis* 1714, *Glyn Walis* 1723, *Clun Wallis* 1830 (***glyn***). It is very likely that we have a local W borrowing from ME *weall(e)s* (OE ***weall***) found in English Pembrokeshire in the form *wall* and *wallis* and in Welsh Pembrokeshire as *walis*. A medieval settlement is recorded at SN 4733-33624 which may account for the 'walls'. The hilly area around the farm and modern village is *Mynydd Gallt Walis* 1729 (***mynydd***) (BowenNSW). The pn. has also been explained, partly on the basis of the first historic form, as ***allt*** with an unrecorded rn. **Dwalais < Dywalais*, 'fierce stream': ***dywal***, dialect *dwal*, and ***glais*** len. *lais*, possibly referring to the stream Nant Alltwalis, but no early historic forms of the latter have been found.

Alltyferin SN 523236
Llanegwad
Uncertain
Tir-Alt-veryn 1669, *Galt y Verin* 1713, *Alltveryn* 1736, *Galltyferin* 1813, *Gallt-y-ferin* 1831, *Allt-y-ferin Farm* 1888-9
The first el. is ***allt*** 'wooded-slope' with def. art. ***y*** but the third el. is uncertain. A rn. Merin is recorded in co. Cardigan and there is a very small stream next to the ho. Plas Alltyferin (***plas***) though there is no evidence that this was ever called Merin or Berin. A pers.n. **Merin** possibly derived from Latin *Marīnus* is noted by R.J. Thomas (EANC 210-1; PNCrd 1262; CO³ no. 123) is doubtful because of the def.art. ***y*** found in most historic forms. This is unlikely to be intrusive because it also appears in *ynys y veryn* 1681 and *Ynysyferin* 1813 (***ynys***), the n. of a former ho. located north of the motte-and-bailey castle (SN 522233; HER 689). Melville Richards (WATU 6) records Dinweilir as an alternative n. for Alltyferin but see Pencader. The n. Alltyferin was also borne by a mansion (SN 518227) built in 1869 and demolished after World War II (HCH 6).

Allt-y-gaer SN 572210
Llangathen
'wooded-slope at the fort', ***allt, y, caer***
Athelgair, Athelgar 1353, *Altegar* 1432, *Althegar* 1532, (queen's manor) *allt-y-gaer or Kethinog* 1596, (manor) *Alltigaer* c.1597, *Alltygare* 1786, *Allt-y-gaer* 1831
A t. and manor and the location of courts for Catheiniog (q.v.). There is a wooded slope immediately north of the ho. The nearest fort shown on OS maps is Gron-gaer (SN 573216) nearly 1 km away.

Aman
rn. Llandeilo Fawr, Llandybïe
'river which roots and dashes like a pig': ***amanw***
Aman 1203, *ammen* 1222, *Amman* 1538, *Aman* 1574, *Amond flu:* 1578, *Amone* 1583, *Ammanfach, fawr* 1831
The n. might describe the nature of the r.; cf. Twrch, cos. Brecon and Montgomery, Hwch, co. Brecon, and especially Banw, co. Montgomery (***twrch, hwch, banw²***). See also Rhydaman/Ammanford, Cwmaman and Glanaman.

Amgoed
cmt. (with Peuliniog)
'(area) around woodland': ***am¹, coed***
(cmt.) *Amgodde* 1282, (cmt.) *Amgodde* 1282, (demesne land) *Amgoyt* 1310, *Amcoyt* 1382, *Comm. Amgoet* c.1538
A former cmt. which was generally linked with Peuliniog which together lay between the r. Cleddau Wen/Eastern Cleddau and r. Cynin and defined on the south by the r. Taf. The two became the basis of the AN lp. of Sanclêr/

St Clears which was later divided into 'thirds' (***traean***) (HCrm I, 11; ETG 49).

Ammanford see **Rhydaman**

Annell, Afon　　　　　　　　　　SN 6537
rn. SN 710426 to Cothi SN 644355 Caeo
?'little silvery river': ***afon, ar(i)an, -ell***
yr Annell 15th cent, *yr anell* 1566, *yr Annell* 1668, *Rhannell river* c.1700, *Rannell Brook* 1729, *River Annell* 1887, *Afon Annell* 1906
Earlier Ar(i)annell > Annell through association of Ar- with the def.art. *yr* (before a vowel). Historic forms of the rn. are uncommon but see Glanyrannell and note Afon Annell (to Tywi SN 465206) in Abergwili recorded in *Tythyn pen pont yr annel* 1760 (***tyddyn, pen¹, pont***) and as *River Annell* 1889 also recalled in Glanyrannell (SN 479228) recorded as *Glanrannell* 1729 (***glan***). Many historic forms bear close comparison with rns. in co. Brecon and elsewhere (EANC 92-96).

Asen, Afon　　　　　　　　　　SN 2628
rn. SN 260315 to Afon Cynin SN 261287
'donkey river, river likened to a donkey': ***afon, asen¹*** or perhaps ***asen²***
Afon Asen, (ho.) *Aber-Asen* 1889
Earlier refs. have not been found apart from in the ho.n. Aberasen (SN 261287) recorded as *Aberasen* 1810 and *Aber-asen* 1831 (***aber***). ***asen²*** has a number of extended meanings including 'rib of a boat or basket' but 'donkey river' seems likelier in view of the numerous rns. containing animal ns. The r. might have been perceived as slow but strong like a donkey.

Ashfield　　　　　　　　　　SN 693286
Llansadwrn
'field or enclosure having ash-trees': E ***ash, field***
Ashfield Farm 1887, (ho.) *Ashfield* 1901, (ht.) *Ashfield* 1996
The n. shifted from the fm. (SN 691287) to a row of hos. shown as *Glansefin Row* 1887 and 1891 (see Glansefin) but *Ashfield Row* 1906, 1977-82.

Babel to Bynea

Brechfa. Forest Arms Hotel and St Teilo's church c.1930.

Babel SN 833356
Llanfair-ar-y-bryn
'(ho. named) Babel': ***Babel***
Babel 1869, (building) *Babel* 1888, 1947
Babel is probably derogatory and now applies to a building located near a small stream and access roads to the hos. Efail-fach and Trallwng. The n. first appears some time between 1839 and 1869 and in 1905 applied to a post office. The n. was also borne by a nearby former school now known as Babel Hall. It is possible that the n. once applied to an unrecorded Nonconformist meeting ho. though Babel as a chp. n. is uncommon. Capel y Babel (***capel***) is recorded at Cwmfelin-fach, near Caerffili, and there was a Babel Calvinistic Methodist chp. at Groes-faen, Glamorgan.

Bacau, Backe SN 260156
Sanclêr/St Clears
'(the) ridges': ***bac*** pl. *bacau* dialect *bace*
Backey 1718, *Baccê* 1807, *Bacce* 1826, (hill) *Waun Bacau*, (area) *Pen-y-bac* 1831, (hos.) *Backe*, (ho.) *Pen-y-back* 1888
The ho. Pen-y-bac, now Penback on OS maps, is recorded as *Pen y back* 1663 (***pen¹***, **y**) with identical examples in Llandysilio, Llansteffan and Pen-bre. The n. describes a scatter of hos. on hill slopes leading up to exposed pasture (***gwaun***) north of Afon Taf. ***bac*** is a local borrowing from E ***back*** cf. the use of ***cefn***.

Bachsylw SN 165257
Cilymaenllwyd p.

'nook of (man called) Sulfyw': ***bach²***, pers.n. ***Sulfyw***
Backsulnew 1325, *Bakesilio* 1590, (ht.) *Bach y Sulw* c.1700, *Bachsylw* 1819
See also Maenor Bachsylw. The pers.n. is attested as that of a saint in 12th cent sources with a ch. dedication at Llancillo (***llan***), co. Hereford (HERPns 145). There is, however, no proof of any connection between the saint and Bachsylw; Sulfyw is more likely to be a secular n. Some later forms have an intrusive -*y*-, probably by association with the def.art. *y* and supposition that -*sylw* is a common n.

Bach-y-ffrainc SN 2222, 2422
Llanboidy
'nook of the Frenchmen': ***bach²***, ***y***, ***ffranc¹*** pl. *ffrainc*
Bâch y ffraingc, Bach y ffraink Issa, ~ ucha c.1700, *Bach y ffraink* 1738
Historic sps. are late and the suggested meaning canot be regarded as certain. See also Maenor Bach-y-ffrainc.

Backe see **Bacau**

Bae Caerfyrddin, Carmarthen Bay
'bay near Carmarthen': ***bae¹***, **Caerfyrddin**, E ***bay***, **Carmarthen**
Carmarthen Bay 1747, 1794, *Caermarthen Bay* 1811, (*yn*) *mau Caerfyrddin* 1888, *Machwy Caerfyrddin* 1898, *Bae Caerfyrddin* 2016
Named simply from its proximity to Caerfyrddin/Carmarthen, a port of considerable importance in both local and international trade down to the 19th century, but unable to cope with large ships owing to sand and mud banks in Afon Tywi. The bay is otherwise recorded as *Tenby Bay* c.1695 from the town in co. Pembroke. The form for 1898 has ***machwy*** 'bay (part of sea)' and may be purely literary.

Bancffosfelen SN 488121
Llangyndeyrn
'bank near Ffosfelen': ***banc¹***, pn. *Ffosfelen*
Bancffosfelen 1876, *Banc-ffos-felin* 1880-7, *Bankffosfelen* 1903, *Banc-ffos-felen* 1906
The n. earlier applied to the hos. around crossroads but now also describes development along the Pontyberem road towards Llwynypiod. Ffosfelen is *Ffos-felyn* 1831 and *Ffosfelin* 1845 referring to a ho. (at or adjoining the former Wesleyan chp.) meaning 'yellow-brown ditch' (***ffos, melen***). Some forms suggest association with ***melin*** but there is no record of a mill here. The land is poor and waterlogged; note nearby Trallwm (SN 486118) (***trallwng***).

Bancycapel SN 430151
Llandyfaelog
'ridge at the chapel': ***banc¹, y, capel***
Bankycapel 1858, *Banc-y-capel* 1888
The n. probably refers to a former parochial chp. Llangynheiddon (q.v.) rather than the Methodist chp. built in 1834.

Bancyfelin SN 323180
Llanfihangel Abercywyn
'ridge at the mill': ***banc¹, y, melin***
Bankyfelin 1767, (meeting ho.) *Bankyfelyn* 1780, *Banc y felin* 1808, *Banc-y-felin* 1831, *Bancyfelin* 1851
banc¹ describes raised ground above Afon Cywyn. The OS plan in 1889 shows a corn mill at the main crossroads (SN 323180) and a mill-race. The location of a Calvinistic Methodist chp. (HMGC II, 536).

Banc-y-ffordd SN 409377
Llangeler
'bank of the road', ***banc¹, y, ffordd***
Bankyffordd 1885, *Banc-y-ffordd* 1889, 1964, *Bancyffordd* 1980-1
Early evidence is lacking but the ht. developed near a ho. Bancyffynnon (*Banc-y-ffynnon* 1831) 'bank at the well' (***ffynnon***), on the unclassified road running south from Llandysul towards Llanpumpsaint.

Barddu, Afon SN 3034, 3134
rn. SN 317340 to Cuch SN 293349
'river dark as soot'?: ***afon***, ?***parddu*** len. *barddu*
Afon Barddu 1889-90
The r. is in a deeply-cut, wooded valley often in shadow and it is possible that ***parddu*** is a

fanciful description for a river as dark as soot but historic forms are too late for certainty.

Bargod see **Nant Bargod**

Bele, **Afon** SN 3632
Cynwyl Elfed rn. SN 336344 to SN 376312
'river in an area notable for pine martens': *afon*, *belau¹*, *bele²*
Afon Bele 1889
An identical rn. is recorded in Llanfihangel-ar-arth as *Blen pele* 1608, *Blan bele* 1632, *Blaen bele* 1780 (*blaen*) and Bele, co. Montgomery, recorded as *Beleu* 1420, *Bele* 1661. Both rs. might describe wooded areas where the pine marten or some similar predatory animal is found or perhaps allude to the nature of these rs. leaping and darting like a pine-marten; cf. Twrch.

Berach (Berrach) SN 580190
'?'pointed like a spear': ?*bêr*, *-ach*
Berrach 1777, 1811, *Berach* 1831
Named from a stream *Berach* 1831 which rises in Blaenberach (*blaen*) and runs through the valley Cwm Berach (*cwm*) to the r. Bythych. Historic evidence is too late for certainty but 'spear, lance' might be a fanciful description of its shape or more likely the manner in which the stream is observed as cutting and piercing the landscape; cf. Rhymni, cos. Glamorgan and Monmouth, with *rhwmp¹* 'auger' (PNGlam 195-6). I think we can rule out **bêr* as perhaps in *llifeiriaint* 'a flowing, flood' since this is such a common one of rs. and streams. The rn. has also been explained as Ir *bearach* 'pointed' describing a r. in a narrow valley tapering to a point at its source or one in which there are sharp pointed rocks (EANC 3). Berach is the name of a stream rising on Mynydd Du/Black Mountain (SN 683162) and joining Aman (SN 673138) in Glanaman but here again the evidence is late, recorded in ho.ns. *Cwmbarach isaf* and *Cwmbarach uchaf* 1831 (*isaf*, *uchaf*). Later OS maps from 1889 refer to this stream as *Nant Berach* (*nant*). Both rs. Berach bear comparison with Barach found as a rn. in co. Caernarfon and probably as the final el. in *Parkey Barrach* and *Parke yr barrach* (*parciau*) in co. Pembroke. For the latter Charles (PNPemb 32, 746) favoured *bar(r)*, ie. *bar²* with the ?Ir suffix *ach* possibly indicating a place where there were lumps or banks. This cannot be ruled out because Barrach > Berach is a possible outcome of dissimilation.

Berth-lwyd SN 6129
Llangathen
'(the) grey bush': (*y*), *perth*, *llwyd*
Berthloet 13th cent (1331), *Berth-lwyd-uchaf*, ~ *-isaf* 1831, (ht.) *Berthlwyd* 1863
A very common ho.n.; cf. Berth-lwyd, in Llandeilo Fawr, recorded as *y Berth lwyd* 1628, *Berth Llwyd* 1734. Berth-lwyd was the location of property of Talyllychau abbey.

Bertwn SN 376069
St Ishmael
'barley farm, demesne farm': ME *barton*
Berton 14th cent, *Bertwn* 1798, 1871, *Bertwin* 1808, *Burton* 1811, 1830
Recorded in some sources as *Burton* 1811-1889 reverting to the cymricised form Bertwn before 1906. The ho. has been replaced by a country club and caravan park. Also the n. of a former quay on the Gwendraeth estuary.

Berwig SN 548988
Llanelli
'outlying farm': ME *berwick*
Berwicke, *-e*, alias *Verwicke* 14th cent, *Burwyk* c.1501, *Burwyg* 1542, *Burwig* 1543, *Burwik* 1545, *Vurwig* 1551, *Burwige* 1598, *Bwrwig* 1646, *Berwick* 1790-1852, (ht.) *Berwicke*, otherwise *Verwicke* 1752, *Berwig* 1833, *Berwic* 1880-1992
A former ht. sometimes described as a manor; see Maenor Berwig. Berwig possessed its own chp. variously known as Capel Berwig and Capel Gwynllyw (q.v.). OE **bere-wīc** earlier meant 'barley farm' but later developed the sense 'outlying grange'. Cf. Ferwig (DPNW 149).

Bethlehem SN 684252
Llangadog
Bethlehem 1851
Named from an Independent chp. (SN 687249) erected 1800, rebuilt 1834 (HEAC III, 563-6; RC I, 294). The village developed around Dolgoy, Bancyfedwen and Groesffordd Inn. A fairly common chp. n.; cf. Bethlehem, at Sanclêr/St Clears, built in 1793.

Betws SN 632117
'house of prayer, chapel of ease': ***betws***[1]
Betwse 1536, *y betws* c.1550, *Bethos alias Striveland* 1609, *Betws* 1767, *Bettws* 1831
The paucity of early refs. may be attributed to it having once been a chp. to Llangyfelach, co. Glamorgan. Betws was co-terminous with Maenor Betws, which earlier formed part of Gower/Gŵyr, and maintained ecclesiastical connections with Llangyfelach and the deanery of Gower down to the 17th cent. ***betws***[1] is ultimately of OE origin but passed into W vocabulary before the 13th cent., often with the specific sense of chapel-of-ease, subordinate church or chapel.

Betws (Llandyfaelog) see **Llangynheiddon**

Betws Ystumgwili SN 409264
Abergwili later Llanpumsaint
'house-of-prayer in Ystumgwili', ***betws***, **Ystumgwili**
Bettus 1758, (fm.) *Bettws* 1811, 1889, *Bettws Ystum Gwili* 1836
The chp. may be identified with the rectangular cropmark near the fm. Bettws. Traces of what were supposed to be those of a chp. were found in a field *Park Penygar* (SN 4086 2667) (***parc***) within which a circular – probably prehistoric – enclosure has been recorded north of the ho. The fns. *Park Penygar* and *Glan Pengare* (SN 4098 2678) (***glan***) together with the adjoining *Pengare* (SN 4099 2667) and *Llether Penygar* (***llethr***), recorded in the p. tithe apportionment 1838, favour a fort. *Pengare, Penygar* clearly stand for Pen-y-gaer, 'top of the fort' or 'hill of the fort' (***pen***[1], ***y***, ***caer***).

Betws see **Capel y Betws** (Tre-lech a'r Betws)

Bigyn SS 511997
Llanelli
'point, peak': (***y***), ***pigyn***[1]
(field) *the biggin* 1735, *Pigyn* 1831, (ho.) *Bigin* 1880, *the Bigyn* 1889, *Bigyn* 1903, 1907
The n. may describe the small hill on which the ho. stood or perhaps some lost mound – natural or artificial. It is noteworthy that Bigyn (SN 443031), in Pen-bre, recorded as *Bigin* 1880, is located near a Bronze Age burial mound. Cf. Castell Pigyn (q.v.).

Black Mountain see **Mynydd Du**

Blaenau[1] SN 600140
Llandybïe
'uplands; upper reaches of rivers': ***blaen*** pl. *blaenau*
y Blaeneu c.1700, *Blayney* 1729, *Blaenau* 1757, *Blaenau, Blaenau-bach* 1831
The upland, hilly part of the p. and one of its eight hts. Also the n. of a ho. that belonged to minor gentry (HCH 12-13).

Blaenau[2]
Llanfihangel-ar-arth
'uplands; upper reaches of rivers': ***blaen*** pl. *blaenau*
Recorded as a t. by Melville Richards (WATU) but historic evidence is lacking. It may have applied to the southern, hilly part of the parish, probably around New Inn and Gwyddgrug where there are a number of ho.ns. including Blaenblodau (SN 469371), Blaen-gwen (SN 460339) and Blaen-gwyddgrug (SN 483351) incorporating the el. ***blaen***.

Blaenau[3] SN 482151
Llangyndeyrn
'uplands; upper reaches of rivers': ***blaen*** pl. *blaenau*
the Blayne 1589, *Y Blayne* 1622, *y Blayney* 1647, *The Blayney* 1657, *Blyney* otherwise *Codwgan* 1704, *Blaeneu* 1811, (ho.) *Blaenau* 1831

The n. applied to a t. and the ho. (grid ref. above) near Gwendraeth Fach upstream of the p.ch. The ho. is otherwise known as Coedwgan, recorded as Cadwgan in Maenor Cadwgan (HCH 20), *Coedwgan* 1716 and 1791, a n. now confined to two hos. Cadwgan Fawr, ~ Fach a little to the west recorded as *Cadwgan-fawr, ~ -fawr* in 1831, meaning 'Gwgan's wood' (**coed**, pers.n. **Gwgan**) qualified as 'great, greater' and 'little, lesser' (**mawr**, **bach**[1]). Late forms indicate association with a distinct pers.n. **Cadwgan**.

Blaenau[4]   Blaenau-bach SN 560107
Llan-non
'uplands; upper reaches of rivers': **blaen** pl. *blaenau*
Blaenau c.1700, 1831, *Bleyne* 1804, *Blaenau, -mawr, -bach* 1831
The n. is applied in 1831 to the area centred on SN 563117 near Pen-twyn but once described the low hills around Bryn-mawr (SN 557111) between Llan-non and Cross Hands. The n. of two fms. distinguished as 'lesser' (**bach**[1]) and 'greater' (**mawr**).

Blaen-nant-llin   SN 313357
Cenarth
'headwaters of Nant-llin': **blaen**, rn. or valley n. *Nant Llin*
(mess.) *blaennant y llin* 1756, (ho.) *Blaen-nant-llin* 1831, 1889
The n. applies to a ho. in Groesffordd and was the name of a t. probably corresponding with the former East Cilrhedyn cp. Nant Llin is not current but may have referred to the short valley (**nant**) of the stream which runs northwestwards to meet Afon Mamog near Abergeletsh (SN 309361). Llin could be **llin**[1] in the specific sense 'line, streak, groove' as a description of the valley or less likely **llin**[2] 'flax'. Abergeletsh is *Aber-gelatch* 1831, *Abergelech* 1840, *Aber-gelech* 1889, 1964, *Abergeletsh* 1979. Geletsh probably refers to the stream rising at SN 320363 and may have **geletsh**, a dialect form of **elestr**, *gelestr* 'sword-flag, fleur-de-lis, iris; lily', with **aber**.

Blaensawdde   SN 784239
Llanddeusant
'headwaters of (r.) Sawdde', **blaen**, rn. **Sawdde**
Blaen Sawdde c.1700, *Blaen Sawddey* 1744, *Blaensawdde* 1788, *Blaen-Sawdde* 1887
The ho. is located on a hill-spur between Afon Sawdde and Afon Mihertach but as the n. of a t. must have once applied to a wider area extending up to the source of Sawdde in Llyn y Fan Fach. Blaensawdde was the legendary home of a widow's only son sent by her into the mountains and who sought out and married the Lady of the Lake in the 'Legend of Llyn-y-Van-Vach'. The story is recorded in 1841. Another ho. Blaenau (SN 793240), so spelt 1831, about a mile further upstream has the looser sense of a ho. near headwaters.

Blaen-waun   SN 236272
Llanwinio
'head of the moor', **blaen**, (**y**), **gwaun**
Blain y Wain 1754, *Blenwaun* 1811, *Blaen-y-waun* 1839, *Blaen-waun* 1843
A small linear ht. along the Meidrum-Tegryn road near another road forking towards Llanglydwen. This was the location of Moriah Independent chp. (recorded 1807). The moorland is also recalled in the ho.ns. *Waun-fawr* (SN 243272) and *Waun-fach* 1843 (**bach**[1], **mawr**).

Blaen-y-coed   SN 3527
Cynwyl Elfed
'head of the wood': **blaen**, **y**, **coed**
Blaenycoed 1755, *Blaen-y-coed* 1831
Located near the head of two wooded valleys leading eastwards down to Cynwyl Elfed. The n. contrasts with the ho.ns. Blaen-bryn (SN 346268), and Blaen-y-cwm (SN 350274) and Blaen-y-waun (SN 343264) (**bryn**, **cwm**, **gwaun**). The location of an Independent chp. in 1807.

Blaenynys   ?SN 5720
Llangathen
'head of the river-meadow': **blaen**, **ynys**
Blaenynys 1851, (ht. Llan and) *Blaenynys* 1863

The n. is not current and its precise location is uncertain though the ht. is usually combined in descriptions with Llan (q.v.) in the southeast of the p. of Llangathen. Its likeliest location was near Afon Tywi. The el. *ynys* occurs in the Ro-fawr (SN 580209) area in the tithe apportionment 1839.

Blotweth, Afon SN 5237
rn. SN 536391 to Clydach SN 521355
Uncertain
Blotweith 1584, (ho.) *Blodwedd* 1831, (stream) *Nant Blodwedd* 1888, *Afon Blodwedd*, (ho.) *Blotweth, Blotweth-fawr* 1905
Historic forms appear to favour **blotwaith**, *blodwaith* nm. 'meal-dust' but literary evidence for this begins only in 1776 (GPC) and it is difficult in any case to understand how this could apply to a r. A likelier explanation is that the rn. contains *blawd*² npl. 'flowers, blooms' and *gwaith*¹ nm. 'work, labour' or 'something which work has produced', a r. perhaps imagined as 'a product of flowers' or simply as one which is attractive and praiseworthy. It is worth noting Blodeuyn, a r. which rises just above a ho. Blodeuen (SN 480362) about 4km west of the ho. Blodweth (SN 528368) containing **blodeuyn**, *blodeuen* 'a flower'. Recorded in *Aber blodoyen* and *Blaen blodoyen* 1633 (**aber, blaen**).

Bocs, Box SN 515008
Llanelli
'box-trees': (*y*), **bocs**¹
Box (and Landshare) 1740, 1772, *gwaith glo y Bocs* 1820, *Box Colliery* 1813, 1830, *Box* 1880
The pn. is first recorded at a time when Llanelli was predominantly W-speaking so that E **box** is unlikely. The el. occurs elsewhere in fns., eg. *Box* 1789 (in Llangennech) and Box (SN 581014), Llandeilo Tal-y-bont, co. Glamorgan, recorded as *tee un e bockes* 1719, (mess. etc) *Box* 1770, *Tuy yn y Box* 1764 (**tyddyn, tyn**). The area developed near Box, Talsarnau and Llannerch collieries and especially after the establishment of the Poor Law union workhouse.

Brân¹**, Afon** SN 8042, 7532
rn. to Tywi SN 753327
'dark river (like a crow)': **afon, brân**
Brane River 1536-9, *the Brane* 1587, *Braen* c.1762
In Llandingad and Llanfair-ar-y-bryn. The n. might also describe its perceived behaviour, dipping and weaving like a crow. Formed by union (SN 807427) of Nant Cynnant Fawr and Nant Llwynor descending by way of Glanbrân (q.v.). The rn. is common throughout Wales.

Brân²**, Afon** SN 8123, 6928
rn. SN 812300 to 696284
'dark river (like a crow)': **afon, brân**
(r.) *Bran, Brane* 1754, *Bran River* 1811, *River Brân* 1887
As Afon Brân¹. In Llangadog. The lower reaches appear as *Sefin River* on the OS map 1831 which may be an inference from the ho.n. Glansefin (q.v.).

Brechfa SN 524302
Llanegwad
'pied, spotted place (near r.) Cothi': **brech, -fa**
Brechva before 1265 (1332), *Brechya Cothy* 1281, *Lanteilan Brechua* after 1271 (1332), *Brechva in Strateuwy* 1288, *Brezka* c.1291, *Brechwacothy* 1301, *Brechfa gothi* c.1566, *Llandilo brechva gothi* 1633, *Brechfa* 1788
The n. might describe an area littered with stones or marked by different types of vegetation and is sometimes applied to turf-cutting operations (CStudies 117). Generally known in full as Brechfa Gothi down to c.1800 with the rn. Cothi as a qualifer distinguishing it from Brechfa, in Llangeitho, co. Cardigan, recorded in fm.ns. *Brechfa Vach* 1664, *Brechfa-fâch, ~ -isaf, ~ -uchaf* 1891 (**bach**¹**, isaf, uchaf**) and Brechfa (Llandyfalle), co. Brecon, recorded as *Brechva* 1466. The ch. is dedicated to St Teilo; cf. Llandeilo (q.v.). For *Strateuwy* see Ystrad Tywi.

Brianne see **Llyn Brianne**

Brisgen SN 5828
Llanfynydd
?'(area of) shrubs': ?*prysgen*
Brisken 1733, 1794, *Briscen* 1851, *Brisken* 1863
prysgen nf. 'shrub, bush' has just one literary ref. dated 1805 (GPC) but nothing else seems likely and it is worth noting that a ho. (HCH 13) in this parish is recorded as *Brisken Helig* 1729, *Briskenhelyg* 1767, *Briskenhelig* 1795 (*helyg*¹ 'willow-trees'). The n. may be used in a collective sense for an area characterised by bushes. The n. survives in two ho.ns. (hos.) *Briscen-isaf*, ~ *-uchaf* 1831 (SN 581284, SN 584287) (*isaf, uchaf*).

Broadlay SN 371092
Llanismel/St Ishmael
'broad clearing', E *broad, ley*
Broadley 1599, 1871, (place) *Brode Lay* 1605, *Broadelay* 1616, *Broadlay* 1889
Possibly *Bradley* c.1447. Cf. Broadlay (SN 346133), in Llansteffan, recorded as *Brodle* 1762, *Broadlay* 1799, *Broadley* 1831, and Broadway (q.v.).

Broad Oak see **Derwen-fawr**

Broadway¹ SN 294101
Talacharn²/Laugharne, Llansadyrnin
'broad road': E *broad, way*
Broadway 1675, 1736, 1809, *Broad-way* 1831
Named from the road running south-westwards and westwards towards Pen-dein/Pendine. Also the n. of a 17th cent mansion ruinous in 1810, replaced by a new ho. before 1878 on a different site (HCH 14). The OS Landranger (2016) applies the n. to an area between Gosport, Talacharn²/Laugharne, and St John's Hill.

Broadway² SN 388088
Llanismel/St Ishmael
'broad road': E *broad, way*
Broadway 1889-90
On the road running north from Llan-saint to the Glanyfferi-Cydweli road south of Pen-y-banc 'top of the bank' recorded as *Pen-y-banc* 1831, *Pen-y-bank* 1889-90: *pen*¹, *y, banc*¹.

Bronwydd SN 417237
Llannewydd/Newchurch
'wooded hillside': *bronwydd*
Bronwydd 1831, the *Bronwydd Arms* 1856
A transferred n. taken directly from the inn Bronwydd Arms which is lower down the r. Gwili. The pn. now also applies to hos. near the former railway station hence the n. Bronwydd Station found on many OS maps. Bronwydd is the n. of a mansion – largely ruinous by 1983 – in Llangynllo, co. Cardigan, belonging to the Lloyd family

Bronysgawen SN 2125
Llanboidy
'hill-side where an elder-tree grow': *bron, ysgawen*
Vronskawen 1692 (1758), 1738, (ht.) *Bronyskawen* c.1700, *Vronskawen* 1738
ysgawen may be used otherwise in a collective sense. A parochial ht. named from a ho. recorded as *Bron yskawen* 1596 and *Bronyscawen* 1620 and the n. of a former ht. merged with Ffynnon-oer to form Kingsland (q.v.) before 1738. Located on slopes of a hill crowned by a prehistoric fort Y Gaer (SN 212263) (*caer*).

Brook SN 267094
Llansadyrnin
'brook, stream': E *brook*
Brooke farm 1673, *Brook* 1799, 1831
Located where a small stream reaches West Marsh behind Pendine Burrows. The n. of a chp.-of- ease to Laugharne from 1865.

Bryn SN 546007
Llanelli, Llangennech
'(the) hill', (*y*), *bryn*
le bryn 1446, *y bryn* 1548, *Bryn* 1717, *Brin* 1740, *Bryn Coal pit* 1833
Located on a low hill on the Llangennech-Llanelli road also recalled in the ho.n. *Brynbach* (SN 542011) 1833 'little hill' (*bach*¹), *Brynmawr* (SN 544004) 1880 'great hill' (*mawr*)

and *Bryn-cornel* (SN 544006) 1880 'hill at the corner' (***cornel***) in Llanelli. Bryn-mawr may be *Bryn mawr* 1645, *Brinmawre* 1707. The n. of a mansion (HCH 14-15).

Brynaman   SN 713143
Llandeilo Fawr, Cwarter Bach
'hill near (r.) Aman': ***bryn***, rn. **Aman**
(fm.) *Bryn-amman* 1831, *Bryn Amman* 1844, *Brynaman* 1872, *Brynaman, neu'r Gwtter Fawr* 1874
For the rn. see also Rhydaman. Named from the fm. which stood near the br. over the r. (SN 713138). The alternative n. Y Gwter Fawr, recorded as *Guter vawr* 1805, (hillside) *Gwtter-fawr* 1831 and *Gwter vawr* 1852, properly refers to the 'great gutter' (***cwter, mawr***) across the valley in Llan-giwg, co. Glamorgan. This is associated with coal-workings including (*the Big Vein at*) *Guterfawr* 1810 and Pwll y Gwter sunk in 1855 (Cwm Aman 50) (***pwll***). The actual village is said to have developed around the ironworks (later belonging to the Aman Iron Co.) begun in 1847 (HMorg 228) but it is evident that there was a settlement on the site of modern Brynaman before that date (PNGlamorgan 22). Rees and Thomas record that the Independent chp. here was opened in 1843 (HEAC II, 320). The name Gwter Fawr was apparently changed to Brynaman in 1864 when the Swansea Vale Railway Station was built (CAS VI, 83-4; HBryn 9, 10, 37). Lower Brynaman appears on OS maps from the 1890s displacing Glyn-y-beudy otherwise Glyn-beudy (***glyn, y, beudy***).

Brynaman-uchaf, Upper Brynaman   SN 718147
Llandeilo Fawr (Cwarter Bach)
'upper part of Brynaman': pn. **Brynaman**, ***uchaf, upper***
Upper Brynamman 1892, *Brynaman Uchaf* 1906
Ribbon development extended along the Llangadog road before 1877 north of Tinworks Row (on Heol Newydd/New Road) but the pn. does not appear until c.1890.

Bryn-du   SN 540094
Llan-non
'dark hill': ***bryn, du***
Brin dû 1626, *Bryn Du* 1745, *Bryndu* 1776, *Bryndû* 1813, *Bryn-du* 1831, 1960
The n. is taken from the ho. later applied to post World War II hos. extending eastwards to the A476.

Bryneglwys see **Henllan** (Caeo)

Bryngwynne   Bryngwynne Uchaf SN 593179
Llanfihangel Aberbythych
?'red-brown hill': ***bryn***, ?***gwinau***
(place) *brin gwiney* 1604, (mess. etc) *Bryngwynne, Bringwynne* 1724, *Bringwine*, (fm.) *Bringwine Issia* 1782, (ht.) *Brynn Gwineu* 1811, (hos.) *Bryn-gwynau-ganol,~ -uchaf, ~ -isaf* 1831, *Bryngwyne* [= Bryngwynne Uchaf], *Bryngwyne-canol, ~ -fâch* 1886-7
The hill is centred around the ho. Bryngwynne-canol (SN 576178) known as Brynawelon since c.1970. Richards (in WATU) uses the form Bryn-gwyn probably in the belief that -gwynne is an affectation comparable with Maesgwynne (SN 20238), in Llanboidy, properly Maes-gwyn, recorded as *y Maesgwyn* 1595, *Maesgwyn* 1736, *Maesgwynne* 1768 (***maes, gwyn***). The hos. are distinguished as 'upper', 'lower' and 'middle' (***uchaf, isaf, canol***).

Brynhafod   Brynhafod-fawr SN 593239
Llangathen
'hill near summer-dwelling': ***bryn, hafod***
Brynhavod 1626, *Brin-Havod* 1678, *Bryn-hafod* 1831, (capital mess.) *Brynhavod* (with water corn grist mill) *Brynhavod Mill* 1835, *Bryn-hafod* (= -fawr), ~ -fâch 1887
The two hos. are distinguished as 'great' (***mawr***) and 'little' (***bach¹***). Both are located on small hills above Pentrefelin.

Brynhyfryd   SN 5500
Llangennech
'pleasant hill': ***bryn, hyfryd***
Brynhyfryd 1959
An area of housing constructed in the 1950s and 1960s. The n. appears to be a new adoption

as it does not appear on earlier OS maps and in the Llangennech tithe apportionment 1844.

Bryn Iwan SN 316314
Aber-nant, Cilrhedyn
'Iwan's hill': ***bryn***, pers.n. ***Iwan***
(chp.) *Brynevan* 1854, *Bryn Iwan* 1869, (ht.) *Bryn Iwan* 1978-9
Named from the Independent chp. opened in 1852 (HEAC III, 410-1; Bryn Iwan 6-16) on land belonging to Crugiwan fm. The chp. is variously recorded as Capel Iwan, Capel Bryn Iwan and Capel Bryn-Evan but there seems to be no precise record of when the n. Bryn Iwan was extended to adjoining hos. It first appears on OS maps in 1978-9 though it was probably in use before 1924. The n. is drawn in part from an antiquity Crug Iwan (SN 312311) meaning 'Iwan's mound' (***crug***) applying to a former heap of stones (SN 312312) marking the p. boundary of Cilrhedyn with Aber-nant and equidistant from the fm.ns. Crugiwan Fawr (SN 321301) and Crugiwan (SN 307321). The first is otherwise recorded as *Crygywan* 1802 and *Crug-ieuan-fawr* 1831 and the second as *Plâs penrhyw yr evel als' Crig Evan* 1721, *Penrhyw yr Efel alias Cr g Evan* 1781, *Crug-ieuan* 1831, *Crug Evan* 1840 and *Crug-Iwan* 1843. Forms for 1721 and 1781 indicate an alternative n. Penrhiw'refail, 'top of the smithy slope' (***pen¹***, ***rhiw***, ***y***, `**r**, ***gefail***). Historical sources are very unclear with regard to the preferred form of the pers.n. varying between colloquial ***Iwan*** and ***Ifan*** (anglicised form *Evan*) and more formal ***Ieuan*** (E *John*); cf. Capel Iwan. Iwan is unidentified. The pers.n. is also found in a ho.n. Maniwan (SN 311307) recorded as *Maenieuan* 1831 and *Maen-Ieuan* 1889 recalling a stone (***maen***) marking the boundary of Aber-nant with Tre-lech a'r Betws.

Bryn Nicol SN 830441
mtn. Llanfair-ar-y-bryn
'hill of (man or family called) Nicol': ***bryn***, surname ***Nicol***
Bryn nichol 1832, *Bryn Nichol* 1887-8
There is also a ho. Gilfach Nichol, in Llanfihangel Aberbythych, recorded as *Kilfach nicols* 1799 and *Gilvach Nichol* 1822 (***cilfach***). A common surname as a variant of Nicholas in south Wales and the borders (WSN 187)

Bryn-y-beirdd Cwrt Bryn-y-beirdd SN 662181
Llandeilo Fawr
'hill of the bards': ***bryn***, ***y***, ***bardd*** pl. ***beirdd***
Bryn y beirdd yngharec kynen c.1550, (*terris vocatis*) *Brynybeyrd* 1609, *Bryn-y-Beyrdd* 1762, *Brynn y Beirdd* 1811
Centred on the ho. Cwrt Bryn-y-beirdd (HCH 17) recorded as *Court Brin y Byrdd* 1761, ('former British residence') *Cwrt Brynn y Beirdd* 1811 (see Fenton 57). The OS map 1831 calls the ho. *Cwrtpenybanc*, 'court (or grange) at the top of the bank' (***cwrt¹***, ***pen¹***, ***y***, ***banc¹***).

Burry Port see **Porth Tywyn**

Bwlchnewydd SN 369248
Aber-nant, Llannewydd/Newchurch
'new pass', ***bwlch***, ***newydd***
(meeting ho.) *Bwlch* 1741, *Bwlch Newydd* 1777, *Bwlchnewydd* 1831, *Bwlch-newydd* 1889
The village developed along the Caerfyrddin/Carmarthen-Cynwyl Elfed road in the 19th cent south of an Independent chp. The earlier n. for the settlement may have been plain Bwlch, a n. confined to a PH here (later Bwlch Stores) in 1889 and 1907; ***newydd*** may have originally referred specifically to the chp. The pass is either that at Plas-bach (SN 374254) on a minor road above the chp. or that near Bryneryl (SN 363254).

Bynea SS 545994
Llanelli
Uncertain
Banie 1597, *Baine* 1607, *Binie* 1772, *y Bynie* 1786, *Binea Farm* 1814, *Binyea* 1813, 1833, *Bynea* 1860
Pronounced [ˈbɪnjə] though this is at odds with the two earliest historic forms. The first one seems to favour *baniau* a pl. of ***ban¹*** perhaps with the specific sense of 'corner, angle' rather than the more common 'summits, peaks' etc. It is possible for *Baniau* > **Beiniau* > *Biniau, Binia,*

etc through analogy with common nouns such as **gwaith** pl. *gweithiau* and **ffair** pl. *ffeiriau* but no instances of *Beiniau* or anything closely similar have been found. Some late forms of Bynea bear comparison with nearby Glynea shown on 20th cent OS maps as (fm.) *Glynea* (SS 553991) and *Glynea Pit* (SS 548991) and earlier as *Glyne* 1546, *Gleine* 1750, *Gleinau* 1811, *Glynea* 1851. This might perhaps stand for *glein(i)au* pl. of **glain**[1] 'gem, precious stone' etc but quite what this might mean is as obscure as Bynea. I think we can reject *glyniau* 'valleys' as a pl. form of **glyn** which is inappropriate for this flat marshy area near Afon Llwchwr. Bynea earlier applied to Bynea fm. (SS 559985), next to part of the r. and was extended to Bynea Colliery (SS 556988) which opened in 1837 (IMLlanelli 48) then to the railway station on the Llanelli-Swansea line and Bynea Brick works (SS 552990). What is now shown on modern OS maps as the village of Bynea is given as *Cwmfelin* 1891-1953 (**cwm**, **melin**) – virtually an extension of Llwynhendy (CMLlanelli 34). Bynea gradually replaced Cwmfelin after 1911 since James Morris of Llansteffan treats Bynea and Cwmfelin as distinct places (HMSG 456).

Caeo to Cywyn

Cydweli/Kidwelly. Castle Street near former school looking towards the castle c.1910.

Caeo[1]
cmt., hd.
?'(land of man called) Caeo': pers.n. ?*Caeo kemmoto ... de Kaeoe* c.1191, *Caeav* 13th cent, (land, cmt. of) *Kayo* 1257-1379, *Kayou* 1282-1309, (land of) *Cayo* 1277, (cmt.) *Cayo* 1302, *Cayo* 1401 (early 15th cent), (cmt.) *Caeau* 1292, (cmt.) *Caio* 1439, *Gayo, Caeaw* c.1440-1, *Comm. Cayau* c.1537
The cmt. (***cwmwd***) covered the ps. of Caeo (otherwise Cynwyl Gaeo) and Llansawel with part of Talyllychau/Talley. Probably a pers.n. (ETG 51). Caeo is clearly a n. of long standing, as Lloyd says (HCrm I, 7), and is otherwise recalled in the n. of an unidentified locality at Pumsaint (q.v.).

Caeo[2] SN 676399
p.
'(church of Cynwyl in) Caeo': (pers.n. **Cynwyl**), **Caeo**[1]
Kenwell Cayo c.1291, *Kynewelle* 1302, *ecclesia Sancti Kynwil* 1331, *Kunwillgaeo* 1396, *kaeo plwyf kynnwyl* c.1500, *Conwyll Gayo* 1535, *Kayo, Kynwylgayo* 1544, *Cynwylgaio* 1724, *Conwil Gaio* 1766, *Cayo Village* 1786
Caeo, as the short form of Cynwyl Gaeo, seems to originate in the 18th cent with specific ref. to

the ht. and in its fuller form distinguishes the p. from Cynwyl Elfed (q.v.).

Cae'r-bryn SN 591138
loc. Llandybïe
'enclosure at the hill', *cae, yr, `r, bryn*
(parcel of lands) *kae yr bryn* 1614, *Caerbryn* 1762, (ho.) *Car-bryn* 1831, (ho.) *Cae'r-bryn* 1876-89, 1906
The n. is taken from an area on the eastern side of Cae'r-bryn Colliery (SN 594135) opened c.1870-5 and hos. now known as Cae'r-bryn Terrace (SN 598137) constructed c.1900. The n. was extended during the 1960s to include hos. built about the same time as Cae'r-bryn Terrace around an older ho. Blaenau-bach (see Blaenau). Cae'r-bryn referred earlier to a ho. (SN 589137) near Pen-y-groes.

Caerfyrddin, Carmarthen SN 4120
mb, cp.
'fort at Moridunum', *caer*, pn. *Moridunum*
Maridounon, Moridounon 2nd cent, *Muridono* ?2nd cent, *Chaermerthin* c.1130, *caer uyrtin* c.1215 (early 14th cent), *yg caer ua6r uyrtin* early 13th cent (14th cent), *(ciuitate) Chermerdi, Carmerdin* c.1170, *Kair Merdin* 1154, *Cairmerdin* 1159, *Kaermerdin* 1184, 1219, *Kaermerdyn* c.1191, *Kaer Uyrdin* 1109 (c.1400), *Kaervirddhyn* 1440, *kaer fyrðin* c.1566, *Kermerdyn* 1309, *Carmarthen* 1546, *Caerfyrthin* 1612, *Caermarthen, or Caer Fyrddin* 1811
Roman *Maridunum, Moridunum* is composed of Brit **mori-* and **dūno-*, **dūnon* meaning 'sea fort' (> *môr¹, din*), apparently influenced in Roman and later Latin sources by *mare* 'sea', producing OW **morddin > merddin, myrddin* (DPNW 71-2). The n. no doubt arose because of its location close to the head of marine navigation. The fort was eventually replaced by a town and in the post-Roman period the earlier pn. was prefixed with *caer* 'fort' (a fairly common descriptor for Roman towns) which accounts for len. *m > f* [v] in standard form. The second part of the n. *myrddin* was then associated by Geoffrey of Monmouth (c.1090-1155) in his Vita Merlini c.1150 with a pers.n. **Myrddin** found in older tales. This was latinised as *Merlinus*, Fr and E *Merlin* in the tales of King Arthur. These led to the false supposition that Caerfyrddin/Carmarthen was the 'town (or fortress) of Merlin' (CCH 33-34). This became a familiar theme in antiquarian discussions of the town right down to the 19th century despite the fact that the scholar William Camden had identified Caerfyrddin/Carmarthen with *Maridunum, Moridounon* in his publication Britannia in 1586. The supposed link with Merlin accounts for Merlin's Gate otherwise Porth Myrddin (SN 450206), Merlin's Grove (in Abergwili) and Bryn Myrddin (SN 449212) (E *gate, porth², bryn*), the n. of a large ho. built c.1858. The latter n. is taken from Merlin's Hill (SN 455205) recorded as *Merling hill* 1586, *Allt virthin* 1722, *Merlin's Hill* 1729, *Gallt Myrddin* 1831 and *Galltyfyrddin* 1838 (*hill, allt*).

Camnant, Afon SN 6720
SN 690219 to Cennen SN 676196
'crooked stream': *cam², nant*
(brook) *Camnant* 1609, 1754, *River Camnant* 1886-7
In Llandeilo Fawr and Llangadog (Gwynfe). The stream rises on slopes southwest of the hill Trichrug twisting southwards to Afon Cennen. A ho. near the stream is recorded as *Tir y Camnant* 1603 (*tir*).

Cantref Bychan
ctf. covering Hirfryn, Is Cennen and Perfedd cmts.
'little cantref': **cantref, bychan**
Canterbochan 1121, *Cantrebachan* c.1145, *Cantrefbochan, id est, kantaredo brevi* c.1191, (land of) *Canterbohan* 1289, *Cantretboghan* 1319-20, *y Cantref Bychan* c.1400, *Cantref Bachan* c.1537
'little' in contrast to Cantref Mawr (q.v.); the two ctfs. are thought to have been subdivisions of the territory of Ystrad Tywi formed before the 12th cent (HCrm I, 6). Centred on the castle Carreg Cennen (q.v.).

Cantref Eginog
ctf. covering Cedweli, Carnwyllion and Gŵyr
'cantref called Eginog': **cantref**, pn. *Eginog*

Cantref Eginoc 1559, *Eginoc* mid 16th cent
The meaning of Eginog is uncertain. Melville Richards (ETG 52) suggested that it might contain **egin** npl. 'shoots, sprouts' with the particular sense of 'descendants' or a possible pers.n. **Egin** with a suffix *-og*. The first might indicate 'territory belonging to descendants (of some unknown person)' and the second 'land associated with (man called) Egin'. The latter is more convincing in view of other territorial ns. such as Pebidiog, co. Pembroke, Cyfeiliog, co. Montgomery, and Brycheiniog. The late and sparse refs. to the n. raise some doubts with regard to its historical authenticity.

Cantref Gwarthaf
ctf.
'uppermost cantref': *cantref, gwarthaf*²
Cantrewartha c.1236, *Canntref Gwarthaf* 1216 (c.1400), *Cantref Guentha* c.1537
'uppermost' because it was a part of the ancient land of Dyfed (less Efelffre) (HCrm I, 6). The ctf. covered the cmts. of Amgoed, Elfed, Talacharn, Derllys, Penrhyn and Ysterlwyf.

Cantref Mawr
ctf. covering cmts. Caeo, Cetheiniog, Gwidigada, Mabelfyw, Mabudrud, Manordeilo, Malláen
'big, greater cantref', *cantref, mawr*
cantref maur c.1145, *Kantrefmaur* c.1191, *Cantremaur* 1222, *Cantrefmaur* late 13th cent, *Kantremaur* 1309, *Cantremawre* 1415
'big, greater' in contrast to Cantref Bychan (q.v.).

Capel SN 522006
Llanelli
'(area near) a chapel': *capel*
Capel 1950
The n. is taken ultimately from a chp. Capel Dewi³. The area was developed from c.1910. Misapplied on the OS Landranger to the area of Tyrfran and Llannerch (SN 512011).

Capel Abergorlech see **Abergorlech**

Capel Bach SN 444243
chp. Abergwili
'little chapel', *capel, bach*¹
(mess.) *Cappel bach* 1722, *Bach Chap* 1742, (ruins of) *Cappel Bâch* 1811, *Capel-bach* 1831
The chp. is shown on the OS 1889 as a chp.-of-ease on the site of an ancient chp and gives its n. to two hos. *Capel-bach-isaf, ~ -uchaf* 1889. Otherwise known as Capel y Groes, 'chapel with a cross' (*capel, y, croes*). A description of the consecration of this chp. in 1888 states that 'the building ... was really the re-erection of one of the old chapels which lay in a central position, with a grave-yard attached to it.... The walls of the old chapel were standing in the memory of some now [1916] living, and there remain in the churchyard two fine yew trees, a stone cross and traces of graves' (CAS XI, 10). An early stone incised with a cross is recorded as leaning against the ch. in 1912 (RCAHM V, 2, 6, 180), now in Carmarthen Museum. This may have functioned as a focus within a cemetery rather than marking a single grave (Corpus II, 201-2).

Capel Bach see **Capel Brynach** (Llanddarog)

Capel Bach y Drindod SN 31044073
chp. Cenarth; Castellnewydd Emlyn
'little chapel dedicated to the Trinity': *capel, bach*¹, *y, trindod*
Y Castel, Chapel to Kenarth 1786
Capel Bach y Drindod was built in 1780 on the site of a medieval chapel close to the outer earthworks of the castle also dedicated to the Holy Trinity and probably to be identified with *Capel y Castell* in 1733 (Cymmrodor 22: 55) referring not to a chp. in the castle (*castell*) but to a chp.-of-ease serving the town Castellnewydd Emlyn/Newcastle Emlyn. The builders of the new chp. were evidently aware of the older chp. and re-adopted its n. but at some point it became more widely known as Capel Bach y Drindod. The chp. was replaced by Holy Trinity ch. (built on a separate site in 1842-3) and pulled down after 1856

Capel Begewdin SN 511148
chp. Llanddarog
'chapel of Begewdin': *capel*, ?pers.n. **Begewdin**
Capel Bigewdyn 1884, *Capel Begewdin* 1974-93
The second el. is probably drawn from a ho. recorded as *Begewdyn* 1743, *y Bigewdyn* 1800, *Bigewdin* 1802, *Begowden* 1831, (ho.) *Bigawdin* 1888. A ho. Cefnbrisgen (SN 584295), in Llanfynydd, also bore the n. *Bigewdin* in 1841. The chp. is thought to be partly late medieval but mainly 16th cent (HER 638). Francis Jones states that the chp. is built over a well reputedly visited for strains (HWW 28). Its meaning remains uncertain but the def.art. in 1800 implies a pn. rather than a pers.n. as John Rhŷs suggested (AC 1894: 166. He was 'not inclined to think it Welsh'). The ruins of the chp. were described in 1884, 1894 and 1950 but no dating evidence was found. Local people believed it to be a Catholic chp., i.e. pre-Reformation. It is similar in character to Capel Erbach (q.v.).

Capel Betws see **Capel y Betws**

Capel Brynach
chp. Llanddarog
'chapel dedicated to Brynach': *capel*, pers.n. **Brynach**
(*Llanddarog* and) *Bronach chapel* 1552, (remains of old chp.) *St. Bernard's Chapel* 1811, *Hen Gapel St Bernard a Chapel Bach* 1898
This may also be Capel Bach listed by Richards (WATU) although Rice Rees (WSS 329) – an unreliable source – and Wade-Evans (Cymmrodor 22: 51) treat them separately. Capel Bach is 'little chapel': *capel*, *bach¹*. It is possible that Capel Brynach was simply a chp. in or attached to the p.ch. or that it is a duplication of Capel Begewdin. Brynach may be the Ir saint, for whom a 'Life' survives dated c.1200 (VSB 2-15; TYP 294), dedicated at Llanfrynach, co. Brecon, and Llanfyrnach, co. Pembroke (DPNW 252, 253).

Capel Cadfan
Llangathen
'chapel dedicated to Cadfan': *capel*, pers.n. **Cadfan**

Melville Richards (WATU) records this as a chp. in Llangathen p. but others refer to it as a chp. within the ch. and further evidence is needed. There is a ho.n. Llethrcadfan (SN 579231) recorded as *Llather Kadvan* 1611 and *Llethercadvan* 1736 (**llethr**), one mile north of the ch., but there seems to be no proof of a chp. there. A St Cadfan reputedly lived in the 5th cent. (LBS II, 1-9).

Capel Cadog see **Llangadog** (Cydweli)

Capel Cefnberach SN 568187
chp. Llanfihangel Aberbythych
'chapel at Cefnberach': *capel*, *Cefnberach*
(society of) *Ceven Berrach* 1791
Cefnberach means 'ridge at Berach': **cefn** and **Berach**. Berach (q.v.) also survives in the ho.n. (SN 580190) and a wooded area Allt Berrach (**allt**) on its northern side. A Calvinistic Methodist chp. was opened here (SN 568187) in 1747 (MethC I, 442; HMGC I, 385) or 1749-50 close to the road.

Capel Ceinwyry approx. SN 612291
chp. Talley
'chapel dedicated to Ceinwyry': *capel*, pers.n. *Ceinwyry*
(ruined chp.) *Cynhwm* 1742, 1786, (~) *Cynhwrn* 1763
The pn. is poorly recorded and the third ref. in 1763 simply repeats the first (with a variation in typesetting). Described in 1917 as Capel Llanceinwyryf and located on Danycapel farm (SN 612291), which is *Dan y Capel* 1729, 'where stone foundations are occasionally met with during ploughing' (RCAHM V, 264-5), meaning '(place) below the chapel' (**dan**). The pers.n. bears comparison with Ceinwyr in Llangeinwyr, co. Glamorgan (PNGlamorgan 123) containing the pers.n. **Cain** and **gwyry**, *gwyryf* 'a virgin' (ETG 175). Cain was a reputed daughter of the semi-legendary Brychan Brycheiniog. The pers.n. is recalled in the ho.n. Llwyncynhwyra (SN 608290), formerly a gr. of Talyllychau/Talley abbey, recorded as *Lankeinwyry* 13th cent (1331), *Llwyn Cyn hurra*

1745, *Llwyncynhwyra* 1828, and *Llwyn-cynhwyra* (SN 608290) 1831, 1889 (***llwyn***).

Capel Coker
chp. Cydweli
'chapel of (man called) Coker': ***capel***, pers.n. ?***Coker***
Historical evidence seems to be lacking though Melville Richards (WATU) records it as a chp. It is more likely to have been a private chp. named from Geoffrey de Coker, prior of Cydweli/ Kidwelly in 1301 (WSS 329). Geoffrey, or a member of his family, probably gave his name to the burgage in Cydweli/Kidwelly known as *Coker is Parke* in 1533.

Capel Crist ?SN 636327
chp. Talyllychau/Talley
'chapel dedicated to (Jesus) Christ': ***capel***, ***Crist***
(ruined chp.) *Crist [Holy Trinity]* 1742, *Capel Crist* 1836
Reputedly located in a small field *Mynwent Capel Crist* (1888) (***mynwent***) between Edwinsford Arms and a ho. Delfryn (RCAHM V, 264) in the village of Talyllychau. An uncommon n. for what was apparently a chp.-of-ease dedicated to the Holy Trinity in 1763 and 1836. All chs. and chps. belong to God or Christ though dedications are usually re-attributed to the Holy Trinity. Identical to Capel Crist, at Llannarth, co. Cardigan, recorded as (place) *Chappell Llangrest* 1592 (***llan***), *Capel Chryt* 1610, *Capel Crist* 1777.

Capel Cynfab ?SN 785388
chp. Llanfair-ar-y-bryn
'chapel dedicated to Cynfab': ***capel***, pers.n. ***Cynfab***
Llange'vab c.1291, *Llangenvab* 1552-3, *Cynfab Ch.* 1742, *Capel Cynfab* 1836
Melville Richards suggests that it has the n. of an obscure saint (ETG 216) presumably ***cynfab*** nm. 'eldest son, first-born' as an epithet. It is not known whether the chp. mentioned in 1742 still held services but it is probably significant that it is omitted in an index for 1763. The chp. has been identified with reputed ruins in a field called Cae Capel (SN 78573889) (***cae***, ***capel***) on the p. tithe map, on the lands of Cefnllan (***cefn***, ***llan***) fm. The latter, recorded as *Kevenllann* 1730, *Kevenllan* 1755, appears to confirm the existence of an ecclesiastical site in the vicinity. A well Ffynnon Gwrfil (***ffynnon***, ?pers.n. *****Gwrfyl***), credited with medicinal properties, was said to have lain in a field next to Cae Capel. This is *Cae gwrfil* 1839 on the north side of Cae Capel but the tithe plan and OS plans do not record a well. The n. Capel Cynfab was transferred to a school licensed by the bishop of St Davids in 1849 where services were held (*Capel Cynfal* 1851, *Capel Cynfab* 1903) and shown on OS maps from 1888. The chp. has also been identified with Capel Newydd (q.v.).

Capel Cynnor ?SN 482021
chp. Pen-bre
'chapel dedicated to (St) Cynnwr', ***capel***, pers.n. ***Cynnwr***
(chp.) *Cappell Kwnnwr* 1591, (ruined chp.) *K. Kynnor* c.1700
Recorded as *Capel Cynnor ym Mhendryn* (AC 1915: 331) which places it at or near Penrhyn (SN 482021). The ch. in Llangynnwr (q.v.) is dedicated to the same saint according to Melville Richards (ETG 216).

Capel Dewi[1] SN 475202
chp. Llanarthne
'chapel dedicated to Dewi', ***capel***, ***Dewi***
Saynt D'D is chaple 1552, *Capel dewye* 1578, *Chappell Dewie* 1586, *Capeldewye* 1607, (place) *Chappell Dewi* 1609, *Cappell dewy* 1710, (ruins of chp.) *Capel Ddewi, ~ Isaf* 1811, (fms.) *Capel Dewi-isaf, ~ -uchaf, ~ ganol, Ffynnon Dewi* 1831
Capel Dewi was a former chapelry of Llanarthne. The well *Ffynnon Ddewi* (***ffynnon***) is first recorded in the p. in 1699. Llanarthne also has another well dedicated to Dewi at Pistyll Dewi with a spring (SN 542196) recorded as *Pistill deivi* 1663 and *Pistilldewy* 1824 (***pistyll***).

Capel Dewi[2] SN 659178
chp. Llandeilo Fawr
'chapel dedicated to Dewi', ***capel***, ***Dewi***
Capel Dewi (Remains of) 1887-8
Traces of a small chp. located in a field recorded as *Cae'r Cappel* 1841, 'the chapel close' (***cae***, ***yr***,

`r, capel`) immediately north of Llwyn Dewi which is *Llwynddwi* 1811, *Llŵyn Dewi Farm* 1887-8, *Llwyn-Dewi* 1906 (***llwyn***).

Capel Dewi³   SN 52010059
chp. Llanelli
'chapel dedicated to Dewi', ***capel***, ***Dewi***
chaple of Saynt D'D 1552, *Capel dewye* 1578, *Capell Dewi* 1598, (chp. yard) *Chappell dewy* 1600, *Cappel Dewi* 1619, *Cappel Ddewy* 1729, *Dewi Chap. (ruined.)* 1742, (small ruin) *Chapel Dewy* 1807
This may also be the chp. of St David recorded between 1223 and 1250 in conjunction with Llanelli and Capel Ifan¹. It also appears to be *Capel (Remains of)* 1880 located near a fm. shown as *Capel-isaf* (recorded thus 1830-1954) and a short distance from *Capel-uchaf* (recorded as *Capeluchaf* 1813 and *Capel-uchaf* 1830-1954). A piece of wall in the yard of Capel-isaf was said in 1917 to be traditionally part of the chp. Investigations in 2007 and 2019 revealed evidence of medieval burials and a possible chapel. The two ns. simply indicate 'lower' (***isaf***) and 'upper' (***uchaf***) parts of hos. known collectively as Capel (q.v.). The site has been confused (LPC l.; RCAHM V, 119, no. 348) with Capel Gwynllyw² (q.v.) otherwise known as Capel Berwig, near Llwynhendy.

Capel Dolhywel see **Dolhywel** (Myddfai)

Capel Dyddgen   SN 465126
Llangyndeyrn
'chapel dedicated to ?Llyddgen': ***capel***, pers.n. ****Llyddgen***
Lanlothegeyn 1358, *Lanlothegeyne* 1493, *llanylydd gain* c.1550, (chp.) *Saynt Dethgen* 1552, *ll. llyðgen* c.1566, *S. Lethgenis Chaple* late 16th cent, *Capel Llanthithagyn* 1607, *Cappell Llanelithan* 1729, *Cappel Dyddgen* 1798, *Capel Duddgen* 1804, *Capel-dyddgen* 1831, *Capel Dyddgen (Ruins of)* 1906
The identification of earlier forms with Capel Dyddgen is made by Melville Richards (ETG 191, 216-7). The n. begins as Llanllyddgen 'church of Llyddgen' (***llan***) and gradually develops by way of Capel Llanllyddgen to Capel Dyddgen possibly through association with later forms of Capel Dyddgu (q.v.). The presumed pers.n. Llyddgen does not seem to be recorded elsewhere. The 1358 and 1552 refs. describe it as a chp. of Llandyfaelog. Llangyndeyrn was formerly under Llandyfaelog. The ch. was in ruins in 1742, its tower still standing but its walls largely disappeared in 1906 (CAS II, 168).

Capel Dyddgu   ?SN 51401 06684
chp. Llanelli
'chapel dedicated to (St) Dyddgu': ***capel***, pers.n. ***Dyddgu***
the chaple of Saynt Diddgye 1552, *Capel duthgye* 1578, *Capel duthgye* 1607
This seems to correspond with the chp. in Maenor Hengoed mentioned by a local historian David Bowen (DLlanelli 34; and Arthur Mee in LPC l). Bowen states that it was quarter of a mile from the farm Capel Sylen (shown as *Blaen-y-ddau-gwm* in 1880 at SN 512070) in the middle of a plantation on the side of Mynydd Sylen but he noted that it had left no surviving traces. The likeliest location is what is now shown as a nearly-complete rectangular enclosure in a former plantation (grid ref. above) south of Sylen crossroads but conclusive evidence is lacking. The plantation was cleared between 1907 and 1915. County mapmakers from Saxton 1574 down to Coltman 1794 show Capel Dyddgu as midway between Llanelli and Llannon which accords well with Capel Sylen and Maenor Hengoed. We cannot be sure when the chp. was abandoned because many later county maps plagiarise Saxton and contain obsolete details. The chp. should not be confused with the former Baptist chp. at *Capel-sulen* 1830 (SN 519067) or the ho. Capel Sylen (SN 511065).

Capel Erbach   SN 529147
chp. Llanarthne
'chapel of adoration': ***capel***, ***erbarch***
(mansion ho.) *Tyr Cappell yrbarch* 1613, *Capel-irbach* 1831, *Capel-hir-bâch* 1889-91, *Capel-erbach* 1906, 1991-2
The second el. is almost certainly ***erbarch*** 'great honour, respect; adoration; object of worship', a very uncommon el., influenced here perhaps by ***bach***¹. The n. might describe a chp.

established by a particular person or family or might be taken directly from the well Ffynnon Capel Herbach (*ffynnon*) which was notable for the treatment of 'spasms' (HWW 28). The chp. has been described as a medieval well chapel possibly constructed in the 13th or 14th cents. The ivy-covered ruins were described in 1894 (AC 1894: 21-22, 166), 1971 and 2003 (DavisSS 37).

Capel Gwenlais SN 600161
Llanfihangel Aberbythych
'chapel by (r.) Gwenlais': *capel*, rn. **Gwenlais**
Kappel Gwenlaish, Gwenlaish springs at Cappel Gwenlais c.1700
The chp. seems to be first mentioned by the scholar Edward Lhuyd c.1700 but it had disappeared before 1809. It may have stood near a yew tree overhanging the r. Gwenlais which rises in the well Ffynnon Gwenlais (*ffynnon*) (Fenton 60) at Llygad Gwenlais, 'source of (r.) Gwenlais' (*llygad*) (HWW 2, 44, 165)

Capel Gwilym Foethus approx. SN 50632181
chp. Llanegwad
'chapel associated with Gwilym Foethus', *capel*, pers.n. **Gwilym**, *moethus*
Capel Gwilim Foethus 1715, (ruined chp.) *Capel. Gwilym-foethus by Cothi-bridge* 1742, (chp. ruins) *Cappel Gwilym Foethus* 1811
The chp. has been identified with 'our Lady Chapell at the Bridge end of [r.] Cothy', i.e. in Pontargothi, in the will of Sir Rhys ap Thomas (1449-1525). It may have been successor to an earlier chp. at Llandeilo-yr-ynys (q.v.) recorded in 1332. All of the cited forms are related and Gabriel suggests (CAS XVIII, 58) that Gwilym, often found in historical sources as *Glm'*, is an error for *Lln'*. That seems unlikely in view of the pn. evidence. He further suggests that it could refer to Llywelyn Foethus of the adjoining p. of Llangathen (recorded in poetry of Lewys Glyn Cothi GLGC², nos. 53, 54, 56 and notes). Llywelyn has been identified as a great-great-grandfather of Sir Rhys (ETG 218). It is possible that *moethus* adj. 'fond of luxury or a life of ease' is a family epithet but any direct link between Gwilym and Llywelyn is so far unproven and the evidence is far too late to support the suggestion. The chp. was occupied by a stable or a hog-sty in 1715 and was described in 1811 as having been in ruins for more than one hundred years. Its precise location is uncertain but was probably in or adjoining *Cae'rcapel* (SN 50632181) (*cae, y, `r*) in the tithe apportionment (no. 1160) 1839 north of the A40 and close to the Cressely Arms. The archaeological record (HER 49279) places it at SN 510220 on the north side of Pontargothi.

Capel Gwyn SN 465225
Abergwili
'white chapel': *capel, gwyn*
Capel gwyn 1811, *Capel Gwyn* 1812
Probably describing its lime-washed walls. Named from a Calvinistic Methodist chp. recorded in 1800 (HMGC II, 538). Located near a smithy and a ho. recorded as *Pant-y-gleien* in 1889 and *Pant-yr-hely'-isaf* 1906, 'hollow in the clay-ground' (*pant, y, cleien*), 'hollow of the lower willow-trees' (*helyg, isaf*).

Capel Gwynfe SN 722220
chp. Llangadog
'chapel-of-ease at Gwynfe': *capel*, pn. **Gwynfe**
Gwynvey alias Wynvey 1316, *Wynvey* 1543, *Capel Gwenuye* 1578, *Gwenwie chappell* 1586, (t.) *Gwynfe* 1596, *Capell Gwenfayr* 1603, *Gwynvaye* 1600-7, *Gwinvey* 1626, 1748, *Capel Gwynvey* 1729, *Gwynfe Chapel* 1757, *Capel Gwynfe* 1831
Gwynfe is 'white *or* fair place': *gwyn, mai* but employed in the less specific sense of 'fertile' or 'favoured place'. The development -ai > -ei > -e -ai reflects dialect. This may also be *Wylfry* 1288 (in erratic sp.). Little is known of the chp. The present late 19th cent ch. dedicated to All Saints replaced an 18th cent structure later used as a hall.

Capel Gwynllyw[1] SN 5324
Llanegwad
'chapel dedicated to Gwynllyw', *capel*, pers.n. **Gwynllyw**

chaple of Gwnllow 1552, *Capel Gwnlliw* 1650
The location of this former chp. is uncertain though it has been identified with a reputed chp. of *Llechgron* 1703 in Llech-gron ht. (RCAHM II, 118; CAS XVIII, 59; JHSCW III, 29). The chp. has also been confused with Llanhernin (q.v.) apparently because of the similarity of a fm.n. Nantergwynllyw (SN 547219) recorded as *Nantarwaynllif* 1804, *Nantarwenllew* 1828 and *Nant-yr-wenlliw* 1831. This may actually take its n. from an adjoining unnamed stream which joins Afon Dulas near Felindre (SN 25142117). The two earliest forms of the rn. appear to contain *Gwenlliw or *Gwynlliw describing one notable for its whiteness or 'brilliance (**nant**, ?**ar**, **gwynlliw**[1]). The similarity to **Gwynllyw** in Capel Gwynllyw[1] is evident and may well have encouraged misidentification with Llanhernin. Llanegwad possessed another chapel for which there is fn. evidence near the fm. Glancapel (SN 523243) in Llech-fraith t. The tithe apportionment 1839 names the ho. as *Llanycapel* (*Glan y capel* on the plan) (covering parcels 3079-3101) including an area north-west of the ho. recorded as *Caecapel* (3079, 3080 at SN 251872443), *Caecapelissa* (3083a), *Caecapelycha* (3082) and *Caercapel* (2895, 2896) (**glan**, **cae**, **isaf**, **uchaf**). David Jones (Haul 23: 176) claims that the chp. was built by the family from nearby Ynys-wen but conclusive evidence is lacking.

Capel Gwynllyw²   SS 53889969
chp. Llanelli, near Berwig
'chapel dedicated to Gwynllyw', **capel**, pers.n. **Gwynllyw**
(p.) *Llanwnlliw* (in Carnwyllion) 1533, (chp.) *llanvnllu* 1590, *capellam gunlliw* 1553, *Capel Gunllo* 1578, *the chaple of S. Gwnlet* late 16th cent, *Capell Gwnlliw* 1605, (chp. of ease) *Chappell Gwnllian* 1612, *Cap: Gwenllwee* 1675, (ruined) *Capel Gwenlhiw ... now called Capel Byrwig* c.1700, *Cappel Gwenllo* 1729, *Dyddwen Chap. (ruined)* 1742, *Burwich or Dyddwen Chap. (ruined.)* 1763, *Capel Berwig* 1796
Gwynllyw is presumably Gwynllyw, confessor, father of St Cadog who has a number of dedications in Carnwyllion covering the Llanelli area. The location of the chp. in some historical sources is not always clear and the ruins were described in 1888 as 'St. David's' (LPC li) and in 1917 as Capel Dewi (RCAHM V, 119, no. 347). Capel Dewi[3], however, was in Westfa ht. and refs. for 1553 and c.1700 state that Capel Gwynllyw was at or near Berwig (q.v.) which is confirmed by *Berwick Chapel* 1836 and *Capel Berwic (Ruin of)* 1880. The confusion with Capel Dewi[3] seems to begin with the local historian David Bowen in 1856 (DLlanelli 34). Iolo Morganwg describes the chp. as a 'ruin' with its churchyard 'open to the high roads' in 1796 (CAS 24: 48). The alternative form *Dyddwen* in 1742 and 1763 is probably through confusion with Capel Dyddgu (q.v.). St David's Church Hall was built on the site and the remains were uncovered when the hall was demolished before 2012. The present ch. of St David, on the opposite side of the road, was built in 1882 and its dedication was probably inspired by the misidentification of Capel Dewi[3].

Capel Hendre   SN 594113
Llandybïe
'chapel (in the) Hendre'; **capel**, pn. *(Yr) Hendre*
Capel-`rhendre 1831, *Capel yr Hendre* 1836, (village) *Hendre*, (village and chp.) *Capel Hendre* 1906, *yr Hendre* 1908
Named from a Calvinistic Methodist chp. built in 1812 (HMGC II, 357, 541; HPLland 41, 152). There is little evidence for any Anglican ch. or chp. which the historian Rice Rees locates here (WSS 330). Hendre, 'winter dwelling, home farm' (**hendref**) is presumably taken from the ho.n. Hendre-siclath (SN 594115) recorded as *Hendre syglaeth the name of a Sydhyn* c.1700 (**syddyn**) and *Hendre-Siclath* 1879 or from Hendre-isaf (**isaf**). The precise sense of Siclath is unclear though the word is apparently composed of **sigl**[1] nm. 'a swinging, a swing' and **llath** nf. 'rod, staff; sail-yard, spar'.

Capel Iago   SN 547423
chp. Llanybydder
'chapel dedicated to (St) Iago (James)', **capel**, **Iago**

(fair at) *a Chapel Iago* 1612, *Cappel Jaco* 1676, *Cappell Jago* 1725, *Capel Iago* 1773, *Capel-iago* 1778, *Capeljago* 1831
The location of a chp., reputedly medieval. A geophysical survey in 2011 found little physical evidence though the survival of a yew-tree (at SN 54764240) seems significant. Otherwise, only the fm.n. Capel Iago and an adjoining fn. *yr hen fynwent* meaning 'the old cemetery' (**hen**, **mynwent**) – not found in the tithe schedule – hint at the existence of an early chp. Francis Jones (HWW 166) mentions a Ffynnon Iago in Cilymaenllwyd near the boundary with Llandysilio East, and another west of Mynydd Ystefflau-garn, Llanllawddog.

Capel Ifan see **Capel Iwan** (Cilrhedyn)

Capel Ifan¹　　　　　　　　　　SN 498104
chp. Llanelli
'chapel dedicated to (St) Ifan (John)': **capel**, pers.n. **Ifan**
the chaple of S. John late 16th cent, *Capel Ivan* 1745, *Ifan St. John Chap. to Llanelly* 1763, *Cappel Evan* 1822, *Capel Evan* 1839, *Capel Ifan church* 1851, *St. John's Church*, (ho.) *Capel-Ifan* 1880
The chp. lay in Glyn ht. and cmt. of Carnwyllion (q.v.) and is likely to be the chp. of St John recorded between 1223 and 1250 (NLWJ III, 136). Described as in ruins in 1735 (LPC l-li) and 1763 (Ecton 1763: 472) but apparently repaired and used by Methodists in 1811. The vicar of Llanelli later re-claimed it and it was reconsecrated in 1837 possessing regular services in 1855. It was disused again in 1906 but later reinstated after 1922 as Eglwys Capel Ifan (SN 498103) (**eglwys**). The dedication was also applied to St John's ch. (SN 50401036) in Pontyberem built 1893-4. The chp. is recalled in (tmt. of land) *Cappell-Evan-Caernawllon* 1740 and *Capel Evan* 1839 which presumably refers to either of two hos. Capel Mawr (SN 494102) and Capel Bach (SN 498102). The first is *Capel mawr* 1813 and the second is *littill Chappell Evan* 1676, *Capel bach* 1813 (**bach¹**, **mawr**).

Capel Ifan²　　　　　　　　　　SN 490154
chp. Llangyndeyrn
'chapel dedicated to (St) Ifan (John)': **capel**, pers.n. **Ieuan**, later **Ifan**
Saynt Johns chaple 1552, *Chappell Jeuan* 1609, *Cappell Evan* 1613, *Cappel Evan* 1674, *Capel Ifan* 1813, (mess.) *Cappel Evan otherwise Tir cappel Evan* 1817, *Capel-Ifan* 1831, *Capel Evan* 1839
The earliest sp. may simply be a clerical translation of an existing Capel Ieuan. Described in 1811 as a fm. and former chp. Also the n. of a former colliery.

Capel Isaac　　　　　　　　　　SN 583269
Llandeilo Fawr
'chapel associated with Isaac': **capel**, *Isaac*
Chapelisac 1742, (dissenting meeting ho.) *Cappell Isaac* 1772, *Capel Isaac* 1831
Reputedly named from Isaac Thomas who granted a cottage and the plot of land on which Capel Isaac Independent chp. was built in 1672 (CAS 39: 151-4). The cause was established in 1650 (Cofiadur 10 and 11: 39), apparently in Y Brynmelyn (SN 585270), moving to a meeting-ho. near Glan-y-nant, in Llangathen, later to land below Tanycapel, in Llandyfeisant (HEAC III, 529-42). Otherwise called Mynydd Bach recorded as *y mynydd Bach* 1715, *mynydd bach* 1758 (**mynydd**, **bach¹**).

Capel Iwan　　　　　　　　　　SN 289363
chp. Cilrhedyn
'chapel dedicated to (St) Iwan (John)': **capel**, *Iwan* variants **Ifan**, **Ioan**
Saynt John is Chaple 1552, *Capeleuan* 1578, (mess.) *tythen cappell Jevan* 1607, *Capeleuen* 1607, (lands) *Cappell Evan* 1680, (mess.) *Tir Capel Evan* 1728, *Cappell Joan* 1757, (mess.) *Cappell Ivan* alias *Cappel-Evan* 1799, *Capel-efan* 1831, *Capel Ifan* 1851, *Capel Iwan* 1869
Sources (*-euan* for *-evan*, *Evan*, *Jevan*) generally favour the form Capel Ifan, with an occasional anglicised form Capel Evan, and this is continued on OS maps. Capel Iwan is the usual form in W newspapers. The chp. has been identified with an Independent chp. (SN 28973628) said to have been erected in 1723 and enlarged in 1795 (RC I, 485; HEAC III, 412-4; HECI 82).

The n. actually derives from that of a chapel-of-ease serving the eastern part of Cilrhedyn in co. Carmarthen. The Independent chp. stands within the cemetery of the older chp. and a field (SN 28813628) to the west is described as *Park dan y capel* in 1841 (*parc*). The land on which the Independent chp. was built is described as near *ffynnon Evan* in 1723 (*ffynnon*).

Capel Llanddu ?SN 474226
chp. Abergwili
'chapel at Llan-ddu': *capel, Llan-ddu*
Capel Llanddu 1836
There seem to be no other refs. to Capel Llanddu and the reliability of the source (WSS 329) is very questionable. The supposition that there was a chp. here may be based on little more than an assumption that the pn. Llan-ddu contains *llan* 'church' with *du* 'dark, black'. That would presumably describe a church which is dark in appearance but it is difficult to think of suitable comparisons. Llan-ddu seems to be first recorded as *Llanddee* in 1740 applying to a messuage located near a watermill on Afon Annell. The mill is *Felin-llan-ddu* 1831, *Velin Llanddu* 1838, *Glan-ddu Corn Mill* 1889, *Glan-ddu Mill* 1964 (*melin*). It is far more likely that Llan-ddu is properly Lan-ddu, containing the lenited form of *glan* 'bank, river bank', and describing a bank which was dark in appearance, perhaps one in shadow. Comparison should be made with Glan-ddu (SN 580251), in Llanfihangel Cilfargen, recorded as *Lan-ddu* 1831-1964 and *Glanddu* 1975-7. The ho. adjoining the mill-site is now Llwyn Howel (*llwyn*, pers.n. *Hywel*).

Capel Llandyri see **Llandyry**

Capel Llanlluan Capel Farm SN 556155
chp. Llanarthne
'chapel called Llanlluan', *capel*, pn. *Llanlluan*
Llanlluan 1326, *Daullians chaple* late 16th cent, *ynghapel llan Llyan* 1741, *Cappel Llanlluan* 1753, *Llanlluan, Chap. to Llanarthney* 1763, *Capel-llanllian, Melin-llanllian* 1831, *Capel-Llan-Lluan* 1887-8

A former chp.-of-ease at what is now Capel Farm. Llanlluan is 'church dedicated to Lluan': *llan*, pers.n. **Lluan** who has been identified with a daughter of the numerous progeny of the semi-legendary Brychan Brycheiniog (AC 1915: 395). Remains of what appeared to be a small burial ground were noted in 1917 (RCAHM V, 71). Llanlluan was also the n. of the t. and manor recorded as *Llanlean* 1535 and *Llanllylan* 1685. The n. is recalled in Melin Llanlluan (SN 540159) referring to a mill (*melin*) located on an unnamed tributary of Gwendraeth Fach at Pont Lan-dwr. The chp. was used by Methodists in the 18th century (HCrm II, 195-6; HMSG 81-86) down to 1830. This was itself replaced in 1839 by the Calvinistic Methodist chp. Capel Llanlluan (SN 556159) near Myrtle Hill.

Capel Maesnonni approx SN 499397
chp. Llanllwni
'chapel at Maesnonni', *capel*, pn. *Maesnonni*
Capel Nonni ruinous time out of mind c.1700, *Capel Noni* 1818, *Capel-maes-nonny* 1831
The former chp.-of-ease was apparently located on the north-east side of Maesnonni fm. where the tithe apportionment records fields called *Caecapel* and *Caedany Capel* in 1843. A 9th cent inscribed stone found in 1907 in the first field has been identified as a grave marker lying in an early medieval cemetery (Corpus II, 258-60). Maesnonni is *Maes none* 1638, *maes Nonny* 1704, (mess.) *Maes tŷ Nonni* 1707, *Maesynonni* 1738/9, *Maesnonny* 1776, apparently meaning 'open-field of (person called) Nonni or None' with **maes** and pers.n. (ETG 219) *None* or *Nonni*. Capel Maesnonni was said to have remains of a nunnery known as *Maes Nonny*, i.e. *The Nun's Field*, with a spring *Ffynnon Noni* 1811 (CarlisleTD) (*ffynnon*) near a castle tumulus called Y Castell (HWW 171) (*castell*) but the explanation is onomastic. References to *Tir blaen nonne* 1600, *Tir blaen nant none* 1615 and 1621 (*blaen*) tend to favour a lost rn. rather than a pers.n. The Independent chp. Capel Nonni (SN 494402) was established in 1810.

Capel Mair[1] SN 404380
chp. Llangeler
'chapel dedicated to (St) Mair (Mary)': *capel, Mair*
Caple Vayre 1553, *Caple Vayre* late 16th cent, *Castell Mair so call'd situated near a Chapel in ye Grange dedicated to B. Virgin* c.1700, *Capel-Mair (School)* 1889, *St. Mary's Church* 1905
A former chp. said to be in ruins in 1811 and 'entirely demolished' in 1844, rebuilt in 1849 (HPLlangeler 177-8), but only a school is shown here in 1889. The ch. of Mair/Mary appears between 1891 and 1905. A well Ffynnon Fair (*ffynnon*) is recorded near the site of the chp. in 1899 on Llwynffynnon land (HWW 165) but it has also been identified with a well within the castle Pencastell (SN 402379). A 6th cent inscribed standing-stone, recorded in the ch. cemetery in 1828, was later moved to Dancapel fm. (SN 404382) and broken up. Two fragments were found in the walls of outbuildings of the fm. in 1900 and a third fragment was in use as a drain cover (AC 1907: 293-310; Corpus II, 251-4) and moved to the modern ch. before 1912 but only one fragment survives. Dancapel is *Dan y Cappel* 1740 and *Danycapel* 1811 '(ho.) below the chapel' (*tan, dan*).

Capel Mair[2] ?SN 555325
chp. Llanybydder
'chapel dedicated to (St) Mair (Mary)': *capel, Mair*
(ruined chp.) *Mair* 1742, 1763, (chp. ruins) *St. Mary* 1811
The two earliest refs. and mention of a chp. *Mynwent Cappel Mair* 1811 (*mynwent*), which draws on them, associate it with Talyllychau p. but this is uncorroborated. The answer is seemingly provided by D. Long Price (AC 1879: 180) who suggests it was at Glan-capel-Mair, a fm. now lost to forestry. The fm. is recorded as *Glan-y-capel* 1831, *Glanycapel* 1840, *Lan-capel* 1888, and *Glan-capel-Mair* 1906 (*glan*). The chp. itself may actually have been at or very close to the fm. Capel Mair on the crown of a hill overlooking Afon Cothi located near several hill-roads. The fm. is *Capelmair* 1811 and *Capel Mair* 1831. The chp. probably served the southernmost part of the large p. of Llanybydder.

Capel Mihangel approx SN 41850645
Cydweli
'church of (St) Mihangel (Michael)': *llan, Mihangel*
(*Butter is Parke near*) *Seynt Mighhell is Chapell* 1505, *Llanfihangel, St. Michael* 1836
One of two unnamed chps. on the Muddlescwm estate in 1720, recorded in 1919 as about 300 yds. north-west of the entrance to the fm. (SN 42070592). This places it north of the B4308 at the above grid ref. The dedication is confirmed by the n. of a well Ffynnon Mihangel (*ffynnon*) recorded in 1908. No traces of the chp. were found in 1836 but it is said to have been located in a field Parc y Maenllwyd. which may be translated as 'enclosure at the grey (or holy) stone' (*parc, y, maen, llwyd*). This tends to favour a standing or boundary stone perhaps identifiable with a 'chamfered stone', reputedly taken from the chp., and laid over the nearby brook Hêd close to Muddlescwm fm. (SN 42070592). It may also be identifed with *Horestonefeld* mentioned with the chp. in 1505 probably meaning 'grey-stone field' (E *hoar, stone, field*) recalling a Bronze Age stone Maenllwyd Mawr (HER 5327) 'greater grey-stone' (*mawr*).

Capel Newydd SN ?8040
chp. Llanfair-ar-y-bryn
'new chapel': *capel, newydd*
Capel newith 1578, *Capelnewith* 1607, *Chapel Newith* 1695, *Newid Chap.* 1742, 1763
Early county maps show the chp. a little to the north of Glanbrân on the west side of r. Brân in the vicinity of Cynghordy in the p. of Llanfair-ar-y-bryn. This is partly confirmed by the chronicler Raphael Holinshed in 1587 (Holinshed I, 131) who states that the r. Brân rises two or three miles north of *Capel Newith*. By contrast, Bowen's map 1729 (BowenNSW) omits the n. and other evidence of a chp. seems to be lacking. The presumed location is remarkably close to the site of Capel Cynfab (q.v.) in the same p. and they may refer to

the same place. It seems significant that Capel Cynfab appears to be missing from historical sources between 1553 and 1742. The refs. to Capel Newydd in 1742 and 1763 need not contradict this suggestion. The source in question Liber Valorum (Ecton) went through several editions between 1711 and 1786 and contains obsolete material taken from older sources.

Capel Penarw SN 58692167
Llangathen
'chapel at Penarw': *capel*, ho.n. *Penarw*
Cappell Pennarwy 1602, (vestiges of old chp.) *Cappel Pen Arw* 1811, *Cappel penarw Cot* 1813, *Capel Penarw* 1836
The tithe map shows a former building recorded in the apportionment in 1839 as *Capel Cot and Garden* above the wooded slopes of Allt y Capel (*allt*) but this had been removed before the publication of the OS map 1887. Jones (HWW 167) adds that there is a holy well Ffynnon Capel Pen Arw (*ffynnon*) within half-a-mile of the p. ch. notable for curing sore eyes and rheumatism. Penarw appears to derive from Penarwy probably with *pen*¹ in the particular sense of 'hill' but *arw(y)* is uncertain. There is no stream here which might justify comparison with Arwy, E form Arrow, in cos. Radnor and Hereford, for which see CDEPN 18-19.

Capel Pencader see **Pencader**

Capel Peulin SN 788470
chp. Llandingad
'chapel dedicated to (St) Peulin': *capel*, *Peulin* (chp.) *Sancti Paulini* 1339, *Capel Pylyr* 1578, *Capel Pylyn* 1607, *Pilin capell* 1587, *Pelyn Chap. (ruined)* 1742, *Capel Pilin* 1798, *Capel Peulin* 1836, *St. Paulinus' Church (Chapel of Ease)* 1887-8
Capel Peulin was a part of the gr. of Nant-bai in Rhandir Abad (q.v.). Little certain is known of Peulin but he may be identified with Paulinus (Paulens) described in a Life of St Dewi/David (11th cent) as a scribe and disciple of St Germanus, the bishop, in the late 5th cent. Paulinus has also been identified as the scholar to whom St Teilo went to perfect his learning (DWB: Paulinus). Peulin/Paulinus was apparently confused by the monk Wrmonoc c.884 with St Paul Aurelian but some of the details in St Paul's Life appear to derive from the earlier tradition of St Peulin. These include a ref. to *Brehant Dincat* (see Llandingad), described as the patrimony of his father. Peulin is likely to be the person dedicated at Llan-gors, co. Brecon, and perhaps on a pillar stone at Llantrisant, Anglesey. He has also been identified with Paulinus described as 'Preserver of the Faith' on a 5th cent stone (HER 9939) found in use as a footbridge on Brondeilo fm. at Pantypolion (Talyllychau) but this may rather refer to a secular person. The chp., otherwise known as Ystrad-ffin (q.v.), is described by the antiquarian Richard Fenton in 1809 as fallen down, with services transferred to 'a miserable cot just above a little stream on the road to the Agent's house' at Nant-bai. The present church, restored in 1900, is said to have replaced 'a larger and more dignified structure to which tradition gives the date 1117'.

Capel Sant Silin SN 529340
Llanfihangel Rhos-y-corn
'chapel dedicated to (man called) Silin': *capel*, *sant*, pers.n. *Silin*
Capel sant Silin 1842
The ref. in 1842 applies to what is now Capel Farm and it is possible that this was the location of a former chp.-of-ease or perhaps a Nonconformist meeting-ho. Little is known of the chp. though the n. was current in the late 18th cent because the writer and Unitarian preacher Tomos Glyn Cothi (1766-1833) was born here (*Capel St. Silin*: HCrm II, 435; DWB). Lewis' Topographical Dictionary (first published 1833) (LewisTD) simply says that its existence had been inferred from the existence of a well Ffynnon Capel 'situated near an ancient yew tree'. Neither the well (*ffynnon*) nor the yew-tree can be identified on OS maps from 1889 but the well was probably in or near a field *Cae ffynon* 1844 (SN 54023470) 'well field' (*cae*) just above the fm. Esgairfynwent. The latter is probably *Esgerfynwent* 1748 and *Esgervynwent*

1752. This is 'ridge of the cemetery' (*esgair, mynwent*) and adds weight to the supposition of some sort of chp. in the neighbourhood. The p.ch. of Llanfihangel Rhos-y-corn is about 1.5km to the east and has its own graveyard. The chp. has been confused (HER 11765) with the Independent chapel on the other side of Afon Clydach in Gwernogle.

Capel Seion SN 517133
Llanddarog
'chapel called Seion': *capel*, chp.n. **Seion**
(*yn*) *Nghapel Seion* 1714 (1873), *Capel Seion* 1867, *Capel Seion (Independent)* 1906
An Independent chp. built in 1712 and rebuilt in 1848 (HEAC III, 507). This is probably *Chappel Sion* 1762. The n. was transferred to the ht. and appears on OS maps after 1991. Seion, Sion or Zion is the biblical Mount Zion, possibly alluding to the chp.'s location on a hill.

Capel Taliaris see **Taliaris**

Capel Teilo[1] SN 435074
chp. Cydweli
'chapel dedicated to (St) Teilo': *capel*, pers.n. **Teilo**
Chappel Tylo 1594, *Cappel Tilo* 1674, *Cappel Tylo* 1811, *Capel Deilo* 1831, *Capel-Teilo* 1888
The chp. was located near a former ho. (SN 433076) and its whereabouts lost until re-discovery in 1966 above the narrow valley of Cwm Teilo (SGStudies 255-9, with plan and description). A spring Pistyll Teilo (*Pistyll Deilo* 1831) (*pistyll*) lay a little to the south of the chapel, its waters used for the treatment of rheumatism in 1811 (CarlisleTD).

Capel Teilo[2] ?SN 651375
chp. Talyllychau/Talley
'chapel dedicated to (St) Teilo': *capel*, pers.n. **Teilo**
Cappel Teillo 1633, *Cappell Teilo* 1725, (ruined chp.) *Trilo* 1742, 1763, *Capel Teilo* 1836
Possibly located at Brondeilo (grid ref. above) which has a Pistyll Teilo (*pistyll*) though no remains of a chp. have been found here. Capel Teilo may also be identical to *Lann Teliau garth teuir uilla tantum super ripam cothi* and *Lann teliav garthteuir* c.1145 (LL 124, 254) which probably stand for Llandeilo Garthtefir or Garthdefir (*garth*[1] with an uncertain el.). Brondeilo is located at the bottom of the south slopes of an unnamed hill partly covered by the woods Allt Pen-y-coed and Allt Ynysau (*allt*) and near a ho. Garth (so recorded 1831). Recorded as *Bronedeilo* 1537-9, *Tyr bron deilo* 1607, *Tir bronn deilo* 1639, *Tir bron deilo* (in gr. of *Traethnelgan*) 1658 (*bron*[1]).

Capel Troed-y-rhiw ?SN 373313
chp. Cynwyl Elfed
'chapel at Troed-y-rhiw': *capel*, pn. Troed-y-rhiw
Capel Troed y Rhiw 1836
A former chp. recorded by Rice Rees (WSS 329) on the fm. Troed-y-rhiw (SN 373313) in the northern part of Cynwyl Elfed. Its walls could be traced and an adjoining field was said to be *Cae'r Hên Eglwys*, 'enclosure of the old church' (*cae, yr, 'r, hen, eglwys*). A correspondent to a newspaper Y Llan in 1898 adds that it was in a place called *Parc yr Hen Eglwys* on the brow of a hill opposite a dwelling ho. called *Troed y Rhyw Chwere* close to a hedge *Clawdd Rhyfel*. The ho. evidently refers to Troed-y-rhiw otherwise recorded as *Troedyrhiw trachwerw* 1788, *Troed-y-rhiw* 1831 and *Troed y Rhyw Chwere* 1898, but *Parc yr Hen Eglwys* and *Clawdd Rhyfel* (*parc, clawdd, rhyfel*) cannot be identified. Troed-y-rhiw is '(place at) foot of the slope' (*troed, y, rhiw*) but *trachwerw* is uncertain. It would appear to stand for *tra chwerw* 'very bitter' figuratively 'very harsh' (*tra*[2], *chwerw*) or a re-interpretation of *achwre*, *achre*[2] 'palisade, fence'. Melville Richards (WATU) places the chp. in Aber-nant because Cynwyl Elfed was a chp. of Aber-nant down to the 19th century.

Capel Tydyst SN 667240
chp. Llangadog
'chapel of (St) Tudwystl', *capel*, pers.n. **Tudwystl**
Merthir Tudhistil 11th cent (c.1200), *Merthyr Tutbystil* 13th cent (16th cent), *Llan Dydystyl*

o vy6n y vaenor Vabon 15th cent, *Cappel Tydyst* 1680, (estate) *Cappeltudis* 1785, *Cappeltydist* 1797, (mess. etc) *Capeltydist* 1811, *Capel-ty-dyst* 1887

The site of the chp. is shown on OS maps on the south side of the ho. The pn. starts in the historical record as Merthyr Tudwystl meaning 'graveyard of Tudwystl' with **merthyr**[2] 'graveyard containing the bones of a saint' and later appears as Llandudwystl 'church dedicated to Tudwystl' with **llan**. The progression from **merthyr**[2] to **capel** may suggest a decline in status to that of a chp.-of-ease within the p. of Llangadog followed by abandonment probably at some date before 1680. Tudwystl was one of the reputed daughters of the semi-legendary Brychan Brycheiniog and a sister of Tudful (HCrm II, 118; and PNGlamorgan 140-1 for Merthyr Tudful) thought to have lived in the 5th or early 6th cents.

Capel y Betws　　　　　　　　SN 279281
Tre-lech ar'r Betws
'chapel (called or at) Y Betws': **capel**, **y**, **betws**[1]
'chapel, house of prayer'
Bettus Capella 1552, *y bettws* c.1566, *Capel bettus* 1578, *Capell Bettus* 1586, *Bettws C.* 1742, *Bettws Chap. to Trelêch* 1763, *Capel y Bettws* 1741-2, *Capel Bettws* 1798, (chp.) *Cappel Bettws* 1811
capel and **betws**[1] are very close in meaning and it possible that the first el. was prefixed at a time when the second el. was falling out of use as a common colloquial n. Described in 1684 as 'but a chapel of ease and in an indifferent repair' (CAS XII, 18). Services at the chp. had ceased before 1710. Located near a fm. Gilfach-y-betws (SN 27882814) (*cilfach*).

Capel y Castell see **Capel Bach y Drindod**

Capel y Croesfeini　　　　　SN 39622387
chp. Llannewydd/Newchurch
'chapel near or possessing stone crosses': **capel**, **y**, **croesfaen** pl. *croesfeini*
Capel y Groesveini 1733, *Capel-groes-feini* 1831, *Capel y Groesfeini* 1836

Sometimes described as Llanfihangel Croesfeini but the evidence for this form is late. The chp. had been converted into a barn before 1811 and was apparently demolished in 1847. Located in a field known as *Lan capel* in 1844 (**llan**). The OS map 1831 places it near a mound Crug (SN 396241) (**crug**) close to Capel-bach (SN 39237) (**bach**[1]), a former Baptist chp., which was a little to the west on the lane leading past Pen-yr-heol. Davies (CAS I, 117-8, 120) states that stones from Capel y Croesfeini were used in the construction of stables at Waunllanau (SN 399230), about 1.2 km to the south, and he records an upright stone, possibly the shaft of an ancient cross. This has been dated 7th-9th century and is said to 'point to the early medieval origins of the chapel of Llanfihangel Croesfeini' (Corpus II, 273).

Capel y Drindod　　　　　　SN 354387
chp. Llangeler, Penboyr
'chapel dedicated to the Trinity': **capel**, **y**, **trindod**
Chapel called Trinity 1713, *Cappel y Drindod* 1753, *Cappel-y-Drindod* 1755, *Ydrindod, Chapel to Penboyr, alias Trinity* 1786
A former chp. generally known as Capel Bach (**bach**[1]) (HPLlangeler 184-5; HCrm II, 196n.) shown on the tithe plan 1840 on the west side of the br. Pont y Capel (**pont**) over Nant Bargod. The chp. survived as a Sunday school down to the opening in 1863 of a new ch. dedicated to St Barnabas (SN 354384) in Felindre.

Capel y Groes see **Capel Bach** (Abergwili)

Capel yr Hendre see **Capel Hendre**

Capel yr Ywen　　　　　　　SN 672266
chp. Llandeilo Fawr
'chapel with a yew-tree': **capel**, **yr**, **ywen**
(ruins) *Cappel yr Ywen* 1811, *Capel yr Ywen* 1836
No traces were discernible in 1913 though graves were disturbed in the 1850s when digging foundations of a building on Ty'r-capel farm (RCAHM V, 89). A second chp. Capel-

isaf (SN 661252), 'lower chapel' (*capel, isaf*), recorded as *Capel-bach* (***bach***[1]) in 1831, lies just 2 km to the south-west.

Cardiff Forest see **Fforest Gaerdydd**

Carmarthen see **Caerfyrddin**

Carmarthen Bay see **Bae Caerfyrddin**

Carmel SN 584165
Llanfihangel Aberbythych
'(place named from chapel called) Carmel: ***Carmel***
Charmel 1851, *Carmel* 1867
A private Baptist chp. erected in 1833. Carmel was a mountain in biblical Canaan.

Carnwyllion (Carnwallan) Carnwallon-fawr CRM SN 490100
cmt., lp., hd.
'land of Carnwall', pers.n. ***Carnwall, -(i)on***
(ch. of) *Carwathlan*, (ch. of) *St. eslini of Carnewarlan* c.1144, *Carnewothan* c.1165, *Cornoguatlaun* c.1200, *Carnwotlhan* c.1236, *carnwaliaun* early 13th cent, (*ecclesia de*) *Carnewaylan* c.1240, *Karnwathlan* 1283, (land) *Kaerwarthlan* 1299, (cmt.) *Karnegwelliawn* 1322, *Carnewathlan* 1361, *Karnywyllawn* 1201 (c.1400), *a Charn6llonn* c.1500, *Comm. Carnwatllaun* c.1537, (place) *Karnewllon* 1550, *Carnowllan* 1609
Carnwyll is the form of the pers.n. favoured by Richards (ETG 152) probably because Carnwyllion is said to have been named after *Cornouguill* in the Life of St Cadog (VSB 24; HCrm I, 9) but most historic forms favour *Carnwall*. The n. of a cmt. (*cwmwd*) which survives as that of two hos. distinguished as *Carnawllon, Carnawllonfach* 1822 'greater' (*mawr*) and 'lesser' (***bach***[1]) located near the B4317 between Pontyberem and Pont Henri. Local pron. at least for Carnwyllion Fawr is 'Carnawllon or Carnwllon' (CHist 17: 70-76) Carnawllon-fach was displaced by Lenham Court some time between 1953 and 1961. Historic forms with *St. eslini* (above) and *St. Elyni* c.1282 refer to Llanelli which lies in this cmt.

Mynydd Carnwyllion recorded as (mountain) *Mynidd Carnalltharn* 1609 (***mynydd***) is not current. The cmt. and hd. covered the ps. of Llanedi, Llanelli, Llangennech and Llan-non (HCrm I, 230-1).

Carreg Cennen SN 668181
cas. Llandeilo Fawr, lp.
'rock near (Afon) Cennen': ***carreg***, rn. **Cennen**
Karrekennen 1257, *Karreckennen* after 1271 (1332), (cas.) *Karakenny* 1280, (cas.) *Carregkennen* late 13th cent, (cas.) *Karregkenyn* 1327, *castell Carec Kennenn* 1248 (c.1400), *Kaerkenyn* c.1456, *Kerikennen* c.1537, *Cast: Carreg* 1574
The rock is the craggy hill on which the castle (HER 3998) is located on the north side of Afon Cennen. Short descriptions of this have been given by the antiquary Richard Fenton (Fenton 58-9) and Lloyd (HCrm I, 283-4); it was destroyed in 1461.

Carwe, Carway SN 464065
Llangyndeyrn
(place named from r.) Carwed: **Carwe**
(land) *Cwyn Carwed* 1565, *Carowed* 1578, (brook, r.) *Carwed* 1609, *Karwed* 1688, *Carwed* 1811, *Carway* 1831
Named directly from Carway fm. but ultimately from the now unnamed stream rising above Blaencarway (*Blaen Carway* 1793, *Blaen-carway* 1831) (SN 472068) (***blaen***) which joins Gwendraeth Fawr (SN 465079) at the place recorded as *Aber Karwey* 1612 (***aber***). Carwed probably consists of ***carw*** with suffix ***-ed***[3] indicating a r. imagined as behaving like a deer, semantically identical to a pers.n. Carwed recorded in Tregarwed (***tref***), in Llangaffo, Anglesey. The village developed near Carway Colliery opened in the 1850s by the Carway and Duffryn Co. *Cwyn* 1565 may be a scribal error for ***twyn***.

Castellcosan SN 203261
manor; gr. Eglwys Fair a Churig
'castle of ?': ***castell***, ?pers.n. ****Cosan, -m***
(manor) *Castell Coslom* 1535, (gr.) *Castell Cossam* 1562, (gr.) *Cosham* 1594, (gr.) *Castell Cossam*,

alias Castle Gosham 1600, *Castle Cossam grange* 1605, *Castell Cosien* c.1700, *Castellgosen* 1843
The second el. could be a pers.n. on the lines of Castelldwyran but earlier evidence is needed. The n. of a diffuse gr. of Hendy-gwyn/Whitland abbey extending into several counties, apparently incorporating the early grs. of Blaenpedran, Cilgryman and Nantweirglodd (WCist 182, 314). The actual gr. appears to be the ho. known as *Fron-uchaf* (**bron**[1], **uchaf**) since 1888 at latest. OS maps record it as simply a 'tumulus' but it is otherwise described as 'a considerable mound', thought to be a castle, located in a field Parc Castell (RCAHM II, 41-42; HER 5073) (**parc**).

Castelldwyran SN 144182
chp. Cilymaenllwyd p.
'castle of (man called) Durant': ***castell***, pers.n. ***Durant*** AFr and Fr
Castle Dornod 1325, *Castrum Duraunt* 1358, *Chastelduront* 1361, *Castrum Durant* 1414, *Castle Durham* 1584, *Cast.deram* 1578, *Castle dyran* 1600, *Castledyra'* 1608, *Castell dwyran* 1752, *Castlederran* 1778, *Castledwyran* 1796
The epithet or pers.n. meaning 'steadfast', variations *Durran, Durrant, -d, Durand*, etc., was re-interpreted by W-speakers as **dwyran** (***dwy***, ***rhan***) 'two-parts'. This was traditionally explained as a ref. to the chp. and castle (HER 3732) which were divided into two parts for the benefit of two sisters. There seems to be no good evidence of a castle though one is reputed to have stood near the chp. at which there is an adjoining ho. *Ty'r-Eglwys* 1890, 1907: 'the church house': ***tŷ, yr, 'r, eglwys***. Castelldwyran is described as a chp. 1630 in Cilymaenllwyd parish. Local pron. was said to be 'Casdiran' in 1934. It has been suggested that the dedication of the chp. to St Teilo supports an early medieval origin though there is no conclusive written evidence. The discovery of a late 5th-early 6th century standing stone forming part of a stile built into the church boundary, later in front of Gwarmacwydd House (SN 161209) (Corpus II, 202-6), tends to support the supposition that there was a much older ch. on the site of the later chp. The stone, known as the 'Vorteporix stone' from its inscription, was once identified with Vortipor or Gwerthefyr (***Gwrthefyr***), a 6th cent king of Dyfed, but this is now thought through linguistic analysis to be doubtful.

Castellheli, Castle Ely Upper Castle Ely SN 197108
manor Eglwys Gymyn
'castle of (man called) Heli': ***castell***, E ***castle***, pers.n. ***Heli***
Castelhely 1307, *Great Castell Elye* 1531, *Gretecastell Ely, Lytylcastell Ely* 1535, (tmt. in) *Castellely, Castelly* 1592, *Castle healye* 1603, *Castelye* 1609, *Castle-heli* 1757, *Castellheli* 1819, *Lower Castle Ely, Upper ~, Castle Ely Mill (Corn), Castle Ely Bridge* 1887-9
Lower Castle Ely (SN 195104) is now plain Castle Ely. There is no record of a castle but traces of a prehistoric enclosure have been found on the nearby hill (SN 20091092) (HER 3891) and a rectangular enclosure (SN 20151085) possibly Iron Age or Roman (HER 9659) close to Upper Castle Ely. Heli is a recorded, if uncommon, pers.n. A rn. is also possible, composed of **heli** 'brine, salt water', but unlikely because we must exclude the closest stream marking the border of the historic cos. Carmarthen and Pembroke. This is now unnamed on OS maps but it is recorded as *the Gwair* 1587 (Holinshed I, 132). There is a second unnamed stream rising at SN 207114 near Rhos-goch/Red Roses which reaches Bae Caerfyrddin/Carmarthen Bay at SN 226075 but no proof that this was ever called Heli. The medieval lp. of Talacharn[1]/Laugharne possessed an E reeve (a manorial official) for Small Castle Ely and a W reeve for Great Castle Ely (HCrm I, 239) reflecting the division between E and W customary practice. Very late forms such as *Castle Ely* may have been influenced by association with the rn. Ely, co. Glamorgan.

Castell Llwyd SN 246094
manor Llanddowror; ho. Laugharne
'grey castle': ***castell, llwyd***
Castelloyt 1207, *Castelloyd* 1307, (manor, etc) *Castle Lloyd* 1617, (lp.) *Castle Lloyd* 1703, *Castle*

Lloyd 1775, 1889, *Castle Llwyd* 1786, *Castell Llwyd* 18th cent
Castle Lloyd on OS maps. There seems to be no mention of a castle here in historical records and the n. now applies to a fm. The 'castle' may actually refer to a prehistoric fort (SN 24080932) described as a 'tumulus' and a 'camp' (RCAHM V, 57; HER 3830). The antiquarian Richard Fenton (Fenton Pemb 473) suggested that Castell Llwyd was the older n. of nearby Llanmilo (q.v.) but there is nothing to substantiate this.

Castell Moel, Green Castle SN 396165
ant. Llan-gain
'bare, exposed castle', 'green castle': *castell, moel¹*, E *green, castle*
(place or cliff) *Grene Castel* c.1537, *Greene Castle* 1599, *Greenecastle* 1609, *kastell moel* c.1550, *Castell Moel* c.1600, *Castell Moel* 1798
Green Castle probably has *castle* in the sense of a mansion and both *moel* and *green* might describe one which was ruinous. The ho. dates from the 15th cent according to Francis Jones (CAS 27, 3-20; HCH 86-87) and was associated with the Reed family. It is possible that both ns. were transferred from an earlier reputed fortification 600 yds to the south (HCrm I, 274, 287) at Hengastell alias Old Castle farm (SN 397162, 'old castle' (*hen, castell*). This has been identified as a motte-and-bailey castle (RCAHM V, 149) but is not in current archaeological databases or shown on maps. The history of the lp. is given by Bruce Coplestone-Crow (CAS 48: 15-16).

Castellnewydd Emlyn, Newcastle Emlyn
 SN 3040
p., town
'new castle in (cantref of) Emlyn': *castell, newydd*, E *new, castle*, Emlyn
(*cum*) *Nouo Castello* 1265, *Newcastle in Emelin* c.1287, (*ville*) *Noef Chastel de E....* after 1287, *Novum Castrum in Nemelyn* 1295, *Noua uilla de Emelyn* 1304, (cas., lp.) *Newcastell Emelyn*, (cas., town, cmt.) *Newecastell Emelyn uchuth* 1435, *Castell neuweydd in Emlyn* 1541, *kastell newyδ yn E.* c.1566, *Castell newydd yn Emlyn* 1612, *Emblyn Castel* 1660
Both E and W forms have lost the unstressed locational preps. *in* and *yn*. The 'new castle' was built c.1240 (HW II, 726, n.51; HCrm I, 285-6) replacing Cilgerran (in the cmt. of Emlyn Is Cuch), co. Pembroke, which may be (*ecclesiam de*) *Castelhan Emelin* c.1180, (ch. of) *Castelhan*, (~) *Castelan Emelin* 1231-2 (DPNW 345, 86). Local pron. is 'Castellnewy'. The Latin *castellum* and *novum* are purely clerical. This is also likely with the Fr forms containing MFr *noef* and *novel* ModFr 'neuf, neuve, nouvelle', *chastel* 'castle'.

Castell Pigyn SN 434221
fm. Abergwili
'castle on a point of land'?: *castell, pigyn¹*
Castlepiggin 1579, *Castelbigin* 1602, *Castle Piggin* c.1643 (late 18th cent), *Castle Pigin* 1661, *Castle piggin alias Castle Biggon alias Place gwynn* 1710, *Castell Piggin* 1729, *Castellpigyn* 1779
The n. of a large ho. recorded in the 16th cent and demolished, apart from the stables, in 1981 (HCH 26) for redevelopment. Located on the south end of a promontory. There seems to be no evidence of a castle as in the case of Castellpigyn (Llannarth), co. Cardigan (PNCrd I, 360 (22(a)), but we know that *castell* (and E *castle*) were sometimes used in an ironic manner for an otherwise unremarkable ho., perhaps one in an elevated place. The n. should be compared with a ho. recorded as *Castell-pigyn* 1831 at Pwll-trap near Sanclêr/ St Clears, Castell Pigyn recorded in Llangeler in 1851, and Castell Pigyn in Llangyfelach, co. Glamorgan. If that is the case here, then it is worth considering that *Pigyn* may actually be *picyn* variant *pigyn* (borrowed from E *piggin*) 'a small pail with one stave extended above the rim to serve as a handle', perhaps a fanciful description of the shape of the ho. or its location.

Castell Toch
manor, lp. Llan-dawg/Llandawke, Llanddowror
'castle of ?': *castell*, uncertain el.

Casteltof 1307, *Castel Toch* 1313, *casteltofe* early 14th cent, *Casteltogh* 1443, *Castell Toghe Ycha* 1540, *the high Castletoe, the lower Casteltoe* c.1592, *Castle Togh Issa* 1603, *Castle-Tough* 1699, *Castle-Dock* 1779, *Castell-toch* 1831

There seems to be no evidence of a castle at the fm. (SN 253113) but it is known that both *castell* and E *castle* can describe a place or a hill thought to resemble a castle (see Castell Pigyn). It is classified as the site of a medieval settlement (HER 9701). The alternative explanation is that it applies to what the OS describes as a Roman fortlet (SN 254121) about 1km north of the ho. Castell Toch is located in Llan-dawg (Llandawke) but any possible etymological relationship between dawg and Toch is unlikely. A case could be made for a pers.n. derived from ME *togh* (OE *tōh*) 'steadfast, tough' cymricised at an early date as *toch* but conclusive evidence is lacking. Late forms suggest E-speakers have probably substituted -*ck* 1779 for W -*ch* (χ). Forms such as *Castell Coch* 16th cent and *Castle Koz'* 1600-7 are probably through association with *coch* 'red, red-brown'.

Castellyrhingyll   SN 577148
loc. Llanfihangel Aberbythych
'the officer's castle': *castell*, **y**, *rhingyll*
(tmt.) *Castellyrringill* 1791, *Castell-y-rhingell* 1831, (ho.) *Castell-y-rhingyll* 1887-9, 1906
The precise significance of the n. is unclear and historic refs. are very late. *rhingyll* applied to an officer in the medieval administrative division known as a cmt. who collected rents and issued summonses. Later sources translate the n. as 'an officer, beadle, serjeant, and sergeant (in the army)'. Faint traces of what appear to be 'a small mound castle' were identified south of the present ho. (RCAHM V, 127; HER 642). The n. was transferred from the ho. (SN 57911467) to the ht. which developed along the Llanelli-Llandeilo road (A476) in the 19th cent near a turnpike gate and smithy.

Castle Ely see **Castellheli**

Cathan   SN 6309
rn. SN 661103 to Llwchwr SN 621103
'r. which behaves like a small cat': *cath*, -*an* (water) *Cathan* 1306-1764, *Kathan Amman* 1541
The rn. is also found in Blaencathan recorded as *Blaen Kathan* 1608, *Blaen Cathan* 1650 and *Blaen Cathan alias Cwm alias Kilvach Llyddon* 1749 (*blaen, cwm*) which appears to be Gilfach (SN 648089) (*cilfach*) and may be compared with Cathan, in Llanddewi and Llangennith, co. Glamorgan. R.J. Thomas (EANC 51-2) also notes an Ir pers.n. Cathan corresponding with *Cadan* (*cad¹*) and drew attention to what he identifies as an Ir pers.n. in Llangathen and Catheiniog (q.v.). This cannot be completely ruled out but the suggested meaning makes better sense and places Cathan in a large group of W rns. containing the ns. of animals.

Cathargoed   Cathargoed Uchaf SN 600188
t. Llanfihangel Aberbythych p.
'woodland inhabited by cats': *cath* nf. 'cat, wild cat', *argoed*
Katheargoed 1550, *Kathargod* 1574, (land) *Cathargod* 1674, *Cathaergoed, Cathaergoed ucha* 1745, (ht.) *Cathar Goed* 1811, *Caeth-ar-goed-isaf, -uchaf* 1831
The area north of Cathargoed-isaf and Cathargoed-uchaf (*isaf, uchaf*) along the slopes overlooking the Tywi valley and Gelli-aur/Golden Grove (q.v.) is heavily-wooded.

Catheiniog (Cetheiniog)
cmt. in Cantref Mawr; hd.
'land of (man called) Cathen': pers.n. **Cathen**, -*iog¹*
(cmt.) *Ketheinauch* 1222, *Ketheinneauc* 1261, (cmt.) *Katheynoc* 1288, (cmt.) *Kethoynok* 1292, (cmt.) *Ketheynok* 1303, *Cathenock* 1532, *Cathinog* c.1550, *Kathynogg Hundred* 1671
The same pers.n. is found in Llangathen (q.v.) and probably Myhathan but they need not refer to the same person (HCrm I, 8). The n. was also borne, for example, by a reputed saint Cathen ap Cawrdaf recorded as *Kathan ab Kowrda* in additions to a pedigree 'Bonedd y Saint' compiled in the 12th cent (EWGT 66).

Cathilas[1] ?SN 588166
Llanfihangel Aberbythych
Uncertain
(?tmt.) *Kithilas* 1609, *Kythylase* 1611, (mess. etc) *Tir Cathillas* 1666, *Tythin Cathilas* alias *Brin y Vedwen* 1705, *Kathilas* 1745, (ht.) *Cathilas* 1811
The first el. appears to be **cath** found in other pns. in this area (see Cathargoed) and favours a topographical origin rather than a pers.n. or pers.n. el. The 1705 ref. identifies it with Brynyfedwen otherwise recorded as (mess. etc) *bryn y weddwen* 1664 and *Bryn fedwen* 1784 (grid ref. cited) (**bryn, y, bedwen**). There is another ho.n. Cathilas (SN 646145) in Llandybïe recorded as *Kethilas* 1601, *Cathilas* 1616, *Kathylas* 1658, *Kathilas* 1784 adjoining a small unnamed stream.

Cathilas[2] SN 5931
Llanfynydd
Uncertain
tir Kethilas 1617, *Kethilas* 1656, (messuages etc) *tire Cathilas* (and *teay Lan vach*) 1733, (ht.) *Cathilas* 1851
Probably tmt. *Cathilas Land* 1785. Cf. Cathilas[1]. The ho. has been identified with Ffosyrhwyad (SN 593319) (PêrG 225). The 1733 ref. associates it with *teay Lan vach* which is probably a former ho. recorded as *Penlanfach* 1851, *Penlan-fâch* 1888 (SN 600315) (**pen**[1], **glan, bach**[1]). Cathilas ht. made up the north-eastern part of Llanfynydd between Afon Melingwm (recalled in Felin-gwm SN 586320), Nant Ffrainc and Brisgen. There is nothing very obvious in the landscape to explain the n.

Cedweli see **Cydweli** (Kidwelly)

Cefn-blaidd SN 647334
gr. Talyllychau/Talley
'wolf's ridge' or perhaps '(hill shaped like) a wolf's back': **cefn, blaidd**
Kevenbleith 13th cent (1332), *Keven Blaith* 1520, (bailiff's fee, gr.) *Kevyn Blayth* 1535, (gr.) *Keven blaidd* 1562, 1715, *the Grainge of keven blaith* alias *keven llith* 1636, (ht.) *Cefn y Blaidd* 1811, *Cefnblaidd* 1831

The n. is probably taken from the hill immediately north of the ho. A former gr. or farm of Talyllychau abbey (HCrm I, 155, 352, 362; CStudies 113).

Cefn-bryn-brain SN 744135
Llangadog p.; Cwarter-bach
'ridge near Bryn-brain': **cefn**, pn. Cefn-brain
Cefn-bryn-brân 1877-8, *Cefn-bryn-brain* 1906
The ho. Bryn-brain or Bryn-y-brain (SN 747135) is recorded as *Bryn-brain* 1831, *Bryn-brân* 1877-8, 1906, meaning 'hill of crows', **bryn, brain** pl. of **brân**. A nearby ho. is *Nant Brân* 1887-8, *Nant-y-brain* 1906 'stream of crows' (**nant**) referring to Nant Llynfell.

Cefncaeau SN 532002
Llanelli p.
'ridge characterised by fields or enclosures': **cefn, cae** pl. **caeau**
(row of hos.) *Cefn-caeau* 1880, 1916, *Cefncaeau* 1882, 1953
The n. was transferred to post-War housing before 1953 by which date the original hos. were re-christened Cefn Row. The **cae** theme was continued in the street-ns. of Cae Canol, Cae Isaf, Cae Uchaf (**canol, isaf, uchaf**).

Cefncerrig SN 775323
ht. Myddfai
'ridge characterised by stones': **cefn, carreg** pl. *cerrig*
Cefn Cerrig 1811, *Cefn carreg* 1831, *Cefn-ceryg* 1886, *Cefn-cerig* 1906
From its location between Cwm Mydan and Llwynwermwd Park.

Cefn-coed SN 8136
fm. Llanfair-ar-y-bryn
'ridge covered with trees': **cefn, coed**
Tyr Keven Coed alias *Tyr Ken Coed* 1606, *Keincod Mayor* 1640, (tmts.) *Cencoed* 1791, *Cefn-y-coed* 1832, *Cin-coed* 1888, *Cefn-coed* 1905
Historic forms are similar to those for Cyncoed, co. Glamorgan (PNGlamorgan 63) and for Cefn-coed (in Caeo) (*Kencoed* 1778), Cefn-coed (in Llandeilo Fawr) (*Kencoyd* 1639), Cefn-coed (in Llanegwad) (*Kefncoed* c.1500, *Kencoed* 1776)

and Cefn-coed (in Llanelli) (*Kevencoed* 1618, *Kencoed* 1695).

Cefneithin SN 554138
ht. Llanarthne
'ridge covered with gorse', *cefn*, *eithin*
Cefn-eithin 1831, 1906, *Cefneithin* 1877, 1961-4
The n. applied in 1831 to a ho. roughly at the location of Tan-y-banc (SN 551140) and was gradually extended to housing which developed from the mid 19th cent along the road leading eastwards from crossroads (SN 551139) towards Gors-las road and southwards along Trefenty Road.

Cefn Llwydlo SN 8441, 8542
'ridge near Llwydlo': *cefn*, *Llwydlo*
Ludlo Hill 1791, *Cefn Llwydlo* 1832
Llwydlo is *Llwydlo vach* 1729, *Ludlow fach* 1794, *Llwydlo fach* 1809, 'lesser Ludlow' (*bach¹*) in full. The antiquarian Richard Fenton describes *Ludlow vach* in 1806 as a ruinous ale-house where suitors to the royal Court of the Marches from south Wales used to meet to settle matters (Fenton 16). There seems to be no evidence to corroborate this and there may have been some confusion with the market town of Ludlow, W form Llwydlo, Shropshire, where the court met regularly. The n. could simply describe a place used by drovers and other travellers and imagined to be as busy as Ludlow, perhaps in a mocking sense. Similar transferred ns. occur elsewhere. Examples include a cottage called Llwydlo-fach, in Llangyndeyrn, recorded as *Ludlow vach* otherwise *Llwydlo* 1824 and a ho. in Llanllwni is *Llwydlo* 1891 (SN 487392). Cf. also *Llunden vach* 1729, *Llundainfechan* 1831 (SN 614406) (*bach¹*, *bechan*) in Llansawel, and Little London otherwise Llundain-fach (SO 044892), at Llandinam, co. Montgomery (PNMont 107).

Cefn Padrig SN 4700, 4800
'bank or ridge (of sand) of (man called) Patrick': *cefn*, pers.n. **Patrick**, **Padrig**
the Cevern Patrick Sand, *Cefen Patrick* 1806, *Cefn Patrick Sands* 1813, *Cefn Patrick* 1830

Patrick or Padrig has not been identified. As a surname Patrick is very uncommon in Wales. No instances have been found among wills of St Davids diocese and just one in Llandaf (Richard Patrick at St Athan 1652).

Cefn Sidan SN 3205
shallows
'ridge or bank (of sand) smooth as silk': *cefn*, *sidan¹*
Cefn Sidan or Silken Bank 1800, *Cefn Sidan* 1811, 1830, (*ar dywod*) *y Cefn Sidan* 1819, *Cefn Sidan Sands* 1903
A large expanse of intertidal sands west of the Gwendraeth estuary extending southwards along the coast of Pen-bre Burrows and once notorious for shipwrecks.

Cefntelych see **Telych**

Cefn-y-pant SN 191254
Llanboidy
'ridge at the hollow': *cefn*, *y*, *pant*
Keven Pant 1816, *Cefnypant* 1848, *Cefn-y-pant* 1888-9
The location of a Congregational chp. built by Glandŵr ch.

Cellifor SN 562177
Llanfihangel Aberbythych p.
'short grove': *celli*, *ber*
Kellwer 1712, *Gelly woer* 1713, *Kelliver*, *Kellivor* 1720, *Callifer* 1748, *Calliver* 1773, *Caellifer*, ~ *Estate* 1797 (1833), (ht.) *Calafyr* 1811, *Callifer* 1822, *Cae-llifer* 1831, *Caellifer* 1953, *Cellifor* 1974-5
The combination of *celli* and *ber* generally produces the form Celli-fer stressed on the first syllable but stress seems to have shifted to -i-, ie. *Cellifer prompting confusion of Ce- with *cae* 'enclosure, field'. The n. survives as that of a fm. Richards (WATU) uses the form Caellifer, presumably based on late evidence and possibly by association with **llif²** and ***bêr** the supposed els. in **llifeiraint** nm. and pl. 'a flowing, flood'.

Cenarth SN 269416
p., ht. Cilrhedyn p.
'lichen ridge', *cen¹*, *garth¹*
Cenarth maur c.1125 (c.1154), *Sancti Leudoci ... in loco qui dicitur Kenarthmaur* c.1191, *ecclesiam Sancti Ludoci et novem sanctorum de Canartmaur'* c.1225, *Kenarthvaour* 1222 (1239), (mill and weir) *Kennarth* 1302, (ch.) *Keniarth Vaur* 1315, *Kenarth* 1551, 1651, *Cenarth, or, Cefn garth, i.e. A Ridge of Land behind the Wear* 1811
The late misassociation with **cefn** may derive from familiarity with the common spoken, sometimes written, variation of *Cen-* for *Cefn-* cf. Cefn-coed. Cenarth was once thought to contain Ir *ceann* 'head' as an equivalent to Penarth (ETG 235; PNGlamorgan 162-3). Cenarth had a well Ffynnon Lawddog, 'well dedicated to (St) Llawddog' and a feast-day recorded as *Ffynnon Lawdhog* and *Dygwl Lawdhog* c.1700 (**ffynnon, Dydd G yl, Llawddog**). There was also a Ffynnon Lawddog near a hill in Penboyr (HPLlangeler 12-13) and see Llanllawddog. The 'nine saints' c.1191 are not named. Cenarth village and ch. occupy the northern end of a narrow promontory with steep slopes leading down to the r. Teifi. Cf. Cenarth (SN 977757), co. Radnor, recorded as *Kenarth* late 15th cent, *Cenarth* 1734.

Cennen SN 6418
rn.
?'river notable for lichen' or 'scaly river': *cen¹*, *-en*
(o) *Gennen* mid 15th cent, (stream called) *Kennayn* 1541, *Kennenn Riveret* c.1537, (r.) *Kennen* 1754
The r. rises on Mynydd Du/Black Mountain (SN 706181) above Blaencennen (**blaen**), passes below Carreg Cennen (q.v.) and meets the r. Tywi at Abercennen (SN 633218) (in Ffairfach) recorded as *Aberkennen* 13th cent (1332) in ref. to the ch. in Llandeilo Fawr and 1609 (**aber**). R.J. Thomas (EANC 108) suggests that it may be a pers.n. possibly taken from Carn Cennen (**carn**) applied on OS maps to an area just below its source. His third suggestion is that Cennen contains the el. *cen¹* found also in *cen y cerrig* 'lichen'. This seems more plausible with the specific sense of 'r. notable for rocks covered with lichen' or 'having a scaly surface', finding a parallel in Cenarth (q.v.).

Cilcarw Cilcarw Uchaf SN 499122
t. Llangyndeyrn
'nook of the deer, nook where deer are found': *cil¹*, (*y*), *carw*
Kilarowe 1532, *Kyl a Caro* 1549, *Kilykarw vach*; *Kilykarw vawr* 1591, (place) *kil y carw Vaur* 1608, *Kilicarw*, (o) *Gil y Carw* 1741, (mess. etc) *Killcarrow vach* 1734
Cilcarw Uchaf occupies southerly slopes of a ridge extending southwestwards to Cilcarw Isaf, now plain Cilcarw (SN 472106). Cilcarw Isaf appears to be the same place as Cilcarw Fawr (*mawr*) in earlier records, described as *Kyl y karw fawr* extending to r. Gwendraeth Fawr on south in 1620. Cilcarw Fach (SN 493114) is *Cil-carw-fâch* 1880-7 (*bach¹*) but takes on the unusual forms *Gwely-Harvey* 1906, and *Gwilihervy* from c.1960. Presumably it has been influenced through false association with *gwely* and the pers.n and surname **Harvey**.

Cilellyn ?SN 1522
ht. Cilymaenllwyd
'nook at ?': *cil¹*, ?
(ht.) *Cîl Ellyn* c.1700
Further evidence is gathered under Maenor Cilellyn (q.v.). The first el. is clearly *cil¹* but *ellyn*, nmf. 'razor' seems to make no obvious sense unless it is reinterpretation of what appears to be a rn. possibly found in the ho.n. Dyffryn Henlyn (SN 153224) recorded as *Dyfryn hilin* 1794, *Dyffrynhelin* 1851, *Dyffryn-Helin* 1889-1964 (*dyffryn*, pers.n. **Heilyn**). The alternative explanation is an earlier Cil-y-llyn > Cilellyn with shift of stress to the middle vowel; cf. Cil-y-llyn (in Llansawel) and Bachellyn (? < Bach-y-llyn), co. Caernarfon (ELISG 102).

Cilfargen SN 5724
Llangathen
'nook of ?(person called) Margain': *cil¹*, pers.n. ?*Margain*
Kylvargeyn 1215, *Kiluargayn* 1301, (land of abbot of Whitland of) *Kiluargan* 1304, (land

called) *Kylvargan* (in cmt. of *Ketheynock*) 1309, *Kilbargon Grange* 1605
The pers.n. does not seem to be on record and its etymology is uncertain though -gain is probably *cain* found in other W pers.ns. such as Ceinwen. The location of a gr. of the abbey of Hendy-gwyn/Whitland. See also Llanfihangel Cilfargen

Cilgannwr SN 5426
t. Llanfynydd
'singer's nook': *cil¹*, **canwr**
Kilgannwr 1741, *KillyGanwr* 1748, (ho.) *Cilgannwr* 1831, *Cilcanwr* 1867
canwr may be a songbird, eg. *canwr y coed* 'wood warbler' and *canwr yr ardd* 'garden warbler', or perhaps *cannwr* nm. 'bleacher'. An association with a songbird is more convincing. The t. was centred on Cilgannwr (SN 543263).

Cilgryman Cilgryman Fawr SN 234252
gr. Llanwinio
?'nook shaped like a reaping-hook': *cil¹*, *cryman*
Kylgreman' 1215, (chs. de Sancto Wynnoco et de) *Kylkemara*, (ch. of) *Kylkemeram* 1260, *Kilcrennan* 1287, *Kilgremman* 1316, (mill) *Kylegryman* 1535, *Cilgrymman* 1744, *Cilgrymman-fawr, -fach* 1843
The n. of a gr. which belonged to Hendy-gwyn/Whitland abbey, presumably located at Cilgryman-fawr or Cilgryman-fach (*bach¹*, *mawr*) and once possessing its own ch. or chp. The mill, known as *Monks' Mill* (SN 228342) in 1889, was located in Cilymaenllwyd, and demolished in 1967. For *Sancto Wynnoco* see Llanwinio.

Cilgwyn SN 745299
ht. Myddfai
'white or fair nook': *cil¹*, *gwyn*
y kilgwyn c.1560, *Kilgwyn* 1627, *Kilgwynne* 1686, *Cîl Gwyn* 1811, *Cilgwyn* 1831
The n. of a large ho., which acquired the addition of *Manor* some time after 1964, adjoining Afon Ydw. The actual nook may have been a little to the north around Llety-Ifan-Ddu where there

is a ho. Pistyll-gwyn (SN 743304) (***pistyll***) and a ho. Cae-gwyn (SN 742310) (***cae***).

Cilhernin CRM SN 179251
ht. Llanboidy p.
'nook of (man called) ?H(i)ernin': *cil¹*, pers.n. ?***Hernin*** or ***Hiernin***
Kilhernin Hamlett 1692 (1758), 1738, (ht.) *Kilhernin* c.1700, *Kilherning* 1793, *Cil-hernin* 1889
Centred on Cilhernin (SN 179251) recorded as *Kilhernin* 1692 (1758), 1738, *Kilherning* 1793 and *Cilehernnin* 1819 and located in Maenor Cilheirnin (q.v.). The pers.n. may be a variant of Hoiernin, the n. of a saint and confessor (LBS III, 281-2). Cf. **Llanhernin**.

Ciliau CRM SN 5012
t. Llanddarog
'nooks': *cil¹* pl. *cil(i)au*
(place called) *Kilay* 1606, 1644, *Kilayth* 1612, *Kilhay* or *Killay* 1726, (ht.) *Cilè* 1811
Probably centred on Cilau-uchaf (SN 506122) and Cilau-isaf (SN 503120). Sps. are too late for certainty but they bear comparison with those for Ciliau (Llandysilio) *Kyle* 1589, *Killey* 1700, *Killa* 1796, Ciliau (Llanelli) *Killey* 1728, and Ciliau (Pen-bre) *Kyla* 1542, *Kila* 1549, *Killey* 1693.

Cil-march SN 404130
t. Llandyfaelog
'horse's nook': *cil¹*, *march*
Kilmargh 1584, *Kilmarch Hamlet* 1792, *Cil-y-march* 1889, 1964, *Cilmarch* 1969-70
The def.art. *y* in some forms is intrusive. A ho. located at the head of a wooded dingle leading down to the Tywi. Identified with *Kellymarch* c.1501 though this would favour *celli*. Misrecorded as *Clun-y-march* 1831.

Cilmaren SN 6537, 6837
Caeo
'nook of ?Maren': *cil¹*, ?pers.n. *****Maren***
Kilmaren 1331, *Kylmaren* c.1538, *Kyllemehayren, alias Killemeharen* 1598-9, *Kilmaharen* 1633, *Kilmaharren alias tir Wennalt* 1650, *Grange of Killmaren alias Kilmaharen* 1668, *Cil Maren* 1831

Some later forms have been influenced through association with **maharen** nm. 'a ram' (CStudies 114).

Cilnawen see **Maenor Cilnawen**

Cilrhedyn   SN 279349
chp. Cenarth
'fern nook': *cil¹*, **rhedyn**
lann teliaui cilretin inemblin c.1135, *Kylredyn* c.1291, 1404, *Kilredyn* 1323, *Keyll Reden* 1532, *kil Redyn* c.1566, *Kilrhedin* 1619, *Kilrhedyn* 1750, *Cilrhedyn* 1831
Other sps. are cited in PNPemb 363. The grid ref. is that of the p. ch. in co. Pembroke dedicated to Teilo; cf. Llandeilo Fawr. This seems to have originated as a chp. of Cenarth (in the ctf. of Emlyn). The division of ecclesiastical and civil administration in the 1890s led to the formation of Cilrhedyn West cp., co. Pembroke, and Cilrhedyn East cp. (later part of Cenarth cp.), co. Carmarthen. As a n. Cilrhedyn has shifted on recent OS maps nearly a mile southwards to crossroads (SN 282339) near a ho. Lan-cwm.

Cilsân   SN 595221
Llangathen
?'nook of (?person called) *Sân*: *cil¹*, ?pers.n. **Sân*
Kylsaen 1289, 1590, *Keleseyn* 1295, (land of) *Kilsaen* 1303, *Kil sant* 1527, *Kylsayne* 1532, *Kilsane* 1594, *Keelsaen* 1687, *Cillsant* c.1735, *Cilsaen* 1831, *Cilsân* 1887
See also Rhiw'radar and Cilsân. The second el. could be **sân**, *saen²* nm. 'seine(-net)' or **saen¹** nf. 'cart, wagon', but the first is thought to be a borrowing from E *seine* and the authenticity for the second is in doubt, The first literary refs. to both els. appear to be c.1588 (GPC).

Cilsant   SN 2623
Llanwinio
?'saint's nook': *cil¹*, ?**sant**
Kilsant 1578-1741, *Killsaint* 1750, *Cilsant* 18th cent, 1831
The evidence is quite late and we cannot rule out a like meaning to Cilsân (q.v.). An identical pn. is recorded in Llangyfelach, co. Glamorgan, as *Kilsant* 1764.

Cil-wr   SN 605328
t. Talyllychau/Talley
Uncertain
Kilwr 13th cent (1332), 1557, *Kyl-wr* 1647, *killoure* 1707, (two tmts. called) *Kilwr* 1734, (ht.) *Cîlwr* 1811, *Cilwr* 1831
There seem to be no close parallels and historic forms lend only uncertain derivation from *ciliwr* nm. 'pursuer, one who puts to flight; recluse; who who goes or falls back', with the sense 'retreat suitable for a recluse', or *ciliwr* as a personal epithet. Identified with *Cynhwm*, a former chp. and gr. of Talyllychau/Talley abbey (WSS 331) but this is better identified with Capel Cain Wyry.

Cil-y-cwm (Llanfihangel Cil-y-cwm)   SN 753401
p., ht.
'nook in the valley': *cil¹*, **y**, **cwm**
Ecclesia de Kilcom 1304, *Thlaunvyhangel Kylcome* 1308, (ch.) *St. Michael, Kilcum* 1347, *Kylcombe* 1415, *Kylecom* 1535, *ll. V'el kil y kwm* c.1566, *Kilycoom* 1595, *Kilycwm, Kilykwm* c.1600, *Llanihangell Kil y Cwm* 1703, *Cilycwm* 1740, *Cilc m, Cilycwm* 1831
The def.art. seems to be intrusive, first unequivocally recorded in 1557. The p. and ch. are Llanfihangel Cil-y-cwm in full referring to the ch. dedication to St Michael (**llan**, pers.n. **Mihangel**). Located in the valley of Afon Gwenlais.

Cil-y-gell   SN 569457
Pencarreg
?'the nook with a cell or bower': *cil¹*, **y**, ?**cell¹**
Killygell 1757, *Cil-y-gell-isaf*, ~ *-ganol*, ~ *-uchaf* 1831
Historic forms are too late for confident analysis and interpretation. The n. applies to three hos., 'lower' (**isaf**), 'middle' (**canol**) and 'upper' (**uchaf**), on the eastern side of a stream Nant Call, so spelt 1888. Early forms of this have not been found but the similarity of *gell* and *Call* is notable and cf. Callan, a rn. in co. Brecon

(*Blain Calln'* 1588, *Blaen Callan* 1651) (***blaen***) thought to contain ***call*** adj. in the specific sense of 'sharp, wily', and suffix ***an*** (EANC 46). Located in Rhandir Cil-y-gell.

Cilygernant SN 5719, 5820
t. Llanfihangel Aberbythych
'nook near a stream' ***cil¹***, ?***ger***, ***nant***
(ht.) *Cil Garnant* 1811, (ht. of) *Llan-Cilgernant* 1851
Modern *y* appears to be intrusive. There is no rn. in modern sources to favour a comparison with Garnant (q.v.). In the second ref. the pn. is combined with Llan (q.v.) as a unit for census purposes.

Cilymaenllwyd SN 152234
p., cp.
'nook at the grey *or* holy stone': ***cil¹***, ***y***, ***maen***, ***llwyd***
(*ecclesiam de*) *Kilmainlor* c.1185 (17th cent), *Kilymaenllwyd* 1231, *Kilymaenllwyd* 1231, *Kylmaynthloid* 1332, (ch.) *Kylmaenlloyd* 1382, *p. kil y maen llwyd* c.1566, *Killmaenlloyd* 1646, *Killymaenllwyd* 1672 *Cil-y-maenllwyd* 1751, *Cilmaenllwyd* 1843
The n. is no longer shown on OS maps. There is a short wooded dingle east of Llandre running down to Afon Wenallt but the stone under which St Byrnach is said to have slept has not been located. He is also dedicated at nearby Llanfyrnach, co. Pembroke (AC 1975: 107). The ch. is located at Llandre (so spelt 1889) and is dedicated to Philip and James. The last pers.n. is recalled in the ho.n. Ffynnon-Iago (SN 144231) (***ffynnon***, pers.n. ***Iago*** = James) west of the ch. Two hos. north-east of the ch. are recorded as *Pumsant* and *Pumsant-uchaf* (***uchaf***) with and adjoining hillside *Allt Pumsant* 1889-90 (***allt***). Pumsant 'five saints' (***pump***, ***sant***) appears to suggest another ecclesiastical site or a holy well; cf. Pumsaint.

Cilyrychen SN 615171
Llandybïe
'nook of the ox': ***cil¹***, ***yr***, ***ychen***
kilyrychen c.1550, *Kilyrychen* c.1700, *Killarychen* 1722, *Cilyrychen* 1790, *Cil-yr-ychen* 1831

There is a small depression immediately around the ho. and a ditch running east to Afon Marlais. The n. was transferred to late 18th-20th cent limeworks and quarries (HER 4858). Cf. Cil-yr-ŷch, in Llanfair Caereinion, co. Montgomery, recorded as *Kyllerech* 1309, *Kyllyrech* 1577, *Kilyrch* (*ynghaer Engion*) 16th cent, *Kilyrych* 1653-4.

Clog-y-frân SN 240160
manor Sanclêr/St Clears lp.
'the crow's cliff, cliff inhabited by crows': ***clog²*** nf. 'rock, cliff, precipice', def.art. ***y***, ***brân***
Glogvranet c.1260, *Clogeuran* 1313, *Clogvraen* 1606, *Clog y Fraen* 1665, *Clog y Vrane* c.1700, *Clogvran* 1727, *Clog-y-frân* 1843
A steep, curving cliff overhangs the r. Taf extending southwards from the ho. Clog-y-frân for half-a-mile to the wood Coed Clog-y-frân. A manor in lp. and p. Sanclêr/St Clears and a fm. (HCH 35-36).

Cloidach SN 5126
rn. SN 513276 Abergwili, Llanegwad
?'the one that washes' or 'the strong-flowing one: Ir ?***cláidigh*** variant ***clóidigh*** or ***cláideach***
Clendach, *Cleudach* 1331, *Cloidach* 1889
The n. was associated by R.J. Thomas (EANC 8-12) with Ir ***cladach*** 'shore; rocky foreshore' (FGB) with the sense 'rocky river' (DPNW 89-90). Historic forms of Cloidach (and Clydach) are late but initial *Cleu-* favours identification with Ir ***cláidigh*** variant ***clóidigh*** or ***cláideach*** found in rns. in Ireland such as Clady and Cladagh/An Chlóideach, co. Fermanagh (www.placenamesni.org; DUPN 41); ***cláideach*** is now translated as 'mountain stream' in modern dictionaries (FGB). The first el. Clai-, Cloi- can be traced back to Celt *$kleu$- meaning 'to wash' and the second el. is the suffix -*ach* found in other rns. in south and west Wales such as Llechach (q.v.). The possibility that the first el. has a common origin with W ***clau*** 'swift', generally derived from Celt *$klou$- (GPC), is unproven. Ir settlement in Dyfed and other parts of south Wales 400 x 600 is attested in bilingual monumental inscriptions (Latin and Ir), Ir pers.ns., medieval pedigrees

and later literary sources though the language was probably extinct here before the ninth century (WB 112-5, 114, 174-181 and refs.). By contrast, few clear instances of Ir pns. survive in this area (PNPemb xviii-xxi.). One literary source Historia Brittonum c.829/830 also refers to Ir settlement in Cedweli and Gŵyr (Gower) and there is evidence of Ir dynastic links with Brycheiniog (WB 20). If we accept that the suffix *-ach* is Ir, then Cloitach, Clydach and other rns. such as Bradach and Caeach would appear to suggest that the language was once spoken as far east as Gwent. This has been questioned (PRIS 46-61) and we need to make allowance for the possibility that some rns. contain or may have been influenced by a distinctive W suffix *-ach* meaning 'little', later used as a pejorative, or that Ir *-ach* migrated as a suffix into areas unsettled or thinly settled by Ir-speakers. Even if we accept that *-ach* in all cases is Ir, it does not mean that this language completely displaced the W language in south-west and south Wales for the period 400 x 600 AD. West Wales is far more likely to have been a land of three languages with W and Ir as co-existing vernacular and Latin as the scholarly language (WB 114). If we accept for the sake of argument that the immediate origin of Cloidach and Clydach is Ir *cláideach*, then this first gave rise to Cleudach and later Cloidach in local dialect in a similar fashion to pns. such as Llanboidy (q.v.) < Llanbeudy. In other parts of south Wales the rn. developed by way of Clidach and Clitach. The form Clydach (q.v.) may in some instances be a modern imposition. The most difficult matter is understanding the etymological and historic relationship of the rns. Cloidach, Clidach and Clydach found in south and south-west Wales with their apparent counterparts the rns. Clywedog in north and mid Wales. Clywedog is generally thought to contain *clywed* (Brit *cluúet*-) 'to hear, to listen' with suffix *-og* and the extended sense of 'noisy river' i.e. 'river which can be heard' (PNGlamorgan 43-44, and Clydach[1,2]). There are apparent comparisons in other rns. such as Afon Llafar (*llafar* 'loud, clear'). R.J. Thomas (EANC 8-12), however, casts doubt on this explanation and with good reason. He notes that historic forms for the r. Clywedog (near Llanidloes, co. Montgomery) include *Clawedauc* 14th cent; an identical form for this also occurs c.1207. Thomas's scepticism finds further support in early forms of Clywedog (Machynlleth, co. Montgomery) such as *Clawedauc* 1201 and *Clawedock* 1558 and for Clywedog (co. Radnor) which include *Claweda6g* c.1400. The modern form Clywedog for all these rns. may simply have been influenced by *clywed*. These matters remain contentious but ultimate shared etymology of Cloidach/Clidach/Clydach with Clywedog is very likely. Cloidach/Clidach/Clydach may be best regarded as substitutions made by Ir-speakers for earlier and unrecorded forms of what developed in other parts of Wales to Clywedog. The similarities of the Ir suffix *-ach* and W suffix *-og* must have been obvious in areas of mixed Ir and W settlement.

Cloigyn SN 433140
ht., t. Llandyfaelog
Uncertain
(lp.) *Clogyn* 1467, *Cloegyn* 1497, *Cloygyn*, *Cloygin* c.1550, *Cligin* 1578, *The Mannour of Cligin* 1609, *Cloigin* 1811
Apparently *cloigyn* a dim. of *cloig, clöig* 'hasp, clasp, spindle', but it is difficult to make sense of its application here unless it refers to the shape of the hill on which the present house Cloigyn-fawr stands. The n. of a t., lp. and manor recalled in the ho.ns. *Cloigan-fawr*, (~) *Cloigan-fach* 1831. The topographer Carlisle in 1811 (CarlisleTD) records that there was an extra-parochial chp. here but confirmatory evidence seems to be lacking. The modern ht. is a largely 20th cent development near a former corn-mill recorded as *Felin-cloigan* 1831, *Felin Cloigyn* 1888, 1964 (**melin**).

Clunderwen Pemb SN 121195
Llandysilio
'meadow possessing an oak-tree': *clun*[2], **derwen** (ho.) *Klyn derwen* c.1634, *Klin Derwidd* 1729, *Clynderwen* 1765, 1775, (ho.) *Clyn-derwen*, (ht.) *Clynderwen* 1890

The 1729 form is through association with *derwydd*². Clunderwen applied solely to a ho. (SN 132199) in Llandysilio East cp., co. Carmarthen, and was transferred to the railway station on the Great Western Railway and adjoining hos. The southwestern part of the village was in the t. of Grondre, co. Pembroke, and p. of Castelldwyran.

Clydach¹, **Afon** SN 5333
SN 510377 to Cothi SN 542306
?'the one that washes' or 'the strong-flowing one: Ir ?*cláidigh* variant *clóidigh* or *cláideach*
Cledagh Ddû 1729, *River Clydach* 1889 OS 1:2,500
Formerly qualified with *du* len. *ddu* 'black, dark, darker' in contrast to *Cledagh Wen* c.1762 (*gwyn, gwen*) now described as Nant y Ffin (*nant, y, ffin*). See further under Cloidach

Clydach², **Afon** SN 7221
rn. SN 757194 to Sawdde SN 737234
Cledagh flu: 1578, *River Clydach* 1886-7
Identical in meaning to Clydach¹.

Clynennos (Clenennos) SN 5429
t. Llanfynydd
'meadow characterised by ash-trees': *celynnen, -os*
Clynynnos 1851, *Clynunnos* 1863
Historic forms are very late but the suffix *-os* appears to confirm this interpretation since it is generally attached to a plant-n indicating an abundance of these; cf. Helygos (**helyg**) and Rhigos (**grug**).

Coedcae SN 523005
Llanelli
'land enclosed with a hedge': *coedcae*
(field) *Coed Cae near Bridge* 1842, *Coed-cae* 1880, 1907, *Coetgae* 1950-3, 1964-5, *Coed Cae* 1968
Current OS maps spell this as two words. An area of housing developed from the late 19th cent next to part of the former main road to Llangennech and Llwynhendy now known as Glyncoed Terrace. The br. recorded in 1842 crosses Dafen near Halfway.

Coed-gain SN 472183
t. Llangynnwr
'wood shaped like wedge': *coed, gaing*
Coedgaing 1619, *Coed gain* 1686, *Coedgainge* 1724, *Coedgain* 1750
The name could allude to the shape of the wood (SN 477187) north-east of Coed-gain Manor (for which see HCH 39-40). The second el. is unlikely to be *cain*¹ (in the sense of 'fair or bright wood') because len. would be irregular after *coed* which is a nm. and n.pl. Association with *cain*¹ might account for Coed-gaing > Coed-gain.

Coedmor SN 5946
t. Pencarreg
'big wood': *coed, mawr*
(place) *Coydemore* c.1557, *Coed mor* 1724, *Coedmore* c.1735, *Coedmor* 1798
Shortening of -*aw*- > -*o*- may be due to stress on the first syllable; cf. Coedmor (SN 1943), co. Cardigan, recorded as *Coytemaur* 1314, with later forms *koetmor, koedmor* c.1560, *Coedmore* 1777, and see Maenor Llangoedmor. We cannot completely rule out the pers.n. *Môr* as in Llandremor, co. Glamorgan (PNBont 49-50) with *llodre*.

Common Church see **Eglwys Gymyn**

Cornorion SN 5425
t. Llanfynydd
?'(land belonging to) the leaders *or* horn-blowers': *cornor*¹ pl. *cornorion*
(ht.) *Cornoyron* 1851, *Cornyron* 1863
Recorded forms are very late but the n. may allude to some tenurial or administrative office. An identical n. in Pennant Melangell, co. Montgomery, is recorded as *Cornorion* 1577 and 1600, and there are comparable ns. elsewhere such as Weirglodd y Telynorion, in Ysbyty Ifan, co. Caernarfon (ELlSG 47) (*gweirglodd, y, telynor* pl. *telynorion*) 'harpist'.

Cothi SN 685480
rn.
'spurting or sweeping (r.)': ***coth-*** and suffix ***-i***
Cothi 13th cent (1332), c.1400, (r.) *Gothi* c.1380 (c.1400), *Cothey* c.1537, *Kothey* c.1577 (1730), *Cothie* 1584, (r.) *Cothy* 1612
coth- is thought to be the el. in *ysgothi*, *sgothi* 'to defecate; eject, squirt' found perhaps in the stream n. Nant Sgothi or Scothi co. Glamorgan (EANC 134-5) though this has also been explained as a misdivision of Ynys Cothi. The sense may be that of a r. which spurts and expels like a r. which floods and destroys rather than a r. which is polluted.

Crachdy SN 5727
t. Llanfynydd
?'area of rough land (likened to scabs) near the house': ***crach***, (***y***), ***tŷ***
(place) *Crach y Tuy* 1605, 1665, *Craig y ty* 1669, 1727, (hos.) *Crach-y-ty-isaf*, ~ *-ganol*, ~ *-uchaf* 1831, (ht.) *Crachyty* 1851, *Crachdy-isaf*, ~ *-uchaf* 1887
Crachdy-isaf (SN 570261) and Crachdy-uchaf (SN 576270) are at opposite ends of a short ridge. Crachdy-uchaf and *Crach-y-ty-uchaf* applied to a former ho. (SN 5772720) (***isaf***, ***canol***, ***uchaf***). The first el. ***crach*** also appears in two other ho.ns. Crach-gelli-fach (SN 569296) and Crach-gelli-fawr (SN 564291) recorded as *Crach-y-gelli-fâch* and *Craig-y-gelli-fawr* 1831, *Crach-gelli-fâch*, *Crach-gelli-fawr* 1888 (***celli***, ***bach***[1], ***mawr***) about two miles to the north. Earlier historic forms of Crachdy and Crach-gelli-fach and Craig-gelli-fawr all contain the def.art. ***y*** which favour the n.pl. possibly with the sense of 'area of rough land (likened to scabs)'. ***crach*** is also known to have developed the additional sense of 'poor patch of land, mean tenement' although instances of this meaning cited in GPC date only from 1778. The late transformation of *Crach-y-ty* as *Crachdy* may be based on the common assumption that ***crach*** qualifies ***tŷ*** and causes the len. *-dy*; cf. Crachdir (in Llansawel) recorded as *Crach Tyr* 1729 (***tir***) and Crachfryn (in Llanfihangel Rhydieithon, co. Radnor) recorded as *Crach Vrin* otherwise called *Lletty hoilow* 1791 (***bryn***, pers.n. ***Hoedlyw***). Historic forms for 1669 and 1727 confuse ***crach*** with the more common pn. el. ***craig***.

Croes see **Capel Bach** (Abergwili)

Croesyceiliog SN 407164
ht. Llandyfaelog
'the cock's crossroads', ***croes***, ***y***, ***ceiliog***
Cross y Kilog 1729, *Croesyceiliog* 1745, *Crose y Ceilog* 1787, *Croes Ceiliog* 1804, *Croes-ceiliog* 1831, (village) *Croes-y-ceiliog*, (fm.) ~ *-fawr* 1891
Cf. Croesyceiliog (SN 713316), in Llanwrda, recorded as *Crossyceilog* 1768, *Croesceilog* 1824. ***croes*** is frequently combined with the ns. of birds, eg. Croes-y-frân, co. Pembroke, (***brân*** 'crow'), and Croes-y-giach, in Penrhos, (***gïach*** 'snipe') and Croes-y-mwyalch (***mwyalch*** npl. 'blackbirds'), in Llanfihangel Llantarnam, both in co. Monmouth. Many similar ns. are taken from inns or former inns but that is not always the case and it is possible that some crossroads possessed signs bearing the images of common birds; cf. Cross Hands.

Crosshands SN 195229
Llanboidy
'(signpost having) crossed hands': E ***cross***, ***hand*** pl. *hands*
Cross Hands 1906, *Crosshands* 1947
The n. first appears in 1906 applied to a small group of hos. and a smithy. Earlier recorded as *Cefnbrighlly* 1819, *Cefn-briallu* 1889, 'primrose ridge': ***cefn***, ***briallu***[1] n.pl. 'primrose'.

Cross Hands SN 562127
Llan-non
'(inn bearing sign of) crossed hands': E ***cross***, ***hand*** pl. *hands*
Crosshand 1815, *Cross Hands* 1831, (P.H.) *Crosshands Inn* 1880, (area) *Cross Hands* 1891
The location of an inn and a former toll gate (recorded 1831, 1844). The area developed near several small coal-pits and Crosshands Colliery (SN 567127) largely removed by the construction of the A48 and a business park.

Cross Inn see **Rhydaman**

Cross Inn SN 293125
Talacharn²/Laugharne
'inn at crossroads': E *cross, inn*
The Cross Inn 1834, (area) *Cross Inn* 1888-9
A ho. is shown on OS maps from 1888-9 between the Talacharn²/Laugharne-Sanclêr/St Clears road and a smaller road branching off for Llanddowror but this is not described as a PH. The Smith's Arms on the site of a ho. Bryntaf (SN 295124) is recorded down to 1964 on OS maps south-east of the fork. The n. is common with examples at Llanfihangel-ar-arth (SN 454393) recorded as *Cross Inn* 1756 and in ho.ns. *Cross Inn Vawre, Cross Inn Vach* 1738 (***mawr, bach***¹) and at Llangathen (SN 556223) recorded as *Cross Inn* 1759.

Cruclas SN 474237
t. Abergwili
'green hillock': ***crug, glas***¹
(place) *y Kryklas* 1575, *Criglas* 1683, *Crug-las* 1724, *Cryclas* 1765, *Crûg Glas* 1811, *Cricklas, Crucklas* 1851, *Crug-lâs* 1889-90
Cruclas is the form favoured by Richards (in WATU) but later historic forms appear to show shift of stress to the second syllable and have been influenced by awareness that the qualifier is ***glas***¹. The meeting of *-g* and *g-* regularly > *-c-*. Cruclas appears on OS maps from 1906 as *Cruglas-isaf* with ***isaf*** 'lower' but there seems to be no comparative 'upper Cruglas'.

Crug-glas
manor Llangynog
'green rocks': ***carreg*** pl. ***cerrig, glas***¹
(mess.) *kerick glaas* 1646
Earlier Cerrig-glas > Crug-glas under the influence of ***crug***. The n. is not current and is poorly-evidenced.

Crug-y-bar SN 658378
ht. Caeo p.
'mound at the hill': ***crug, y, bar***²
Crucbar 1331, (land) *Greke Bar'* 1503, (tmt.) *y Tyr yngchrygbarr* 1564, *tir Crigg barr* 1628, *Cryg y Bar* 1739, *Crugybar* 1750-1

The def.art. is first recorded in 1739 and appears to be intrusive unless it is represented by *-e* in *Greke* 1503. The hill may be that immediately west of the ht. between r. Cothi and Annell.

Crwbin SN 471132
ht. Llangyndeyrn
?'little heap': ***crwbyn***
Crwbin 1793, *Crwban* 1831, *Crwbyn* 1875
The suggested meaning cannot be regarded as certain as ***crwbyn*** is given as a dim. of ***crwb***¹ nm. 'hump, lump' and said to be more common in the north (GPC). The n. could apply to one of the spoil tips associated with quarries along the limestone ridge.

Crychan SN 8239
rn. SN 852411 to Tywi SN 789371
'rippling, choppy river': ***crych, -an***
River Crychan 1888
The n. would describe a rough torrent rippling over stones. R.J. Thomas (EANC 61) compares *Crychddwr* (***dŵr***), co. Caernarfon, *Crychnant* (***nant***), co. Merioneth, and other rns. The n. also survives in Abercrychan (SN 790370) recorded as *aber krychan* c.1550, *Aber Krychan* before 1569, *Abercrychan* 1767-1859, *Aber-Crychan* 1886-8 and *Glancrychan* 1760 (***aber, glan***).

Crychan Forest see **Fforest Crychan**

Cryngae SN 3539
fm. Penboyr
?'round enclosure': ***crwn***, ?***cae***
Cryn-gae 1831, *Cryngae* 1888
Historic forms are late though families occupying Cryngae ho. and estate have been traced back to the 14th cent (HCH 44-45). It is a little difficult relating the suggested meaning to topography and the pn. needs further research. Cryngae farm appears within an uneven triangle of small enclosures and gardens on the OS map 1888.

Cuch (Cych), **Afon** SN 2538
rn. SN 290326 to Teifi SN 245415
?'scornful (r.): ***cuwch¹***, *cuch*
Keach flu: 1578, *Kych* 1585, *the Cheach* 1587, (r.) *Kŷch* 1609, (r.) *Cŷch* c.1700
The precise meaning is unclear though it could describe an unpleasant r. notable for causing floods and damage. The r. rises at Blaen-cuch (SN 290325) recorded as *Blaenkych* 1603 (***blaen***) and reaches Teifi at Aber-cuch (SN 248410), co. Pembroke, recorded in *Tythyn Aberkych* 1588, *Aberkeach* 1603, *Abercych* 1791 (***tyddyn***, ***aber***). See also Cwm-cuch and Maenor Blaen-cuch.

Cwarter-bach SN 7216
t. Llangadog p.; cp.
'(the) little quarter': ***cwarter***, ***bach¹***
y Cwarter Bach 1884, (p.) *the Quarter Bach* 1888
cwarter is not used literally since Llangadog had eight sub-divisions or townships (WATU); contrast Cwarter-mawr (q.v.).

Cwarter-mawr
t. Llanddeusant
'(the) big quarter': ***cwarter***, ***mawr***
(ht.) *Quarter-mawr* 1804, *Quarter Mawr* 1811
One of four ts. of Llanddeusant. ***cwarter*** is fairly common in fns. and the ns. of administrative divisions in south Wales.

Cwarter Trysgyrch see **Maenor Cwarter Trysgyrch**

Cwmaman SN 6714
Llandeilo Fawr p.; cp.
'valley of (r.) Aman': ***cwm***, rn. **Aman**
Cwmamman 1742, *Cwmamman* 1757, 1831
Cf. Cwmaman, co. Glamorgan (PNGlamorgan 54-5). The ecclesiastical district was formed out of the ps. Llandeilo Fawr, Llandybïe, Betws, Llan-giwg and Llangadog and the ch. consecrated in 1842 (RC I, 319).

Cwm-ann SN 583472
ht. Pencarreg
?'valley of (woman called) Ann': ***cwm***, ?pers.n. **Ann**
Cwmann 1740, *Tyr Cwm Ann* 1747, *Cwm Ann* 1817, *Cwman* 1834, *Cwmanne* 1859
There is no conclusive proof of anyone bearing the pers.n. The village was centred on the junction of the Llanybydder, Llanbedr Pont Steffan/Lampeter and Pumsaint roads which meet at Pen-y-bont and the former turnpike gate Gwar Gate (SN 584473) developing along the A482 towards *Tre-Herbert* (1888-9).

Cwm-bach¹ SN 488020
ht. Llanelli
'little valley': ***cwm***, ***bach¹***
Cumbach 1787, *Cwmbach* 1830
Located at the south end of a small, narrow valley Cwm Bach close to its junction with Cwm Mawr. The stream in Cwm Bach is shown as Afon Dulais on OS maps since 1910; earlier maps apply the rn. to the stream in Cwm Mawr now known as Afon Cwm-mawr.

Cwm-bach² SN 2525
ht. Llanwinio
'little valley': ***cwm***, ***bach¹***
Cwm-bach 1831, *a'r Cwmbach* 1867
A small side-valley to that of Afon Sïen. The location of a Calvinistic Methodist chp. built in 1765.

Cwm-ban Cwm-ban-fawr SN 569280
t. Llanfynydd
'valley of ?(r.) Ban': ***cwm***, ?rn. **Ban**
Cwmban 1683, 1715, *Cwm-ban-fach*, ~ -*fawr* 1831, *Cwm-y-ban*, *Cwm-ban-bâch* 1887-9
Ban may refer to the stream which rises above Cwm-ban-fach (SN 567283) (***bach¹***) and flows southwards to Afon Felindre (SN 569270) but there seem to be no historic instances and it is possible that the second el. is ***ban¹*** with the particular sense of 'arm, branch' describing a side-valley or a stream in a side-valley. The stream is very small and I think we can reject ***ban⁵*** adj. 'loud, noisy' which would be more appropriate to a torrent.

Cwmblewog see **Cwmliog**

Cwm-byr SN 633321
Talyllychau/Talley
'short valley': *cwm*, ***byr***[1]
Cumbyr 1331, *Cwm-byr* 1831, 1888
A former possession of Talyllychau/Talley abbey. Other examples include Cwm-byr (Llanegwad) recorded as *Cwm Bur* 1733 and Cwm-byr (Llanfihangel-ar-arth) recorded as *Cwmbyr* 1740.

Cwm Capel SN 455022
Pen-bre
'valley with a chapel': ***cwm***, ***capel***
Cwm Capel 1835, (valley) *Cwm Capel, ~ Colliery* 1880, (valley) *Cwm Capel* 1969
The valley of Nant Dyfaty, apparently transferred to hos. on its north side in 1990s. The valley is presumably named from Carmel (SN 458022) erected 1827 (RC I 288) or Jerusalem Independent chps. (SN 446016) erected 1812 (RC I, 287) between Craig Capel (Graig) and Achddu.

Cwmcarnhywel SN 533003
Llanelli
'valley near Hywel's cairn': ***cwm***, ***carn***, pers.n. ***Hywel***
Cwm-carn-Hywel 1880, 1921, *Cwmcarnhywel* 1954-1993
The name applied in 1880 to hos. along the road on either side of the White Lion PH and shifted up to the junction with Llandafen Road at what is now called Pemberton. The name was extended during the 1950s to describe new housing constructed around Brynsierfel 'hill bearing 'chervil' (***bryn***, ***serfel***[1], ***sierfel***[1]) and towards Trallwm (***trallwng***).

Cwm Cathan SN 6409
Betws
'valley of (r.) Cathan': ***cwm***, rn. ***Cathan***
(lands) *Cwm Cathen* 1727, *Cwm Cathan* 1745, (valley) *Cwm Cathan*, (hos.) *Cwm-cathan-isaf, ~-uchaf* 1878
The n. also survives in that of two hos. Cwmcathan-isaf (SN 644094) and Cwmcathan-uchaf (SN 648094) (***isaf***, ***uchaf***).

Cwmcawlwyd SN 6429
t. Llandeilo Fawr
'valley of (r.) Cawlwyd': ***cwm***, rn. ***Cawlwyd***
Cumcawlud 1685, *Cwm Cawlwyd* 1746, *Cwm Carw Llwyd* 1811, (ht.) *Cwmcawrllwyd* 1851
This may be Cwm Caflwyd (in editorial form) which survives in a 15th cent poem (GyN 4.94n) addressed to Siancyn ap Llywelyn whose family associations were with south Wales. This has otherwise been identified with Cwm Cowlyd a n. inferred from the n. of the lake Llyn Cowlyd, co. Caernarfon (ELISG 93), and with a ref. to the Owl of Cwmcawlwyd (*Cuan Cwm Kawlwyt*) in the medieval tale of Culhwch and Olwen (CO[3] nos. 871-2 and notes). We cannot be absolutely sure on the correct identification of these pns. but it is probable that they all contain the authenticated pers.n. ***Caw*** (TYP 306-8; CO3[3] index for examples) with ***llwyd*** probably in the sense 'holy' rather than 'grey'. Cwmcawlyd has also been interpreted as a ref. to a giant apparently known as Cawr Llwyd (GW 43 misplacing 'Cwm Cawr Llwyd' in Talyllychau/Talley) with ***cawr*** but this seems to be a reinterpretation. In the case of Cwmcawlwyd we cannot completely rule out *cawl* 'soup, gruel' or *caw* 'band, bandage, knot' as the qualifier. The first might conceivably describe a r. which was soup-like in appearance (perhaps characterised by rocks or vegetation in clearer water), the second a r. with a notably twisted course. The r. in Cwmcawlwyd may be identified with Nant Llwyd (rising near Nant-y-ffin SN 604285) which meets Afon Dulais (SN 637301) near Cwm-du and which defines the parish boundary between Talyllychau (on the north) and Llandyfeisant and Llandeilo Fawr (on the south). This is *Nant Llŵyd* 1888 (***nant***) but earlier evidence is needed. It is possible that it is simply a misdivided from of Cawlyd.

PLACE-NAMES OF CARMARTHENSHIRE

Cwmcothi SN 6846, 6944, 6946
t. Caeo
'valley of (r.) Cothi': *cwm*, rn. **Cothi**
Cwm Cothy late 16th cent, *Cwm Cothey* 1610, *Cwmcothy* 1754
The north-eastern part of Caeo p., largely east of Afon Cothi and Afon Fanagoed.

Cwm-cuch SN 2735
t. Cenarth
'(place in) valley of (r.) Cuch': *cwm*, rn. **Cuch**
Cwm Cych 1787
The valley is Glyn Cuch (*glyn*) in earlier records: *y glyn Kuch yn Emlyn* 12th cent; *Glynn Cuch* 13th cent, later Cwm Cuch.

Cwm Cynnen SN 3622
valley Llannewydd/Newchurch, Merthyr
'valley of (Nant) Cynnen': *cwm*, rn. **Cynnen**
Cwmcunnen 1764, (ho.) *Cwm-cynnen* 1831, (hos.) *Cwm-Cynnen, ~ -fâch* 1889
Nant Cynnen rises near Blaencynnen (SN 378227) and joins Afon Cywyn (SN 335211) near Felin Ricert. The rn. is *Nant Cynnen* 1889 recalled in *Godre-cynnen* and *Blaen-cynnen* 1831 (*godre*, *blaen*) probably meaning 'battling (r.)': *cynnen* nf. 'contention, strife, battle'

Cwm-du SN 634302
ht. Talyllychau/Talley p.
'dark valley': *cwm*, *du*
(water grist mill at a place called) *Cwm Duy* 1647, *Cwm Duy* 1664, *Cwmdy* 1740, (mess. at) *Cwmdu* 1790, *Cwm-du* 1831
Located below slopes on the north side of Afon Dulas.

Cwmduad SN 3731
ht. Cynwyl Elfed p.
'valley of (r.) Duad': *cwm*, rn. **Duad**
Cwm-Duad 1769, *Cwm-duad* 1831
The r. rises near Blaenduad (Tre-lech a'r Betws) recorded as *Blaen Duad* 1731. Duad may be 'dark (r.)' with *du* adj. 'black, dark and adj. suffix -*ad*¹; *duad*¹ nm. 'a blackening, a darkening' not recorded before the 18th cent in literary sources (GPC). Duad is also found in (place called) *dyad* 1646 (in Brecon) and a holding *Dyad Mawr* 1813 (in Defynnog), co. Brecon.

Cwm-dŵr SN 708327
Llansadwrn, Llanwrda
'watery valley': *cwm*, *dŵr*, *dwfr*
Cwmdwr 1820, 1831, *Cwm-dŵr* 1887, 1964, *Cwm-dwr* 1948
Located in the valley of Afon Dulais.

Cwmdwyfran SN 4024
Llannewydd
'valley of two (?rs.) Brân': *cwm*, *dwy*, ?rns. **Brân**
Cwm Dwfran 1729, *Cwm-dwyfran* 1758, *Cwmdyffrane* 1798, *Cwm-dwy-frân* 1831
The stream is *Nant Cwm-dwyfran* 1889 but may be Dwyfran properly. It has two main headwaters and these probably once bore the n. Brân with **brân** 'crow', probably alluding either to their darkness or their rapid, swooping flight (like a crow). Dwyfran could have referred to the r. below their confluence.

Cwmfelin-boeth SN 192191
ht. Llan-gan
'valley at the hot mill': *cwm*, *melin*, *poeth*
Cwmfelinboeth 1819, *Cwm-felin-boeth* 1889
poeth may also have the sense 'scorched, burnt' perhaps describing a mill which had burned down. The mill leat and sluice are shown on the OS 1:2,500 plan in 1889. The settlement or its valley gives its n. to the stream Nant Cwmfelin-boeth. A mill Y Felin-boeth is recorded in Llandingat as (water corn grist mill) *Y Velyn boeth* 1740.

Cwmfelinmynach SN 229248
ht. Llanwinio
'valley at the monk's mill': *cwm*, (*y*), *melin*, *mynach*¹
Cwm-y-Felin-monach c.1720, (*Tŷ-gwyn*,) *yng nghwm melin-mynych* 1778, *Cwmfelinmynach* 1819, *Cwmfelin-monach* 1839, *Cwm-felin-mynach* 1843
The *Monks' Mill* is recorded in 1889 in the middle of the village on the OS 1:2,500 plan. The mill may have formed part of the gr.

Cilgryman (q.v.) which belonged to Hendy-gwyn/Whitland abbey.

Cwm-ffrwd SN 424171
ch. Llandyfaelog, Llangynnwr
'valley of the stream': *cwm, ffrwd*
(brook) *Cwm y ffroode, Com y Ffrowd* 1609, *Cwmffrwd* 1811, *Cwm-ffrwd* 1831
The stream is now called Nant Cwmffrwd (*nant*) (SN 461168 to Nant Pibwr 415180) but this seems to be a substitution for Cyfor recorded in ho.ns. Abercyfor (*Aberkyvor* 1596), Abercyfor Uchaf (SN 431169), Abercyfor Fawr (*Aber-cyfar-fawr* 1831) and Abercyfor Fach (*Aber-cyfar-fach* 1831) (***mawr, bach***[1]) and a hill Mynydd Cyfor (SN 449169) (***mynydd***) near its source. A ho. Cwm-ffrwd Uchaf (SN 430167) is *Cwm-ffrwd-uchaf* 1831 (***uchaf***). Cyfor probably means 'r. which floods or surges' with ***cyfor***[1].

Cwmgwili[1] SN 423232
mansion Abergwili
'valley of (r.) Gwili': ***cwm***, rn. **Gwili**
Cwmgwilly 1649, 1786, *Comgwylly* 1652, *Cwm-Gwili* 1889
From its location in the valley of r. Gwili; see Abergwili. The families of the ho. have been traced back to the fifteenth century (HCH 48).

Cwmgwili[2] SN 576108
ht. Llandybïe
'crooked, bent (r. called) Gwili': ***cam***[2], rn. **Gwili**
camguili c.1140, (r.) *Combenylye* 1574, *the Combwilie, Comwilie* 1586, (r.) *Kwmwyli* 1592, *river of Comwili* 1609, (brook) *Kwm wilie* 1620, (r.) *Cum willye* c.1675-9, *Cwm-Gwili* 1906, *Cwmgwili* 1953
The first el. has been re-interpreted as ***cwm*** and Cwm Gwili applied to the valley. The short form Gwili was transferred to the ht. in Llanedi recorded as *Gwyly* 1804 and *Gwili* 1811 and later to hos. between Ysgoldy Gibea and Pont Rhydybiswail over Afon Gwili. The n. was extended on later OS maps to hos. built along Thornhill Road and Lotwen Road leading to Morfa and Capel Hendre. The poet John Gwili Jenkins (1872-1936) took his bardic n. from the short form of the rn.

Cwmhiraeth SN 3437
Penboyr
'(place in) valley of long gorse': ***cwm, hir, aith***
Cwmhiraeth 1819, (area) *Hiraeth*, (ho.) *Cwm hiraeth* 1831, *Cwm-hiraeth* 1888-9
This is the likeliest meaning though historic forms are late. The n. has probably been influenced by ***hiraeth*** nm. 'longing, nostalgia; grief or sadness'. Cf. Hiraeth (q.v.) in Llanfallteg and Mynydd Hiraethog (DPNW 333).

Cwm Hwplyn SN 439356
valley Llanfihangel-ar-arth
'valley of (man called) ?Hwplyn': ***cwm***, pers.n. or surname *****Hwplyn**
(mess. etc) *tîr kwm hypplyn* 1684, *Cwmhwplyn* 1786, *Cwmhupplin* 1794, *Cwm-hwplyn* 1831
Hobling, Hoblyn is a recorded E surname as a diminutive of Hobb, a pet form of Robert (ONC 298).

Cwmhywel SN 541076
mansion Llan-non
'valley of (man called) Hywel': ***cwm***, pers.n. ***Hywel***
(ho.) *Cwmhywel* 1813, *Cwm-Howell* 1878, 1990-3
The n. may once have applied to the valley of the unnamed stream north and west of the ho. Other examples of Cwmhywel occur at Abergwili recorded as *Cwmhowel* 1739 and Llansawel recorded as *Cwm Howell* 1687 and *Cwm-howel* 1720.

Cwmifor SN 659256
ht. Llandeilo Fawr
'valley of (man called) Ifor': ***cwm***, pers.n. ***Ifor***
Cwmyfor 1799, *Cwmifor* 1831, *Cwmivor* 1852-3, *Cwm-Ifor* 1885-1974
Ifor has not been identified. The ht. developed near a Baptist chp. There is an identical pn. in the parish of Talyllychau/Talley recorded as *Cwmivor* 1715 and *Cwm Ivor* 1722.

Cwmisfael SN 493158
ht. Llanddarog
'valley of (stream called) Ysfael': ***cwm***, pers.n. ***Ysfael***

(cottage etc at) *Cwmysfail* 1808, *Cwmismel* 1813, *Cwm-ysfael* 1831, 1891
The r. (a tributary of Gwendraeth Fach) rises at Blaenisfael (SN 466171) recorded as *Blaenysvail* (in Llangynnwr) 1701, *blayne ys vaile* 1718, and *Blaen-ysfael* 1831 (**blaen**) but is not named on OS plans. The pers.n. Ysfael (deriving from OW *Osmail* AC 1915: 396) is found also in anglicised form in Llanismael alias St Ishmael (q.v.).

Cwmliog (Cwmblewog)   SN 656349
Caeo, Talyllychau/Talley
'valley characterised by ?bushy vegetation': ***cwm, blewog***
(*terra*) *Cvmblewauc* 1303, (*terra*) *Cumbleauc* 1331, *Cwm blewoge* 1574, 1601, *Coom bleog* 1595, *Tir y Cwmbleoge* 1601, *Cwmlyog, Cwym luog* 1798, *Cwmluog* 1841, *Cwm-lïog* 1906
Cwmliog is a late development deriving from Cwmblewog by way of an intermediate Cwmbleog (with assimilation of -b-). The precise meaning of Cwmblewog is unclear because **blewog** generally means 'hairy, furry'. The n. is applied in 1841 to a small detached part of Talyllychau/Talley p. The n. also survived as an area *Rhos-cwmlluog* in 1831 near Rhos (SN 658366) (**rhos**[2]) 2km north in Caeo p.

Cwmllethryd   SN 496117
ht. Llangyndeyrn
'valley of (stream called) Llethryd': ***cwm***, rn. *Llethryd*
Cwm-llethrid 1831, (ht.) *Cwm-llethryd*, (ho.) *Cwm-llethryd-uchaf* 1878
Llethryd or Llethrid appears to be the former n. of an unnamed tributary of Gwendraeth Fawr. The closest comparison is Cwmllethryd (SN 519046) in Llanelli recorded in ho.ns. *Cum' lethrid* 1614, *Cwm Llethryd* 1710, *Cwmlethrid* 1717, *Cwm llethriad fach* 1831 adjoining an unnamed stream which rises above Cwmllethryd-fawr (SN 526060) flowing southwards to Afon Lliedi. In both cases the n. probably describes a 'ford having a slope *or* slopes' down to it (**llethr, rhyd**).

Cwm-marles   SN 5033
t. Llanfihangel-ar-arth
'valley of (r.) Marlais': ***cwm***, rn. **Marlais**[1]
Cum marlos 1650, (ht.) *Marles* 1706, *Cwm Arlloes* 1811, *Cwmarles* 1845, *Cwmmarles* 1881
The rn. is now Marlais but older forms waver between *Marlos* and *Marles*. That for 1811 seems to be influenced by unrelated **arllwys**[2] 'to pour out'. The r. forms the border between Llanfihangel-ar-arth and Llanfihangel Rhos-y-corn.

Cwm-mawr   SN 532126
ht. Llanarthne
'big valley': ***cwm, mawr***
Coome Mawr 1650, *Cwm mawr* 1800, 1813, *Cwm-mawr* 1831
The n. is taken from a ho. (see Drefach[1]) describing the valley of Gwendraeth Fawr. The focus was on SN 529128 but has shifted to the area around *Melin-y-cwm* 1831 (**melin**).

Cwm-miles   SN 1622
Cilymaenllwyd
'valley of (man called) Miles': ***cwm***, surname *Miles*
Cwmmiles 1819, 1872, *Cwm-Miles* 1889
The ht. developed around the Independent chp. Nebo erected in 1836 (RC I, 362).

Cwm-morgan (Cwmorgan)   SN 291349
t. Cenarth
'valley of (man called) Morgan': ***cwm***, pers.n. *Morgan*
Cwmforgan 1749, *Cwmvorgan* 1758, *Cwm-morgan* 1831, *Cwmforgan* 1851, *Cwmorgan* 1875
Cwmorgan is the current form with assimilation of repeated -m-. Cwm-morgan and *Pant-morgan* 1831 reputedly recall Robert Morgan or one of his brothers, ironmasters and cannon founders, and owners of tinplate works, Caerfyrddin/Carmarthen. They may have owned woodland at Cwm-morgan providing fuel for their works.

Cwmoernant SN 416212
Caerfyrddin/Carmarthen
'valley of Oernant', *cwm*, ?rn. *Oernant*
(place called) *Comeornant* 1633, *Cumoyrnant* (in *the uper ffranches*) 1664, *Cwm orenant* 1725, *Cwmoernant* 1750, *Cwm-oer-nant* 1831
The valley is that of a small stream probably once known as Oernant, 'cold stream': *oer, nant*, flowing eastwards through two reservoirs to Afon Tywi at Tanerdy.

Cwm-pen-graig SN 3536
ht. Llangeler
'valley near Pen-graig': *cwm*, ho.n. *Pen-graig* (ht.) *Cwm-pencraig* 1889, 1906
Located below a ho. Pen-graig recorded as *Pen-y-graig* 1831 and *Pen-graig* with a nearby ho. *Pen-graig-fâch* 1889 (*bach¹*) and also in *Llayn pen y graige* 1634 and (*a*) *pharc pen y graig* 1770 (*llain, parc*). The ho. is '(ho. at) end of the rock': *cwm, pen¹, (y), craig*. The cliff refers to the steep, wooded slopes recalled in Cil-graig (*cil¹*) and Twll-y-graig (*twll¹*) east of the two hos.

Cwm-twrch SN 6541, 6644, 6745, 6845
t. Caeo
'valley of (r.) Twrch': *cwm*, rn. **Twrch¹**
Kwm twrche 1544, *Cwm Twrche* late 16th cent, *yng Nghwm-twrch* 1778, *Hamlet of Cwm Twrch* 1752
Afon Twrch rises above Blaen-twrch-uchaf in the parish of Llanddewibrefi, co. Cardigan, and flows southwestwards and southwards to meet Afon Cothi near Pumpsaint. See also Gwestfa Blaen-twrch. The t. made up that part of the north of Caeo p. adjoining Twrch and largely west of Afon Fanagoed.

Cwm-waun-gron, Afon SN 1814
rn. SN 182127 to Taf SN 200161
'river in (valley called) Cwm-waun-gron': *afon*, pn. *Cwm-waun-gron*
Afon Cwm-waun-gron 1889
Cwm-waun-gron is 'valley near Waun-gron' (*cwm*). The hos. Waun-gron (SN 192159), Waun-gron-isaf (SN 190157) and Waun-gron-uchaf (SN 191153) recorded as *Waingronissa* and *Waingronucha* 1819 (*isaf, uchaf*) probably take their n. 'round moor' (*gwaun, cron*) from a small, oval-shape hill (SN 184154). The rn. has displaced Carfan recorded as *Caier* 1578, *Caruan* c.1603 and recalled in ho.ns. Upper Carvan (SN 178133) and Lower Carvan (SN 177140) recorded as *Carvan-uchaf* and *Carvan-isaf* 1889 (*isaf, uchaf*). Probably **carfan** 'weaver's beam, rail; ridge, boundary' used in the sense of 'boundary stream' (PNPemb 5, also 497).

Cwm-y-glo SN 557135
ht. Llanarthne
'the coal valley', *cwm*, def.art. *y, glo*
(ho.) *Cwmglo* 1813, *Cwm-glo* 1831, *Cwm-y-glo* 1880-9, 1964, *Cwm-y-glô* 1906
Or perhaps a ref. to 'charcoal' (*glo*); cf. Cwm-y-glo, co. Caernarfon (DPNW 114; HEALl 131). The valley is that of an unnamed tributary of Gwendraeth Fawr rising near Blaen-y-cwm (SN 558141). The n. applied in 1880-9 to hos. (SN 557136) on Cwm-y-glo Road but now includes hos. constructed in the late 19th cent along the Dre-fach to Cross Hands road around Carreg-hollt, the Farmers Arms and Capel Tabor Baptist chp. Several small coal-pits are shown on older OS maps.

Cwmysgyfarnog SN 789739
t. Llangathen
'valley of the hare': *cwm, ysgyfarnog*
Cwmsquarnog 1763, 1841, *Cwmysgyfarnog* 1841, 1975 (ho.) *Cwm-ysgyfarnog* 1887
The n. survives in that of a ho. and nearby wooded slopes recorded as *Allt Cwm-ysgyfarnog* 1887 of a small valley (*allt*). Colloquial *sgwarnog* is reflected in *Cwmsquarnog*.

Cwrt SN 144215
ht. Cilymaenllwyd
'court, grange': *cwrt¹*
(ht.) *Cwrt* c.1700, *Melincwrt* 1819, *Cwrt, Felin Cwrt (Woollen)* 1889-90
Possibly once part of the diffuse gr. of Castell Cosan belonging to Hendy-gwyn/Whitland abbey (WCist 314). OS maps from 1977 apply Felin Cwrt to what is plain Cwrt on earlier OS maps. Felin Cwrt properly applies to a row of

hos. now described as Dolycwrt (*dôl*) on and adjoining the former woollen mill (*Felincwrt Woollen Factory* 1895) (*melin*).

Cwrt Henri SN 5522
Llangathen
'Henry's court', *cwrt¹*, *Henri*
Courte Henrye 1613, *Court Henry* 1715, 1734, *Cwrt-henry* 1754, *Court Hendry* 1762, *Cwrt Henry* 1831
The ho. (SN 55662256) was reputedly built by Henri ap Gwilym, father-in-law of of Sir Rhys ap Thomas (executed 1551) (Malkin II, 444). The core of the ho. dates c.1460 (HCH 43). Eurig Davies adds (CAS 51: 10-11) that Henri was brother of Llywelyn ap Gwilym of Brynhafod (q.v.).

Cwrtycadno SN 6944
chp. Caeo
'court of the fox': *cwrt¹*, *y*, *cadno*
(ho., etc) *Clwtt y geinoge* aliter *Cwrt y Cadno* 1660, (ho. etc) *Clwtt y Geynog* otherwise *Curt y Cadno* 1663, *Cwrt-y-Cadno* 1750, (chp.) *Court y Cadno, i.e. Fox Hall* 1811, *Cwrt Cadno* 1831
Probably an elliptical description of a secluded place in woodland. The earlier n. was evidently Clwtygeinog, 'patch of land worth a penny', with *clwt* 'piece, patch of land', *y*, *cein(i)og* 'a penny', literally land which was worth only a penny in value or rent or a disparaging description for land regarded as of little value. The chp. is recorded in 1811 (CarlisleTD) and 1836 (WSS 329).

Cydblwyf SN 4010-4210
Cydweli, Llandyfaelog
'district belonging to or forming part of two or more parishes': *cydblwyf* variant *cydblwydd*
Kydplwyf 1590, *the Kydployf* 1606, *the kydeployth* 1607, *Kidplwydd of Kidwely & llandeveylogg* 1672, *Kidwely & llandeveylogg* 1699, *Kydplywdd Hamlet* 1792, *Cŷd Plwyf* 1811
Earlier refs. indicate that Cydblwyf was an area held in common by Llandyfaelog and Cydweli ps. Precise evidence for the geographical area of Cydblwyf does not seem to be available until 1840 when it is shown on the Cydweli tithe plan as an area extending from Cilfeithy (SN 400104) to Gelli-deg. Cytblwyf was also the n. of a division shared by Llanina and Llanllwchaearn ps., co. Cardigan (PNCrd 1312-3)

Cydweli, Kidwelly SN 409068
'land of Cadwal': pers.n. *Cadwal*, suffix *-i*
Cetgueli c.1090 (c.1200), *Kydwely, Kitweli* 1130 (c.1200), *Chedueli, Cedgueli* c.1170, *Kedweli* c.1191, *Kedewely'* 1229, *Cadwely* c.1236, *Kydwelli* 1319, *Kidwelly* c.1537, *Cedwelli* 1612, *Cidwelly* 1800
The n. derives from Cedweli (with vowel affection *Cad-* > *Ced-*) and the great majority of historic forms from both W and E sources have *Ced-*, *Ked-* down to the 16th cent. *Kyd-* and ModW *Cyd-* [kəd] is probably a later development. *Kid-* does not become the dominant form in E sources until the 18th cent. Cedweli is variously described as *prouincia* c.1090, land 1114, barony 1276, cmt. c.1537, and lp. *Kydwelly, otherwise Kydwellyslande* 1581. The n. was transferred to the castle and town recorded from the 12th cent. The town is thought to lie on the site of a Roman maritime fort. E settlement here in the Middle Ages introduced E fns. and street ns.

Cyffig SN 208139
Talacharn²/Laugharne
'(ch. dedicated to) Cyffig': pers.n. *Cyffig*
Lann ceffic intalacharn c.1170, *Egluskeffeg, -keffig* 1307, *Eglus Kyffig* 1384, *kyffic* c.1566, (p.) *Kiffigge* 1569, *Kyffigg* 1614, *Ciffyg* 1778, *Cyffig* 1843
Medieval forms show that prefixed *llan* and later *eglwys* have been lost, a phenomenon found also in Ceidio and Gwytherin, co. Denbigh, Llywel, co. Brecon, and Baglan, co. Glamorgan. The pers.n. is thought to contain *cyff* nm. 'trunk (of a tree or body), coffer' (ETG 182) apparently used to describe a man strong in appearance or character.

Cynghordy SN 807396
loc. Llanfair-ar-y-bryn
'dog-kennel, (ho. possessing a) dog-kennel' or 'porch or gateway, (ho. possessing a) porch or gateway': *cynhordy*

Kynghordy 1611, *Kynhordy* 1615, *Kinhordy* 1682, *Cunhordy* 1719, *Cynghordy* c.1735, *Cynhordy* 1832

Cynghordy originally applied to what is now Cynghordy Hall (SN 809408) but the n. was transferred to the small settlement near Bethel chp., St Mary's ch. and a former tollgate collectively described on OS maps 1889-1964 as Llanfair-ar-y-bryn (q.v.). The last n. properly applies to the ch. and village near Llanymddyfri/Llandovery. Another Cynghordy (at SN 688146) recorded as *Cynhordy* 1831, *Cynghordy* 1878 is probably identical in meaning. Other instances of Cynghordy in Wales include Cynghordy (SN 663030), in Llangyfelach, co. Glamorgan, recorded as *Cunghordy* 1721, *Cynhordy* 1709, *Cynhordy Ycha*, *Cynhorde* 1764. The modern form appears to favour Cynghordy meaning 'house of counsel' (**cyngor**, **tŷ**) but it is now thought to be a reinterpretation of **cynhordy** explained as 'dog-kennel, dog-house; gate-house, gateway, porch' (GPC). The suggested meanings are so different, however, that they raise legitimate questions on which is applicable here and it is possible that two words of distinct origin have become confused. On balance, 'dog-kennel' seems the likeliest interpretation particularly in view of the importance of hunting and hounds in medieval society so notable in the Welsh laws. It is significant that all cited examples adjoin mountain land where dogs may have been used for hunting or the gathering of domestic animals.

Cynheidre SN 490074
fm. Llanelli
?'settlement *or* farm notable for the first swarm of bees': ?**cyntaid** variants *cynhaid*, *cynnaid*, **tref** *Kynhaydre* 1583, *y Gynheydrey* 1596, (place) *Kynhaydrey* 1610, *Ginhaydre* 1648, *Gronhidref*; *Gunhidref ucha* 1787, *Cynheidre ycha* 1798, (fms.) *Cynhidref, -fach* 1831, (ho.) *Cynheidre* 1920
Possibly a farm or settlement which specialised in bee-keeping, notable for propagating first swarms of bees from older hives. This suggestion should be regarded as tentative because the juxtaposition of -d and d- (as a len. of **tref**) would regularly produce the form *Cynaitre(f) and Cynheitre(f) and similar forms. It may be partly on these grounds that Melville Richards preferred **cynhaeaf**, *cynhaef* nm. 'harvest, autumn' as the first el. with the collective sense of 'harvest farm'. This might describe a settlement or farm which was used specifically for gathering and storage of the harvest. It would certainly be appropriate for Gwenhawdre, co. Cardigan, with somewhat different historic forms including *Tal-y-gynavdre* 1614 (**tâl²**), *Kanavdrey* 1660, *Gynhawdref* and *Gynhafdre* 1690, and similar forms (PNCrd 32). Historic forms of Cynheidre are strikingly similar to an identical pn., now known as Llwyn-y-brain (SN 522152), in Llanddarog, (**llwyn**, **y**, **brân** pl. **brain**). These include *Kenhedre* 1605, *Geneithdre* 1717, and *Llwyn y Brain* alias *Gynheidre* 1732. We can be fairly certain that **cyntaid** is the first el. coupled with **-fa** in Tir y Gelli Gynheidfa, in Llanddeusant, recorded as *Tir yr gelli gunheyfa* 1652 (**tir**, **y**, **celli**).

Cynin, Afon SN 2629
Trelech a'r Betws p.
rn. SN 290319 and SN 270314 uniting at SN 280308 to Afon Taf SN 281152
Uncertain
Garthkiny River c.1537, *Carth keny flu:* 1578, *the Carthkinnie* 1587, *Carthkeny* 1602, (r.) *Carthginning*, (~) *Garthginning* c.1700, *Kathgenny R.* 1729, *Nant Garth Ginning, neu Gathgenni* 1870
Locally pronounced 'Ginning' (OPemb IV, 375). The rn. is also recorded in *Dyffryn Karthgini* 1634 (**dyffryn**). R.J. Thomas (EANC 200) suggests that the r. draws its n. from a specific place Garthgynin which he identifies with the ho. Llan Arthgynin shown as *Llangarthginning* on current OS maps (SN 274213). Other historic forms include *Llancarthginin* 1761 and *Llan-garth Gynin* 1889. He explains Llan- as a reinterpretation of *Lan* the lenited form of **glan** 'bank, river bank'. There seems to be no evidence, however, to prove the existence of any place called Garthcynin. It is far more likely that Garthcynin (or some very similar n.) applied specifically to the r. and that this gave rise

to Llangarthginning on its banks. The meaning remains problematic. Thomas suggested that Garthgynin contains the authenticated pers.n. *Cynin* (cf. Llangynin), presumably qualifying *garth*[1] (though he does not actually say so). The persistence of *Carth-* in early sps. of the rn., however, casts considerable doubt on his analysis. It is also worth noting that terminal -n is not found in historical evidence before c.1700 which casts doubt on the validity of *Cynin* here. Early historic evidence for the rn. seems to be missing but what survives appears to represent what would now be spelt as 'Carthcyny' or 'Carthcyni' or something very similar. This could perhaps contain *carth* 'hemp; hards, oakum' and 'offscourings, sweepings, offal' found in compounds such as *carthbwll* (with *pwll*) 'cesspool' but the remainder of the rn. is unexplained.

Cynnull-bach Cynyll SN 718301
ht. Myddfai
?'lesser place for collection (of tithe, rents, etc)': ***cynnull***[2], **bach**[1]
Kynnill bach 1649, *Cunnill Bach, Tyr y Cunnill Bach* 1710, *Cynall* 1831, (hos.) *Cynull-isaf, ~ -uchaf* 1887-1964
Described as a ht. (WATU) but the n. applied specifically to two neighbouring hos., distinguished as 'upper' (*uchaf*) and 'lower' (*isaf*), down to the 1970s when Cynnull-uchaf was abandoned. The n. may be compared with Cynnull Mawr (SN 654873), co. Cardigan, which Iwan Wmffre identifies with ***cynnull***[2] and suggests that it may refer to the gathering of tithe, particularly since this was a *parsel* of the parish of Llanfihangel Genau'r-glyn. Both fms. lie close to Gwaun-gyd Common, tautological for 'joint moorland, moorland held in common' (***gwaun, cyd***[1], E ***common***).

Cynnull-mawr SN 5116
t. Llanddarog
?'greater place for collection (of tithe, rents, etc)': ***cynnull***[2], **mawr**
kynyll mawr 1602, *Kynill mawr* 1612, (ht.) *y Kynyll-mawr* 1703, (ht.) *Cynill-Mawr* 1804, *Cynnull Bâch, ~ Mawr* 1811

The n. does not appear on OS maps but it appears to have been in the centre of Llanddarog around Pant-gwyn (SN 506135).

Cynwyl Elfed SN 373275
p., ht.
'(ch. of man called) Cynwyl in (cmt. of) Elfed': pers.n. ***Cynwyl***, territorial n. **Elfed**
(chp.) *Kenewell* 1290, *Conwyl* 1395, *Conwil-in-Elvet* c.1550, *Conwell* c.1542, *kynwyl elfed* c.1566, *Conwyl Elvet* 1596, *Kunwylelvet* 1675, *Conwyl-elfed* 1778, *Cynwyl Elfed* 1831
See also Caeo. Little is known of Cynwyl apart from a statement in the medieval tale Culhwch and Olwen that he was a saint (*a Chynwyl Sant* c.1350) (*sant*) and one of the three men that escaped from the legendary King Arthur's final battle at Camlan (CO3[3] no. 230 and n.; Mab[2] 102). He may have had a particular veneration in the cantrefi of Elfed and Caeo. Cynwyl was also described as both a giant and a holy man by the grammarian Siôn Dafydd Rhys (Mab[3] 88; GW 162, 306). The ch. – once a chp. to Aber-nant – is dedicated to St Michael which displaced an earlier dedication to Cynwyl (described as *St. Gonvil* in 1763). The ch. once possessed its own festival Dydd Gŵyl Gynwyl c.1700 (Fenton 341).

Cynwyl Gaeo see **Caeo**

Cywyn (Cowen) SN 3320
rn. SN 315299 to Taf SN 308130
?'(r.) notable for pestilence': ***cowyn***, *cywyn*
Couin c.1145, *Cowen flu:* 1578, *the Gowen or Gow streame* 1586, *Avon Kowin rhwng Kynwyl ag Ebernant* c.1700, *Cowen R.,* (br.) *Pont-Cowen* 1729, *Cowyn* 1814
Its lower reaches are notably marshy and it is possible that the r. and valley were notable for insect infestation. One local historian Conrad Evans (SPLlA) suggests that the r. may have contained 'injurious mineral deposits' but this is based on nothing more than a statement in the Life of St Teilo that the saint revived a man by the name of *Distinnic* found dead at the r.

Dafen to Dynevor

Dinefwr Castle. Viewed from the south, 1740.

Dafen SN 5201, 5301
ht. Llanelli; ep. 1874
?'river which behaves like a tame animal': *dof-, daf-*, suffix *-en*
(r.) *Davan* 1544, (r.) *Daven* 1552, 1609, *Dafen River* 1833, (place) *Dafen* 1871
The n. is taken from the stream otherwise recorded in *Glan Daven* 1488, *blaen Daven* 1583 and (marsh) *Morva Maes ar Daven* 1609 (*glan, blaen, morfa, maes, ar*). Tomos Roberts (ADG¹ 40) suggests that the first el. might be *dof-* found also in *dof¹* adj. 'tame, domesticated' and *dafad* 'sheep'. This might describe a r. which behaves like a calm domesticated animal such as a sheep in contrast to a wild mountain stream; cf. Hwch and Twrch. There are comparable rns. in England such as Dane, cos. Stafford and Chester. See further R.J. Thomas (EANC 111).

Derllys
cmt.
'oak court' or perhaps 'great court': *dâr, llys¹*
Derclis 1276, *Derthles* 1284, *Dercles* 1329, *Comm. Deilis à Penryn* c.1537, *hwndrwd derllysc* before 1569, *Kantref derllysc* c.1606, *Derllysg* 1710, (hd.) *Derllys* 1804
Melville Richards (ETG 50) notes that *dâr* nf. 'oak-tree' is also used as an intensifying prefix. The n. survives in that of Derllys Court otherwise Cwrt Derllys (SN 355201) recorded as *Court Derllis* 1658, *Cort' Derllis* 1682, *Derllys* 1695 (*cwrt¹*). The cmt. of Derllys covered St Peter's Caerfyrddin/Carmarthen, Llan-gain, Llan-llwch, and Merthyr (q.v.) (*Merthier in Derthles* 1399) later in the hd. of Elfed (*Merthir Elved* 1600) (HCrm I, 11) (*merthyr²*).

Derwen-fawr, Broad Oak SN 578228
Llangathen
'broad oak-tree', E *broad, oak* and *derwen, mawr*
an Oake 1675, *Broad Oak* 1760, *Broad Oak Inn* 1839, *Derwen fawr ... a place so called for a large oak there* 1802, *Dderwenfawr* 1813, *Dderwen-fawr* 1831
Located at the meeting place of two lanes with the old Caerfyrddin/Carmarthen-Llandeilo road – now a minor road running parallel with the A40.

Derwydd SN 613178
t. Llandybïe
'oak-trees': **derwydd**²
(lands) *derwyth* 1609, *Derwith* 1660, 1725, *Derwydh* c.1700, *Derwith demesne* 1745
Historic forms are ambiguous with regard to the final consonant. Only that for c.1700 points unequivocally to **derwydd**². The n. survives as that of an historic mansion. Gomer Morgan Roberts – a native of the p. – uses the def.art. *y* (HPLland *passim*) in 1939.

Dewi Fawr, Afon SN 2927, 2921
rn. SN 303314 to Afon Cynin SN 282160
?'greater dark-river': ?**du, duf**, suffix **-i, mawr**
Duddey c.1537, *Towa flu:* 1578, *the Gow or Tow* 1587, (r.) *the Dowy* 1632, *Dewi-fawr* 1831
The suggested meaning is based on comparison with some early forms of the rn. Dyfi in cos. Cardigan, Merioneth and Montgomery for which see EANC 139-40. Other forms vary so much that it is difficult to know how reliable they are and what they are intended to represent. *Gow* 1587 is clearly corrupt since the same source refers to Cywyn (q.v.) as *the Gowen or Gow streame* and *Duddey* c.1587 may be a scribal error for *Duwey* or a very similar form. Late forms show association with the pers.n. **Dewi**. Dewi Fawr properly applies to the r.'s main headwater and to the ho. Dewi-fawr (SN 301301) 'greater Dewi' (**mawr** len. *fawr*) in contrast to Dewi-fach (SN 301298) 'little Dewi' (**bach**¹ len. *fach*) recorded as *Blaendewyfawr* and *Blaendewyfach* 1810 (**blaen**).

Dinas SN 274301
ht. Tre-lech a'r Betws p.
'fortress, stronghold': **dinas**
Dinas, Dinas-fach 1831, (hos.) *Dinas,* (ho.) *Dinas-fâch* 1889
Traces of earthworks were apparently found here on a rock outcrop (SN 27352994) near Dinas-foel (**moel**) c.1887 and recorded in the RCAHM inventory. This may have been an Iron Age defended enclosure or medieval motte (HER 3947). The meaning of **dinas** is sometimes hard to determine and may have been used fancifully for a natural feature as in the case of Dinas (SN 612166), a fortress-like hill, at Llandybïe, partly destroyed by quarrying. Dinas-fach is 'lesser Dinas' (**bach**¹ len. *fach*).

Dinefwr (Dynevor) SN 6121
cas. Llandyfeisant p.
'?defensive hill of (man called) Efwr', **din**, pers.n. **Efwr**
gueithtineuur c.1170, *castrum Dynewr, Dynevur* c.1191, (cas.) *Dinevor* 1257, (cas.) *Dinevor* 1257, *Dynevor* 1280, *llys dinefór* c.1176 (1300), *dinefwr* c.1400, *Dineuer Castel* c.1537, (t.) *Denevour* 1651, *Castell Dinefwr, Dynevor Castle* 1831, *Dynevor Castle* 1889
Regular pron. is [dɪːnˈɛvuːr] with an anglicised form Dynevor [ˈdɪnəvɔr] for the personal title. The mansion Dynevor Castle (SN 614225), now shown as Newton Ho. on the OS Landranger, is at the centre of a park and its home farm *Newton Farm* 1889 is recorded as *Dynevor Farm* on OS maps dating from 1964. The medieval castle (SN 61142173) (HER 882) is located on the edge of the park but there seems to be no evidence suggesting any pre-existing fort here to explain **din**. There are traces of two overlapping Roman forts in the eastern part of the park near Llandeilo (RFWM 251-3) but these are unlikely to account for the n. It is more likely that **din** means simply 'defensive hill, hill which resembles a fortified hill' in ref. to the hill on which the castle was built; cf. the use of **dinas** and **castell**. The second el. **efwr** has been identified as a derivation from Brit **eburo-* supposedly meaning 'yew-tree' on the basis of reputed cognates Scots Gaelic *ibhar* and Ir *ibar* 'yew' and Bret *evor* 'blackthorn'. W **efwr**, however, means 'hogweed, cow-parsnip' and the three tree species have little in common though they may have a distant common etymological origin in Proto-Celtic (Schrijver in MPL 65-76). It is possible that Dinefwr means simply a defensive hill notable for the abundance of this plant. It was, however, a common plant and an unattested W pers.n. Efwr, derived from Celtic and British **eburo-*

seems more likely. This is best known in Britain from its identification in Eburacum, the Romano-British n. for York, composed of *eburo*- and suffix -*aco*- (PNRB 355-7; CAS 29: 4-11; Andrew Breeze in VO 16: 205-211) and in many Continental pers.ns. and tribal ns. and that of a god. The etymological relationship between pers.n. **Efwr** and **efwr** is not certain, however, and seems unlikely in view of the meaning of the latter in MW and ModW. Plant-ns. do occur in other pers. ns., usually alluding to their properties, and that may seem appropriate here because hogweed grows to a height of as much as two metres – metaphorically describing a tall hero – but it is unlikely in every other respect. The first ref. occurs in a description of the bounds of Llandeilo Fawr and probably contains what is now ModW **gwaith** possibly in the obsolete sense of an earthwork. Other historic forms have **llys** and **castell**. Dinefwr possessed a settlement Trefysgolheigion (see Gwestfa Ysgolheigion) at or close to the castle meaning 'settlement of the clerks' possibly referring to the centre of administration of Ystrad Tywi under its Welsh lords. The settlement acquired a weekly market and annual fair in 1280 which was replaced by an expanded, closely adjoining, settlement in the 1290s, later known as Newton (see Drenewydd) probably centred on what later became Newton House.

Dinweilir see **Alltyferin**

Dolaucothi SN 664406
mansion Caeo
'water-meadows near (r.) Cothi': *dôl* pl. *dolau*, rn. **Cothi**
Dole Cothi 1603, *Dole kothi* 1606, *Dole Cothy* 1623, (capital mess.) *Tyddyn Dole Cothy* 1704, *Dolau Cothi* 1831
The ho. was demolished before 1952 (HCH 56) and the site is now Dolaucothi Farm located on the north side of Afon Cothi. The n. is sometimes applied to the Roman gold-workings (see RFWM 280) at Ogofau (SN 664402) meaning 'caves' with **ogof**, *gogof* pl. *gogofau*. Recorded as *Gogoveu*; (lake) *Llyn y Gogove* c.1700, *Gogove a medicinal spring* 1729, *Gogofau* 1831, *Ogofau Gold Mines (Disused)* 1878, and probably to be identified with *tir y gogove* 1614(***tir***). Ogofau is the form on OS maps from 1878. The fort (***caer***) in *Caer y Gogovau* c.1700 probably refers to the mound immediately west of the stone Carreg Pumpsaint near Ogofau Lodge recorded as *Careg Pumpsaint* 1878, *Carreg Pumsaint* 1973 (***carreg***) and see Pumsaint.

Dol-gran SN 432343
Pencader
'water-meadow of (stream called) Grân': *dôl*, rn. *Grân*
Dol-grain 1831, *Dolgran* 1875, *Dol-graean* 1891, *Dol-grân* 1947
The stream is *Nant Graean* 1889, *Nant Grân* 1905, 1980-1. Earlier forms of the rn. favour **graean** npl. and coll.n. 'gravel, coarse sand', possibly in contrast to the fm.n. Nantygragen (SN 430346) recorded as *Nant-y-gregyn* 1831, 1889, *Nant-y-gragen* 1905, and *Nantygragen* 1953: **nant**, **y**, **cragen** nf. 'a shell' indicating a stream where shells were found.

Dol-gwm Dolgwm-isaf SN 554458
t. Pencarreg
'water-meadow valley': *dôl¹*, **cwm**
Dol gwm 1664, *Tir dolgwm* 1695, *Dolegwm* 1708, *Dol-Gwm* 1723, *Dolegwm* 1772, *Dolgwmycha* 1786, *Dôl-gwm*, *Dôl-gwm-uchaf* 1834
Dolgwm-isaf and Dol-gwm-uchaf (SN 558454) (***isaf***, ***uchaf***) are located next to and near to a small stream Nant Dolgwm.

Dolhywel SN 818289
chp., ht. Myddfai
'water-meadow of (man called) Hywel': *dôl¹*, pers.n. **Hywel**
(place called) *Dol howell* 1660-1740, (ruins of chp.) *Dôl Hywel* 1811
This has been identified (MyddfaiLP 225) with a gr. of Talyllychau/Talley abbey recorded as *Dolhenwel* c.1291, *ecclesia Sancti David de Dolhowel* 1331, *Dole Hole* 1537-9, in Cwm Wysg.

The chp. was located on Tir-cŷd fm. ('land held in common': *tir*, *cyd¹*) now among forestry above Cronfa Wysg/Usk Reservoir. Hywel has not been positively identified.

Dol-wyrdd SN 490243
?chp. Llanegwad
'green *or* verdant water-meadow': *dôl¹*, *gwyrdd*
Dolewirth 1729, *Dôl Wyrdd* 1761, *Dolewyrdd* 1775, (ho.) *Dôl-wyrdd* 1831, *Dolwyrdd Chapel* 1836, *Dol-wyrdd* 1889
Historic forms show that the adj. took the masc. form *gwyrdd*; the fem form *gwerdd* appear on OS maps from 1906. The ref. to an ancient chp. occurs in an unreliable source (WSS 330, iii) and seems to be otherwise unrecorded.

Dol-y-bryn (Tal-y-bryn) SN 2738, 2838, 3138
t. Cenarth
'end of the hill', *tâl²* displaced by *dôl¹*, *y*, *bryn*
(*terra de*) *Talebrin* 1222 (1239), *Tale y brynne* 1552, (mill) *Mellyne Tallabryne* 1574, (ht.) *Talybrin* 1713, (ht.) *Dolbryn* 1851, *Ddol-bryn* 1889, *Dol-y-bryn* 1906, *Dolybryn* 1979-80
There are traces of what appears to be a mill (*melin*) on Afon Dwrog close to the ho. Dol-y-bryn (SN 296383) and there was a second downstream at Rhydyfelin (SN 282379) (*rhyd*). A ho. known as Plas Tal-y-bryn, recorded from the 16th down to the mid 18th cent, is apparently unlocated (HCH 179).

Dre-fach¹ SN 529133
ht. Llanarthne, Llanddarog
'little settlement': *tref*, *bach¹*
Trefach 1813, *Tre-fâch* 1831, 1880-9, *Dre-fâch* 1906, *Drefach* 1953
The n. applied 1880-9 to hos. on the B4310 and later shifted southwards to include the area leading down to r. Gwendraeth Fawr and Cwm-mawr. The village owed its initial growth due to the development of several coal-mines including Clos-yr-yn (SN 535132), New Cwm-mawr and Dynant. The village expanded after World War I with the building of Gwendraeth Grammar School in 1927.

Dre-fach² SN 651162
Llandeilo Fawr, Llandyfân
'little settlement': *tref*, *bach¹*
Drefach 1841, *Dre-fâch* 1877-89, 1891
Shown but not named on OS 1831. The ht. developed on the south side of a br. over Afon Llwchwr.

Dre-fach³ SN 354387
ht. Llangeler
'little settlement': *tref*, *bach¹*
tre vach in ... Penboyr 1738, *Trefâch* 1739, *Tref-fâch* 1831, *Drefach* 1789, 1851, *Trefach* 1875
A small village which developed near woollen mills and Capel Pen-rhiw an early Unitarian chp. Penboyr (q.v.) is immediately across Nant Bargod.

Dre-fach⁴ SN 296211
Meidrum
'little settlement': *tref*, *bach¹*
Dre-fâch 1889, 1906, *Dre-fach* 1953
A small collection of hos. in the 19th century later merged with the village of Meidrum.

Drefelin SN 3637
ht. Llangeler
'settlement near a mill': *tref*, *melin*
Pentre-felin 1831, *Pentre-dre-felin* 1891, 1947, *Drefelin* 1914, 1964
Earlier prefixed by Pentre, 'village': *pentref*. The village was notable for its woollen industry, possessing at least one woollen mill, an aqueduct (SN 36053797) supplying water-power, and weavers' cottages. There is no support in historic forms to support the suggestion that the pn. contains *bilain* (as HPLlangeler 109-110).

Drenewydd, Newton¹ SN 616225
Llandeilo Fawr, Llandyfeisant
'new settlement': (*y*), *tref*, *newydd*; E *new*, ME *-ton*,
(*villa*) *Nove ville de Dynevor* 1302, *Neuton in Cantremaur* 1316, *Neweton* 1392, *Newton* 1532,

1586, *Newton near Dinevor* 1538, *Newtowne* 1559, *y dref newydd* c.1500, *Trenewydd* 1809
Newton was established as a borough apparently in 1298 and gradually replaced the older settlement of Dinefwr (q.v.). The latter is described as the 'upper town' (*villa superiori*) and Newton as the 'lower town' (*villa inferiori*) in 1303. Newton itself declined during the 15th century castle partly owing to its proximity to nearby Llandeilo Fawr which was located more conveniently in the valley of the r. Tywi. Newton became the location of Newton House occupied by the family of Rhys ap Gruffudd (attainted for treason in 1531) (SGStudies 214-220).

Dryslwyn SN 5520
t., cas. Llangathen
'bramble-brake', ***dryslwyn***
Droysloyn 1257, *Drusselan* 1287, *Drosselan* 1291, *Deresloyn* late 13th cent, *Droslon* 1327, *y Dryslwyn* 1271 (c.1400), *drysl6yn* c.1550, *Dryslwyn* 1753-4, 1779
The site of a medieval castle recorded from 1257 and town, and a farm (SN 553203). The hos. Dryslwyn-fach (SN 550207), Dryslwyn-fawr (SN 554199) and Dryslwyn-uchaf (SN 559208) are *Little Drusloyne* 1703 and *Dryslwynfach* 1827, *Drusluyn Faur* 1793, *Dryslwyn-fawr* 1831, and *Dryslwyn-uchaf* 1831, *Dryslwyn-ucha* 1887-8 (***bach***[1], ***mawr***, ***isaf***, ***uchaf***). Dryslwyn is also said to have been a manor; see Allt-y-gaer.

Duar, Afon SN 554400, 525445
rn. to Teifi
'noisy (r.)': ***dyar***
River Duar 1889, *Afon Duar* 1905
Later development has been influenced by *du*. The rn. is found in Aberduar (SN 526442) recorded as *Aberdyar* late 15th cent-1831, *Aberduar* 18th cent (***aber***) and Glanduar (q.v.).

Dugoedydd SN 771418
fm. Llanfair-ar-y-bryn
'dark woods': ***du***, ***coed*** double pl. *coedydd*
digoydydd 1708, *Tegoydû* 1720, *Dygoedydd* 1741, 1784, *Dugoedydd* 1812, *Dolgoedydd* 1832
There is scattered woodland east of the ho. and on slopes to the west on the other side of Afon Tywi. Dugoedydd was the location of an early Calvinistic Methodist meeting (HCrm II, 190, 203, 260).

Dulais[1]**, Afon** SN 7033
rn. SN 700419 to Tywi SN 718312
'dark stream': ***du***, ***glais***
the first Dulesse 1587, *River Dulais* 1887
Typically a r. in the shade of overhanging trees or enclosing hills, and often one which flows north or south so that it is in darkness at dawn and dusk. Cf. Dulais in PNGlamorgan 68.

Dulais[2]**, Afon** SN 6031, 6430
Llanfynydd, Talyllychau/Talley
rn. SN 599309 to Tywi SN 647240
'dark stream': ***du***, ***glais***
Dineleis, *Dyneleys* 1222, *Duleis* 1331, *the second Dulesse* 1587, *Dules R.* 1729, *Afon dulas* 1831
The first two historic forms probably have *Dine-*, *Dyne-* for *Dive-*, *Dyve-* for MW ***duf*** (OW *dub*).

Dulais[3]**, Afon** SN 4803
rn. SN 490051 to Afon Llwchwr SN 485005
'dark stream': ***du***, ***glais***
(r.) *Dwles* 1562, (stream) *Dwles* 1574, *Dulas flu:* 1578, (rill) *the Dulesse* 1587, *Diwlais Brook* c.1700
See also Pontarddulais.

Dulas, Afon SN 5930
rn. SN 592304 to Tywi SN 544206
'dark stream': ***du***, ***glais***
Dulashe 1578, *the third Dulesse* 1587, *Dulas* 1831
A variant of Dulais[1-3].

Dunant, Afon SN 7338
rn. SN 713390 to Tywi SN 759384
'black, dark stream': ***du***, ***nant***
River Dynant 1726, (brook) *Dunnant Brook* 1816, *River Dunant* 1888
The rn. is also recorded in ho.ns. Cwmdynant, Glandyvnant 1767 and Cwm-dunant, Glan Dunant and Aber Dunant 1831 (***cwm***, ***glan***, ***aber***).

Dyfatty SN 452009
Pen-bre
'sheepfold, sheepcote': ***dafaty***

Dafadty 1813, *Dafad-ty* 1830, *Dafatty* 1880, 1921, *Dyfatty* 1953

The current form with *Dyf-* for *Daf-* is paralleled in some forms of Dyfatty (SS 6594), Swansea, with historic forms including *Tafatty* 1709, *the Davaty* 1762, *Dyfatty* 1763, *the Davadty* 1773, *Dyvatty* 1852 with -a- [a] > to a middle vowel -y- [ə] in the unstressed first syllable.

Dyfed
gwlad
'territory of the Demet-': tribal n. ***Demet-***
Dimet c.954 (1100), *Dewet* c.1100, *Dyuet* 13th cent, *Devet* 1298

Dyfed is derived from a tribal n. Demetae a latinised form of Br **Demet*-, OW *Dimet* (PNPemb 1-2), recorded as *Demetae* 2cent, *Demetarum* 6cent, *Demetarum patriæ* 884 (10th cent), *Deomodum* 918, *Demetie* early 14th cent but its etymology (PNRB 333), and that of an apparently related pers.n. on an early Christian inscription *FILII DEMETI* c.500 (ECMW 390; Corpus II, 119) at St Dogwells, co. Pembroke, remains uncertain. Patrizia Stempel suggests a derivation from Br **Dam-ét-* meaning perhaps 'the tamed region' or 'the taming people' in the context of horse-breeding (quoted in PtolemyLA 9) and Andrew Breeze argues (CAS 41: 175-6) that it may be a Brit compound of an intensive prefix **do-* and **met-* found in, for example, *medaf* 'I reap', and in Elfed (q.v.). That might be translated as 'expert cutters-down (of enemies)' and can be compared with the tribal n. Ordovices (in north Wales) meaning 'hammer-fighters' (PNRB 434) containing Brit **ordo-* which gives ModW *gordd*, *ordd* nf. 'hammer, mallet'. Neither of the suggested etymologies can be regarded as conclusive, however, because both require vowel changes from **Da-* or **Do-* to produce Demetae. The area covered the west of Carmarthenshire and the whole of Pembrokeshire covering seven cantrefs (WB 18-20). The n. survived as a geographical qualifier of Llandysilio (q.v.) and was revived as the n. of the administrative county of Dyfed (1974-96).

Dyffryn SN 313225
Trelech a'r Betws
Dyffryn 1972
'valley (of r. Selach)': ***dyffryn***, (rn. *Selach*)

The n. applies to a small number of hos. which developed around Ffynnon Bedr Independent chp. founded in 1805 (CYB 447) and *Dyffryn Inn* (recorded 1889), and is drawn immediately from a ho. known in earlier sources as *Rochwîth* 1889 and 1891 (*gro*, *chwith*) and *Ogwydd* 1906, 1953. A neighbouring ho. is recorded as *Dyffryn-uchaf* 1889 and *Dyffryn Sellach* 1744, *Dyfryn selach* 1755, *Dyffryn-selach* 1906 and 1953. Selach may refer to the now unnamed stream which flows westwards to Afon Dewi Fawr possibly deriving from Ir **saileach*** 'salty'; cf. Seilach, co. Cardigan (EANC 2).

Dyffryn Ceidrych SN 6925
t. Llangadog
'valley of (stream) Ceidrych': ***dyffryn***, rn. Ceidrych
Dyffryn keydrych 1549, *Dyffryn Keyrdrych* 1555, *Deferin-Kerdrich* 1562, *Diffryn Kyrdrich* 1653, *Duffryn Keidrich* 1800, *Dyffryn Caead Rhŷch* 1811, (ht.) *Dyffrin Cidrich* 1851

The rn. is independently recorded as *Kydryche* 1552 and *Kyrdrych* 1591. Ceidrych occurs as a pers.n. in *Fforest Llen ap Kydryche* 1688 (*fforest*, pers.n. **Llywelyn***), in Betws or Llangynwyd, co. Glamorgan, and as *Cydrich filius edrit* c.1075 (c.1145) and *cidrich filii gunncu* c.1075 (c.1180) in the Book of Llandaf (LL 272, 212, 276).

Dynevor see **Dinefwr**

East Marsh to Esgob

Egrmwnt/Egremont. Taken from Thomas Kitchin's 'Accurate Map of Carmarthen Shire' c.1762.

East Marsh SN 2808, 3008
Talacharn²/Laugharne
'eastern marsh': E *east, marsh*
ye East Marshe 1592, *East Marsh* 1811, 1891
'East' in contrast to West Marsh (SN 2508, 2608) recorded as *West Marsh* 1811 (**west**) west of Witchett Brook marking the parish boundary of Llansadyrnin and Talacharn²/Laugharne (SGStudies 146-151 with plan). The whole area is *le Mareis* 1307, *the Marsh* 1437, *the Marshes* c.1622-4 and *Marsh* 1811.

Edwinsford see **Rhydodyn**

Efail-wen SN 134253
Cilymaenllwyd, Llandysilio (East)
'white or fair smithy': *gefail*¹ nf. 'smithy, forge', *gwyn*, gwen
Gefailwen 1819, *Efailwen* 1874, *Efail-wen* 1889

The smithy is shown on the OS 1889 and 1907 opposite Rhos Inn but the n. Efail-wen earlier applied to the ho., now Efail-wen Isaf (SN 132249), a little to the south, and was transferred after 1964 to the small settlement that developed around the former smithy. The whiteness may refer to the soil, also recalled in the ho.n. Bryn-gwyn (**bryn, gwyn**). Efail-wen is best known as the location of a turnpike toll gate destroyed in 1839 during the Rebecca disturbances.

Eglwys Fair a Churig SN 202263
Cilymaenllwyd
'church dedicated to (saints) Mary and Curig': *eglwys*, **Mair**, *a*, **Curig**
Capel Vayre 1552, *p. mair a chirig* c.1566, *Eglesuaier achirig* 1578, *the Chappell of Eglwys Vair achirrick* 1618, *Eglwys vair y Cheiricke* 1622,

59

Cappell mair achirrick 1625, *Eglwys Fair a Chyrig* 1710, *Eglwysvair a Churig* 1819
The apparent absence of early refs. may be ascribed to its former status as a chp. (***capel***) to Henllan Amgoed. Curig was identified by M.H. Jones with the saint dedicated at Llangurig, co. Montgomery, and Capel Curig, co. Caernarfon (AC 1915: 397) but this is not certain. The scholar Edward Lhuyd c.1700 (Paroch III, 111) records that some call it *E[glwys] V[air] Vathared* but this looks like poor transcription and is unsubstantiated. His other statement that common pron. is *E[glwys] V[air] Ychyrig* is, however, confirmed by a number of historic forms. The ch. is ruinous.

Eglwys Fair Glan-taf   SN 201162
Llanboidy, Hendy-gwyn/Whitland
'church dedicated to (St) Mary on banks of (r.) Taf': ***eglwys***, **Mair**, ***glan***, rn. **Taf**
(chp. of) *Beate Marie de Glantave* 1535, *Eglosuaier* 1578, (chp. of) *St Mary of Llantave* 1590, *Egloes vair glantave* 1613, *the Chapple of Egloysevaire* 1618, *Eglwisvaire glantaf* 1631, (chp.) *Eglwys fair ar Lan-taf* 1680, *Eglwys Fair llan Tâf* 1710, *Eglwys Fair Glann Tâf* 1811, *Eglwysfair* 1819
Eglwys Fair ar lan Taf (***ar***) is the form used by Richards (WATU) but historic forms generally favour the entry form. Eglwys Fair was a chp. of Llanboidy but served as a p. ch. and perpetual curacy to Hendy-gwyn/Whitland and was probably on the site of the present ch.

Eglwys Gymyn (Eglwys-Cymmyn, Common Church)   SN 239107
'church belonging to ?Cynin': ***eglwys***, pers. ?***Cynin***
Egluscumin c.1250, *Egluskymin* 1307, *Egloiscymyn* 1488, *eglwys gymyn* c.1566, *Egloskemyn* 1595, *Eglwys-Cymmin* 1624, *Eglwyscymmyn* 1715, *Eglwys Cymmyn*, i.e. *The Communion Church* 1811, *Common Church Farm*, ~ *Cottage*, *Eglwyscymmyn* 1843, *Eglwyscummon* 1878-1915
Melville Richards (ETG 212) identified the qualifier as ***cymyn***[1] 'bequest', i.e. a church bequeathed by a secular person, and the majority of written forms support him. The only form which casts serious doubt on this is *Eglusgluneyn* c.1291 taken from a papal taxation notorious for its inconsistent spellings. Nancy Edwards argues that this is a slip for **Eglusguneyn* (Corpus II, 214-7), ie. 'ch. dedicated to Cynin' and it has support in the discovery c. 1855 of a 5th/early 6th cent stone in the church, inscribed (in translation) as 'of Avittorix daughter of Cunignus'. The latter might ultimately derive from an OIr pers.n. **Cunegna(s)* nominative of *Cunigni* developing by way of OW *Cunein* to *Cynin* (see also AC 1907: 20-21, 273, and 1915: 331). Substitution of ***cymyn*** must have taken place before the first historical ref. the pn. c.1250. Eglwys Gymyn appears in late sources misinterpreted as Common Church (*cirice*) and the current OS map has Common Church Fm as if it contains E ***common*** either in the sense of 'church which is commonly held' or 'church on common land'. The ch. has been dedicated to St Margaret (of Scotland) since the 14th cent (AC 1907: 266). Parc-cymyn (SN 228107) recorded as *Parc-gymmyn* 1843 and *Parc-y-cymmyn* 1891 is elliptical for 'enclosure *or* park near Eglwys Gymyn' (***parc***).

Eglwys Trefwenyn see **Pentre-wyn**

Egrmwnt, Egremont   PEM SN 0920
'sharp-pointed hill': OFr ***aigre***, ***mont***
Egremount 1513, *Egermont* 1535, *Egremont* 1545, *egyr mwnt* c.1566, *Egermunte* 1581, *Egremond* 1586, 1710, *Egermount* 1691, *Egremwnt* 1888
Identical in meaning to Egremont, Cumberland (ODEPN 174) which may be a transferred n. from Aigremont, south-east Normandy. This is likely to be the case here though historic forms are very late. The n. is a good description for the small castle here shown as an earthwork on current OS maps. The W form with metathesis -*re*- > -*er*-, -*yr*- is poorly recorded though *Egermunt* 1613 may be an attempt at representing it. The actual village is Llandre Egremont recorded as *Llandre* 1733 with ***llandref*** 'church village, township containing the church' and a ho. here is *Place keven y Lan*

otherwise *Place y Llandre* 1727 (**plas, cefn**). The church of St Michael was roofless in 1889 and is described as 'forsaken and ruinated' in 1915. Early medieval origins for a ch. here are indicated by the existence of a 6th cent stone, first recorded c.1745 in the churchyard, now set in the floor of St Tysilio's ch. in Llandysilio (Corpus II, 217-8).

Egwad SN 5121
Llanegwad
'(ht. of) Egwad': pers.n. **Egwad**
(ht.) *Egwod* 1710, *Egwad Hamlet* 1724, (ht.) *Egwad* (and *Hiernin*) 1804
A ht. or t. centred on the village and p.ch. A reduced form of Llanegwad (q.v.).

Elfed
cmt.; hd. Elvet
?pers.n. **Elfed**
Elfet late 12th cent (c.1300), *Elfed* c.1250, *hwndrwd elfed* c.1566, *kantref elfed* 1606, *Elved* c.1275-1340, (cmt.) *Elved* 1292, *Eluede* 1289, *De comoto de Elueth* 1301-2, (seneschal of *Widigada* and) *Elvet* 1431-3
See also Cynwyl Elfed. A territorial n. which has been compared with Elmet, a British kingdom in West Yorkshire (CDEPN 213), possibly dating to the 5th cent. Regarded as obscure by Richard Coates (CV 345) implicitly rejecting the suggestion by Hamp (BBCS 30: 42-45) that it means 'elm wood'. The fullest discussion is by Andrew Breeze (NH 39.2: 157-171) who favours derivation from a tribal n. meaning '(people or leaders) who cut down many' containing Brit roots *el-* 'fully, many' and *met-* 'mow' found in ModW *medaf* 'I mow' and Medi 'September' a month for harvesting. An alternative explanation is that Elfed derives from a pers.n. (containing the same els.) with parallels in pers.ns. such as Elfedd, Elgi, etc. Elfed might have the sense of 'a great reaper', a suitable n. for a mighty warrior. It is not entirely convincing because one would have expected addition of a territorial suffix such as *-ion* in Carnwyllion (q.v.) and Ceredigion or *-iog¹* as in Pebidiog, co. Pembroke, and Ffestiniog, co. Merioneth.

Emlyn
Cenarth
'(land, area) around a valley': **am, glyn**
emlin c.1140, *emelinn, emblin* c.1170, (cas., land) *Emelin* 1178, *Emlyn* 1257, (cmt.) *Emlyn* 1325, (*a*) *chantref Emlyn* c.1400, (manor) *Emlyn* 1551, (deanery) *Hemelyn* c.1291, (cmt.) *Emelyn Uchuth* 1429, *Emlyn ywch kych* c.1566
Emlyn was one of the seven ancient ctfs. of Dyfed but c.1240 Gilbert Marshall, earl of Pembroke, granted the eastern part known as Emlyn Uwch Cuch (**uwch, Cuch**) to Maredudd ap Rhys who built a castle at Cilgerran. The western part Emlyn Is Cuch (**is**) became part of co. Pembroke in 1536-42 (HCrm I,10).

Esgair¹ SN 374288
Cynwyl Elfed
'ridge': **esgair**
Esgair 1889
Located next to a br. over Afon Duad and on the south-western slopes of an unnamed hill extending from a ho. Brynffwlbart (SN 376291) up towards Penffynnon (SN 290295) (**bryn, ffwlbart, ffynnon, pen¹**).

Esgair² SN 279198
Llanfihangel Abercywyn
'ridge': **esgair**
Esgair 1889
Applied to hos. near Salem Baptist chp. located on a ridge and mountain road leading north from Sanclêr/St Clears.

Esgairdawe SN 612410
Caeo, Pencader
'ridge near (Nant) Tawe': **esgair**, rn. **Tawe**
Usker Doweth c.1762, (Independent chp.) *Escerdowe, Escair Dowie Meeting-house* 1765, *Esgerdawe* 1826, *Esgair Ddewi* 1831, *Esgair-Ddewi* 1887-8
The n. is taken from a fm. (SN 602423) near the headwaters of Nant Tawe (**nant**) and was applied to the nearby meeting-ho. (SN 602430)

established c.1755 (HCrm II, 219) known as *Old Chapel* in 1887-8 and *Esgair-Dawe-fâch* 1905 (***bach**¹*). A new chp., 2km south of the older one, adopted the n. *Capel Esgair-Dawe* in 1844.

Esgair Ferchon SN 716427
mtn. Cil-y-cwm
'ridge near (Afon) Ferchon': ***esgair***, rn. **Merchon**
Esgair Dderchon 1891, *Esgair Ferchon* 1907
A prominent oval-shaped ridge at the head of Afon Merchon and other rivers extending south-westwards to Cefn Bach.

Esgair-gam SN 593326
Llanfynydd
'crooked ridge': ***esgair***, ***cam²***
(mess. etc) *tir Esgeir kam* 1671, *Tyr Esgyr cam* 1795, *Eskirgam* 1841, *Esgair-cam* 1888-9
The ho. is on the south-east slopes of a short, curved ridge extending north from Beili-bedw (SN 586323) ***beili***, ***bedwen*** pl. ***bedw***) in the far north of Llanfynydd parish.

Esgair-garn SN 5630, 5731
Llanfynydd
'ridge at the cairn': ***esgair***, (***y***), ***carn***
(ho.) *Esgair Cam, Eskirgam* 1841
Probably centred on Y Garn (SN 561309) in the north of the parish extending to include Pen-y-garn (SN 574318), Rhyd-y-garn (SN 560301), Nant-y-garn (SN 562320) and Blaen-nant-y-garn (SN 570316) (***blaen***, ***nant***, ***pen¹***, ***rhyd***).

Esgairifan SN 5528
Llanfynydd
'ridge of (man called) Ifan': ***esgair***, pers.n. ***Ifan***
(ht.) *Esgair Evan* 1851, 1863
Evan is an anglicised form of the pers.n. ***Ifan*** but he is unidentified. Llanfynydd also has Maes Ifan (SN 574304) (***maes***) recorded as *Maes Evan* 1841 and *Maes-Ifan* 1888-9 and Tir Ifan Ddu (SN 578281), 'land of Ifan the dark' (***tir***, pers.n. ***Ifan***, ***du***) recorded as *Tyr Evanddu, Tir Evan ddu* 1841 and *Tir-Ifan-ddu* 1888-9.

Esgeirnant SN 6431, 6432
Talyllychau/Talley
'ridge near a stream': ***esgair***, ***nant***
Oskernant c.1291, *Eskeirnant* after 1271 (1332), (p.) *Eskernat'* 1519, *Esgeir nant* 1535 ~ 840, (mess., etc) *Eskyrnant* 1765, *Escer Nant* 1811, *Esgair-nant* 1831
The ridge may be that north of Pantyresgair (SN 646319) centred on SN 651329 and the stream reaches Dulais (SN 649303) below Halfway.

Esgob see **Tiresgob**²

Faenor to Furnace

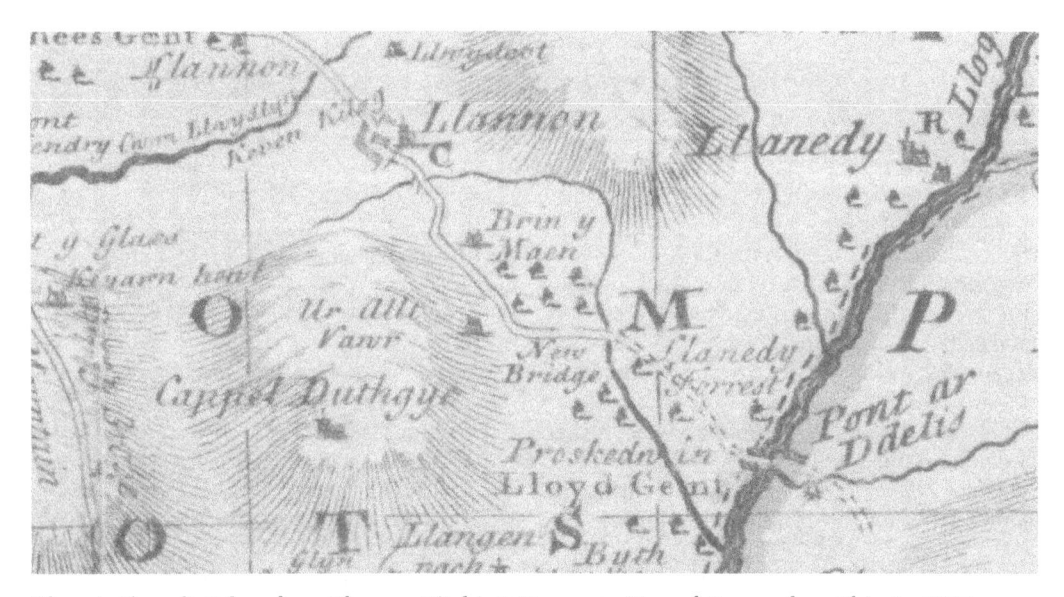

Fforest, Llanedi. Taken from Thomas Kitchin's 'Accurate Map of Carmarthen Shire' c.1762.

Fabon see **Maenorfabon**

Faenor SN 244140
Llanddowror
'(the) administrative unit': (*y*), ***maenor***
Veynor 1785, *Feynor* 1831, *Vaynor Farm* 1845, *Vaynor* 1906-7, *Vaynor Farm* 1971
The late appearance of this pn. in historical records casts some doubt on this explanation. An identical pn. Maenor or Manor Court (SN 236107) occurs in Eglwys Gymyn and is recorded as *Maynor, Vaynor* 1655 (confused with Faenor, in Llanddowror), *Manor-court* 1819 and *Manor Court* 1889 (***court***).

Faenor Isaf SN 6736, 6838, 6939
Caeo
'(the) lower administrative division': (*y*), ***maenor, isaf***
Vynor Yssa 1574-1609, (ht.) *Y Faenor Isaf, or, the Lower Hamlet* 1811
This was the southern and south-eastern part of Caeo. The 'upper hamlet' may correspond with Maenor Rhwng Twrch a Chothi. According to Glanville R. Jones (StC 28: 89-90) Faenor Isaf was divided into five subdivisions or *gwestfâu* (*gwestfa*) including Maestreuddyn, Penarth, Gwestfa Cwmblewog, Tre Cynwyl Gaeo and Gwestfa Y Faenor Isaf.

Faerdref Ferdre-fawr SN 673198
Llandeilo Fawr
'(the) hamlet attached to lord's court': (*y*), ***maerdref***
Vaerdre 1535, *the Maynor of Mayredryffe, (firma Manerii de) Mairdriffe* 1609, (hos.) *Ferdre-isaf, Ferdref-fâch* 1886-7, *Ferdref-fawr* 1907, *Ferdre* 1919

63

Len. (M- > F-) of **maerdref** as a nf. is probably caused by use of def.art. *y*. This is the area around Carreg Cennen castle and formerly the n. of three hos. in what must have been the centre of an administrative unit (**maenor**) in Is Cennen. Ferdre-isaf drops off OS maps between 1890 and 1907.

Faerdref and Fforest Bethel see **Faerdref and Fforest Bedol**

Fanafas, Afon SN 6544
rn. SN 680465 to Twrch SN 650440
Uncertain
River Fanafas 1888, *Afon Fanafas* 1905
Historic forms are too late for reliable interpretation. The r. evidently bore an alternative n. Banw because a ho. (SN 656450) on the north side of the stream is recorded as *Glan-Fano* 1888, *Glanfano* 1978 (**glan**). Afon Fanagoed (q.v.) must also be relevant in this context raising the possibility that *Fanafas* < **Banawes, Banafaes* < **Banwfaes* deriving from **banw²** and ?**maes** with the meaning of '(r.) Banw in open country' (in contrast to '(r.) Banw in woodland' for Fanagoed) or less likely **banwes** nf. 'a young sow, a gilt', with the meaning 'r. which behaves like a young pig'. Both suggestions require shift of stress to the middle vowel to produce the modern form Fanafas [van`avas] as shown in historic forms of Fanagoed. It can hardly be coincidental that Fanafas is a tributary of Twrch (q.v.).

Fanagoed, Afon SN 6844
rn. SN 701468 to Cothi SN 670429 Caeo
?(r.) Banw in woodland': **banw²**, **coed**
(r.) *Banowegoed* 1561, (r. or brook) *Manwgoed* 1724, *River Fanagoed* 1888
The r. joins Afon Cothi near Abermangoed recorded as *Abermangoyd* 1559, *Tir Aber Manwgoed* (and water corn grist mill) *Mellyn Aber Man goed* 1661, (mill) *Aberbenacoid mill* 1728, *Abermaen Goid* 1729, (capital mess.) *Tyr-abermangoed, Abermangoed mills* 1746 and *Aber-Mangoed* 1888 (**aber**) and the unidentified *Tir y Banw Goed* 1613 (**tir**). Historic forms suggest Banwgoed, Manwgoed > *Fanwgoed (with len.) > Fanagoed with regular stress on the middle vowel -w-, later -a-. Forms such as *Abermangoed* show loss of -w-. If the arguments with regard to Afon Fanafas are valid, then this would be a r. called Banw flowing through woodland in contrast to Fanafas flowing through open country. Abermangoed, by contrast, is clearly stressed on -man-, possibly through association with **mangoed** 'small trees'. This is *Abermangoyd* 1559 recalled also in ho.ns. *Brynymangoed* 1841 and *Pant-y-mangoed* 1887-9 (SN 657163) (**bryn, pant**) near Forge Llandyfân (SN 657169).

Felindre¹ (Felindre Sawdde) SN 704275
Llangadog
'settlement with a mill near (r.) Sawdde': **melindref, Sawdde**
Melindreth 1284, *Melindressathney* 1317, *Melynaresauey* 1383, *Velindrefe Sawddey* 1592, *y velindre sowthey* 1626, *Velindre Sawthey* 1636, *Velindre* 1777, *Felindre Sawthey* 1779, *Felindre-sawdde* 1831, *Felindre* 1887
The qualifier Sawdde is dropped in the 19th cent. There was a widespread and mistaken supposition that pns. such as Felindre¹⁻⁶ contain **bilain** nm. 'villein, serf' or **milain** 'villein, peasant' (see Felindre³). A water-mill (SN 6994327790), later a woollen mill, to the north-west may be on the site of a medieval mill (HER 4882).

Felindre² SN 551212
Llangathen
'(the) settlement with a mill': (*y*), **melindref**
Velindre Mill 1752, (village of) *Velindre* 1788, *Felindre* 1831
Adjoining Afon Dulas, a tributary of Tywi. A mill leat runs from Nant Lash, north of Felindre, to a corn mill shown on the OS 1887-8.

Felindre³ SN 354383
Llangeler
'(the) settlement with a mill': (*y*) **melindref**
Y Velyndree 1545, *Velindre Jenkin* 1587, *Velindre yssa* 1662, *Velindre* 1752, (tmt.) *Felin dre* 1773, *Velindre Shinkin* 1799, *Velindre Jenkyn* 1811,

Felindre Siencyn 1831, (toll-house) *Melindre Siencyn* 1843
Jenkin otherwise Siencyn, was identified by Daniel E. Jones (HPLlangeler 106-8) with Siencyn ap Llywelyn, 'lord of Emlyn'. Siencyn was constable of the castle of Castellnewydd Emlyn/Newcastle Emlyn and most notable for transferring his allegiance to Owain Glyndŵr in 1403. Jones also suggests that the pn. contained **bilain** rather than **melin** because he was unable to identify a mill here. The difficulty with this argument is that it fails to match historical forms and ignores the possibility of lost mills, horse-mills, etc.

Felindre[4] SN 460122
Llangyndeyrn
'(the) settlement with a mill': (*y*) ***melindref***
Velindre 1590-1804, *Felindre* 1811, *Felin-dref* 1831
Drysgeirch Mill otherwise Melin *Trysgyrch* is *Triskirth Mylle, Tryscyrch Mille* 1609, *Treskerch Mill* 1644, *Melin Treskir* 1662, *Truscyrch mill* 1716, *Felindrisgill* 1811 (***melin, mill***). See further Maenor Cwarter Trysgyrch. A small tributary of Gwendraeth Fach is recorded as *Ffrwd triscyrch* 1597 powering Drysgeirch corn mill in 1880-90 as well as two woollen factories. The n. survives in truncated form as Ffrwd (SN 455127) recorded as (hos.) *Ffrwd* 1811, 1831, *Ffrŵd* 1880-90 (***ffrwd***).

Felindre[5] SN 4218
Llangynnwr
As Felindre[2]
'(the) settlement with a mill': (*y*), ***melindref***
Velindre 1796-7, (ht. *Llandre* and) *Velindre* 1851
Not shown on OS maps or the tithe plan but centred on Felin Pibwr (SN 421181) for which see Pibwrlwyd.

Felindre[6] SN 687304
Llansadwrn
'(the) settlement with a mill': (*y*) ***melindref***
Tyfelindre 1762-3, *Melindre* 1831, (area) *Felindre* 1887

Shown with a corn mill 1887 and 1906 near Afon Marlais. The first ref. has ***tŷ*** 'house, dwelling'.

Felindre, Afon SN 5728
SN 563305 to Sannan SN 561262
(r.) *Velindre* 1851, *Afon Felindre* 1887-9
Named from Felindre (SN 570275) a former corn mill recorded as *Velindre* 1772. Cf. Felindre[1-6]

Felindre Sawdde see **Felindre**[1]

Felin-foel SN 5102, 5202
Llanelli
'(the) bare mill', (*y*), ***melin, moel***
mill voele c.1500, (mill) *Y Velyn voel* 1542, *Melyn Voiell* 1565, *Melin Voel* 1587, *Velin Vole* 1735, *y Felinfoel* 1778, *Felin-foel* 1833
On Afon Lliedi. Probably a mill which was exposed, perhaps by decay or destruction, or one without any ancillary buildings. A Baptist meeting-ho. built here in 1709 is recorded as *y Tŷ newydd* 1839, 'the new house', ***tŷ, newydd***, but was otherwise known as Y Felin-foel (HECS 1, 3). Local pron. of -oel- as -ô- [ō] is reflected in sps. such as *Velin Vole*.

Felin-gwm Isaf SN 507239
Llanegwad
'lower mill-valley': ***melin, cwm, isaf***
Felin-gwm-isaf 1888-9, *Felinycwm* 1813, *Felingwmisaf* 1884, 1972
'Lower' in contrast to Felin-gwm-uchaf. Recorded as *Penybontbren* 1813 and *Pen-y-bont-pren* 1831, 'end of the footbridge', ***pen***[1], *y*, ***pontbren*** describing a br. over Afon Cloidach. The mill may be *Glandŵr Fulling Mill* 1888-9 (***glan, dŵr***) shown as an empty building in 1972 or a predecessor.

Felin-gwm Uchaf SN 508247
Llanegwad
'upper mill-valley': ***melin, cwm, uchaf***
Felin gwm 1752, *Velin-Gwm* 1753, *Felingwm* 1839, 1884

'Upper' in comparison with Felingwmisaf (q.v.) 1km downstream. Located near *Melin-y-cwm* 1831. What appears to have been an old mill-leat leaves Afon Cloidach and runs in parallel with the latter into the village.

Felin-wen, White Mill SN 463215
Abergwili
'(the) white or pale mill, mill painted white': (*y*), *melin, gwen*, E *white, mill*
the White Mill 1675, (water corn grist mill) *White Mill* 1710, 1786, *Felinwen* 1811, *Felin-wen* 1831
The mill, on the site of Whitemill Farm, was probably whitewashed. The village is Felin-wen 1831-1964 and the fm. is *Felinwen* otherwise *White Mill* 1871, *Felin-wen-isaf* 1906, 1964 (*isaf* 'lower'). The colour theme was transferred to the White Horse Inn, now White Mill Inn.

Fenai SN 4521
Abergwili
Uncertain
Fyneu 1811, (ht.) *Veney* 1851
The apparent absence of early forms makes this very difficult to interpret and there is no suitable r. to justify easy comparison with Afon Menai/Menai Strait (DPNW 317; EANC 29-30) or Fenni (q.v.). If Fenai is based on Brit **men-* signifying ?'flow, current' then it is possible that it may allude to an area subject to flooding of Afon Tywi. The ht. occupied an area extending from Castell Pigyn (SN 435221) to Bwlch-bach (SN 449211) and it is worth observing that the western and southern parts of the ht. are flood-plain.

Fenni, Afon SN 2421
rn. SN 247251 to Taf SN 236163
Uncertain
Venny c.1537, (rill) *Venni ... that cometh through Cardith forest* 1586, *Venny Rivulet*, (brook) *Fenni fâch* c.1700
The final ref. has *bach* len. *fach* 'little' and may properly apply to a specific tributary of Afon Fenni. The rn. is also recorded in Pontyfenni (q.v.) and it is possible that -y- represents a lost initial E-, i.e. Efenni > Fenni. If that is the case then the rn. bears comparison with Y Fenni, the modern W form of anglicised Abergavenny 'mouth of (r.) Gafenni' (*aber*) (DPNW 6-7; EANC 143-5), co. Monmouth. Gafenni is thought to be derived from Gefenni with dissimilation of the first -e- > -a- and is interpreted as 'river of the blacksmiths' or 'river of the ironworks' (Brit **gobann-* and suffix *-i*). Whether that is the correct interpretation of Afon Fenni is very uncertain and there is scope for further research. It must be significant that Fenni or Y Fenni is also recorded as that of a now unnamed brook in Llangeler recorded as *Nant y vennie* 1643 (*nant*) and in the ho.n. Abergavenny (SN 363401) in the same parish recorded as (land) *tir Abergevenni* 1567, *Abergevenny* 1662, *Tir Abergafenny* 1709 (*tir*).

Ferryside see **Glanyfferi**

Ffair-fach SN 629215
Llandeilo Fawr
'little fair': *ffair, bach*¹
Ffair fach yn Llandilo 1612, (water grist mill) *Ffaire Vach Mill* 1662, *fayre vach* 1696, *Fair vach* 1726, *Ffair-fâch* 1757, *Ffair Fach* 1765, *Ffair-fach* 1831
The fair is recorded in 1609 and was 'little' perhaps in contrast to Ffair Maes Abercennen (see Abercennen) at Tre-gib recorded as *ffair Maes Aberkennen* 1609 (*maes*). The fair took place on 11 November.

Ffaldybrenin SN 637446
Llan-y-crwys
'the king's fold': *ffald, y, brenin*
Ffald y Brenin 1831, *Ffaldybrenin* 1851, 1863, *Ffaldybrenin Inn* 1875
It is likely that this was a fold for animals sold at the fairs held here mentioned in 1831 or perhaps one used for stray or working animals in association with the royal forest of Pennant which lay to the north around Blaenaufforest (SN 618451). An identical n. is recorded in Meidrum as *Ffald-y-brenin* 1765.

Ffarmers   SN 650448
Llan-y-crwys
'(village named from) Farmers' Arms
(village) *Farmers* 1888
A cymricised form of E *farmer* referring to an inn allegedly located in the village square. The 1851 Census and OS map 1891 mention only the *Drovers' Arms*.

Fferem-fawr   SN 6414
Llandybïe
'big ?farm': *fferm*, *fferem*, *mawr*
Fferen vawr c.1700, *Ferem Fawr hamlet* 1734, *Fferm Fawr* 1811, *Feremfawr* 1851
Or 'bigger, larger' in contrast to Garn (q.v.) in the same p. which was otherwise known as Fferem y garn (*carn*). The use of *fferm*, *fferem* here is a little unusual because Fferm-fawr was one of the eight hts. of Llandybïe (HPLland 49) and it is possible that it may be better understood as 'large hamlet assessed for tax' rather than a specific agricultural holding. Fferem-fawr is located near Myddynfych which preserves the n. of the ancient administrative division of Maenor Meddyfnych (q.v.) and it is possible that Fferem-fawr corresponds with a much older subdivision.

Fforest[1]   SN 7638, 7738
ht. Llandingad
'(the) forest': (*y*), *fforest*
Y Fforest c.1600, *the fforest* (called *y gilfach Goe*) 1603, (ht.) *Forest*, *Forest y Moswm* 1668, *fforest Hamlett* 1671, *Tyr y fforest* 1693, *Fforest* 1710, *the Hamblett of fforest* 1740, (hill) *Forest* 1831
Presumably a hunting forest in cmt. Hirfryn. *Moswm* appears to stand for *mwswm*, *mwswn* 'moss, lichen' describing a damp forest characterised by moss.

Fforest[2]   SN 5804
Llanedi
'(the) forest': (*y*), *fforest*
the fforreste Llangorath, *fforesta de Kevengorath* 1609, (land) *Forest Keven Gwrath* 1627, *Llanedy Forrest* 1729, *Fforrest* 1745, *Fforest-hall* 1831
The 1609 survey (SDL 282) describes *the fforreste of Kevengorath* as covering 50a. and bounded on its west side by a r. *Come Wyli* now known as Gwili (see Cwmgwili[1]). This also records a second, much smaller, *fforeste de Ketheleuysach* in the same p. next to the r. Llwchwr. Cellifeusach, now Gelli Fm (SN 612104), is recorded as *Kelly veysach* 1640, *Gelly Visach* 1746, *Gellyveysod* 1680, *Gellifaesach* 1831 (*celli*). *Kevengorath* 1357 and 1609 refer to the ridge (*cefn*) on which the former forest and modern village are located but it is not current and its proper form is uncertain. Sir John E. Lloyd (HCrm I, 230-1) uses the form Cefngorach in 1935 but there is little to favour it in historic forms. Confusion of *t*- and -*c*- is fairly common in medieval and early modern script but the same survey which provides us with -*gorath* also carefully distinguishes it from Gellifeusach as *Ketheleuysach*. This suggests that the form Cefngorach may be a late imposition and casts doubt on the suggestion (PNBont 38) that it is connected with Ir *corach* 'marsh'. It is possible that -*gorath*, *Gwrath* are composed of *gor-*, variants *gwor-*, *gwar-* and *garth*[1] in the sense of 'high or prominent ridge'.

Fforest[3]   SN 6243, 6443
Llan-y-crwys
'forest, unenclosed woodland': *fforest*
Y Fforest c.1600, *Fforest* 1811, (ht.) *Fforest* 1851
One of the two hts. of Llan-y-crwys extending southwards from the p. ch. and including Blaenaufforest (SN 618451) (*blaen* pl. *blaenau*). See Pennant (Llan-y-crwys).

Fforest Bedol   SN 6915, 7017
Llangadog
'forest near (r.) Pedol': *fforest*, rn. *Pedol*
(mess. in) *Forest Bedwl* 1610, *forest Bethel* 1772, *fforest Bethel*, *Forest Bethel* 1813
Richards (WATU) describes Faerdref and Fforest Bethel (*sic*) as a manor in Cwarter Bach, in the p. of Llangadog, but Maerdref (q.v.) and the greater part of Fforest Bedol lay in Is Cennen and the p. of Llandeilo Fawr (HCrm I, 233). Nant Pedol (q.v.) rises at Blaen Pedol (SN 717181) (*blaen*) on Mynydd Du/Black Mountain and flows southwestwards through

Cwm Pedol (*cwm*) to meet the r. Aman (SN 688134). The rn. is probably **pedol** 'horseshoe' describing the shape of the valley in its middle reaches above Pantyffynnon (SN 698153). A ho. Cefnyfforest (SN 686148) is located a little to the west of Cwm Pedol near Cynghordy. Pedol occurs in a ho.n. in Caeo recorded as *Ty blaen Cwm Pedol* 1662 and *Tyr Blaen Cwm pedol*) 1733 (*tŷ*, **tir**).

Fforest Crychan SN 82378339, 8440
Llanfair-ar-y-bryn
'forest adjoining (r.) Crychan': **fforest**, rn. **Crychan**
CRYCHAN FOREST 1953
Largely plantations begun by the Forestry Commission after c.1950.

Fforest Gaerdydd, Cardiff Forest SN 2216
'forest called Cardiff': *Cardiff*, **fforest**, E **forest**, *Caerdydd*
silvam de Gardif' 1215, *Cardyth forest* 1578, (forest of) *Cardiffe* 1582, *Cardith forest* 1587, *Cardiff Forrest* 1590, (forest) *Caerdyth'* 1602
Cardiff or Caerdydd has not been precisely located but the pn. may be centred on the ho.n. Fforest (SN 224155), just north of the r. Taf (q.v.), in Llanboidy p., recorded as *Forrest* 1729. The n. appears to be identical in meaning to our capital city Caerdydd/Cardiff, 'fort by (r.) Taf' (**caer**, **Taf**) (see PNGlamorgan 33). Fforest was part of the gr. of Iscoed once held by Hendygwyn/Whitland abbey (WCist 182, 313) but as yet no fort has been located. Terrence James suggests a Roman connection (CAS 34).

Ffwrnes, Furnace¹ SN 503014
Llanelli
'(place named from) a furnace': *ffwrnais*, *ffwrnes*, **furnace** E
1830, 1880, *Ffwrnes* 1892, *y Ffwrnes* 1905
This is also likely to be *the Furnace* 1685 and *Furnace* 1697 and is testimony to the long history of ironworking and industrial activity in the Llanelli area expanding with the construction of a blast furnace in 1793 bought by Alexander Raby, a London ironmaster and cannon founder, in 1796. Locally pronounced 'Ffwrnesh' [ˈfuːrnɛʃ].

Ffynnon-ddrain SN 401219
Caerfyrddin/Carmarthen
'well near thorn-bushes': **ffynnon**, **draen**, *drain Fonon Ddrayne* 1575, *Ffynon ddrain* 1626, *Funnon ddrain* 1657, *Tythin ffynnon ddrain* 1667, *Tyr finon Draine alias Lloyn Tege* 1710, *Ffynnon-ddrain* 1831
A fairly common n.; cf. Ffynnon-ddrain (Llannewydd) recorded as *ffynon draine* 1662 and *Ffunnonddrain* 1799 describing a well located in an area notable for thorn-bushes.

Ffynnonhenri SN 3930
Cynwyl Elfed
'Henry's well', **ffynnon**, pers.n. **Henri**
Funnon Henry 1752, *ffynnon Henry* 1773, *Funnon Hendry* 1795, *Ffynnon Henry* 1831
Henry/Henri is unidentified. The location of a Baptist chp. with the dates 1774 and 1823 above the door but there was an older chp., dating from c.1730, simply described as a dissenting congregation of Protestants in 1752, located apparently in a ho. at the edge of the graveyard (CrSG 233-4). There was another Ffynnonhenri at Llandyfaelog recorded as *ffynnon Henry* 1655.

Ffynnon-oer SN 2025
Llanboidy
'cold well': **ffynnon**, **oer**
Hamlett of ffynnon oer, finnon oer c.1700
Otherwise unrecorded but apparently centred on Cwrtau-mawr (SN 205253) which is recorded as *Cwrtemawr* or *Courtyfynnon Oer* 1874 and as *Court y ffynnon Oer* 1668 and 1699, otherwise *Cwrte mawr* 1738 (1758), with **cwrt** pl. *cwrtau, cyrtiau* 'monastic grange' and **mawr**, belonging to the monastery of Hendy-gwyn/Whitland. The n. may refer to the spring at the head of Afon Tigen (q.v.). Ffynnon-oer was later part of Kingsland (q.v.).

Five Roads see **Pump-heol**

Foelgastell　　　　　　　　SN 545148
Llanarthne
'(the) bare *or* exposed castle': (***y***), ***moel***, ***castell***
y vole gastell 1693, *Foelgastell* 1813, *Foel-gastell* 1831
Also recorded in *Waynvole castell* and a smith's forge *Evel vole Castell* 1745 (***gwaun***, ***gefail***). No traces of any castle are shown on OS maps here but ***castell*** was sometimes used to describe a prominent, steep-sided hill. That might apply to the hill west of the village near a house Pen-y-foel recorded as *Pen y vole* 1851 and *Pen-y-foel* 1880-9 'top of the bare-hill' (***pen***[1], ***y***, ***moel***). Foelgastell lay within the t. of Trecastell[1].

Four Roads see **Pedair-heol**

Fro　　　　　　　　　　　SN 4539
Llanfihangel-ar-arth
'(the) vale, lowland': (***y***), ***bro***
Frô 1811
Poorly evidenced. The ht. appears to have been that part of the parish in the valley of the r. Teifi in contrast to Blaenau[2] (q.v.).

Furnace[1] see **Ffwrnes** (Llanelli)

Furnace[2]　　　　　　　　SN 438011
Pen-bre
'(place near a) furnace: E ***furnace***
Furnace 1830, 1880
The n. probably recalls an ironworks established in 1810 but closed by 1816. The site was re-developed by the Pembrey Iron and Coal Company in 1824 (PBPH). Located near a small coal-pit in 1880. Cf. Ffwrnes in the adjoining p. of Llanelli.

Ganol to Gwynfe

Gwynfe. Pont Glan-rhyd, over Afon Clydach, and Capel Jerusalem c.1900.

Ganol　　　　　　　　　　SN 5936, 6136
Llansawel
'middle (ht.)': *canol*
(ht.) *Genol* 1757, *Genol Hamlet* 1851
The central part of the p. extending from the village of Llansawel a short distance up Afon Melinddwr and Afon Marlais.

Gardd Botaneg Genedlaethol Cymru see **Neuadd Middleton**

Garn　　　　　　　　　　SN 624148
t. Llandybïe
'(the) cairn': (*y*), ***carn***
Fferen y garn c.1700, *Garn* 1734, 1804, 1831, (*Dosbarth*) *y Garn* 1803
There are no obvious physical traces of any cairn here and no clues in field-names in the tithe map and apportionment 1843. G.M. Roberts uses the form 'Fferen y Garn' (***fferm***, ***fferem***); cf. Fferm-fawr and Pistyll. ***carn*** is a common el. in this p. (HPLand 23); cf. Garnbica (SN 637169) (***pica^1***) recorded as *Carnbicca* 1831 but the scholar Edward Lhuyd noted only a limestone quarry c.1700 (Fenton 337).

Garnant SN 689131
Betws
'rough brook': ***garw, nant***
(brook) *Garnant* 1609, (brook) *y Garnant* 1610, 1764, (tmt.) *Teer blan y garnant* 1614, (brook) *Gar'nant, Cwm-gar'nant, Garnant Colliery* 1831
Named from a stream which rises at SN 699099, marking the border of Glamorgan (now Neath Port Talbot) with co. Carmarthen, and flowing northwards to meet Aman (SN 689133). The village developed in the early-mid 19th century near a railway station and coal-workings along the valley Cwm Garnant recorded in (mess.) *Cwm y garnant* 1808 and (valley) *Cwm Garnant* 1878 (***cwm***). A mess. *Glan y garnant* (***glan, y***) is recorded in the parish of Llan-giwg, Glamorgan, in 1610. This cannot be the ho. *Glan Garnant* 1898 which is earlier *Tir-bach* 1831, *Tir-bâch* 1878 (***tir, bach¹***).

Gelli-aur, Golden Grove SN 590197
Llanfihangel Aberbythych
'(the) grove of trees likened to gold': def.art. (***y***), ***celli, aur***; E ***golden, grove***;
Goldengroue 1578, *Golden Grove* 1578-1831 (frequent), *Goulden Grove* 1600-7, *Goldengrove* 1683, *y Gelli Aur* 1596
Golden Grove has been explained as a translation of Y Gelli-aur probably coined soon after the first recorded ho. was built here in 1560 (HCH 83-85) or soon afterwards. The W form is not formally recorded before 1596 but there are mentions in earlier literary sources dating to the 1560s. It is possible that Gelli-aur is in actuality a reinterpretation of an unrecorded Gelli-oer 'cold grove' (***oer***) describing the location of the earlier ho. (burnt down 1729) further north on the damp floor of Afon Tywi. This cannot be proven but it is worth noting local pronunciation of *aur* as *oer* (Eurig Davies in WPNS Bulletin 20). The E n. is found elsewhere, eg. Gelli-aur (SN 950800), in Llangurig, co. Montgomery, is *Gelly Aur* 1765, *Gellyaur* 1787, and there other late instances in Bugeildy, co. Radnor, and Cadoxton-juxta-Barry, co. Glamorgan.

Gelli-ddu SN 5216
Llanddarog
?'dense grove': ***celli, ?tyn¹***
(place) *tre gelli ddyn* (called *gweyn y felin*) 1612, (ht.) *Kellyr dyn* 1670, ht.) *Kelly-dyn* 1703, (ht.) *Gelluddie* 1804, *Gelli-ddu-fach* [= Gelli-ddu], ~ -*fawr* 1831
A ht. centred on a fm. Gelli-ddu Fawr and Gelli-ddu Fach (SN 527166) (***mawr, bach¹***). The modern form is evidently a re-interpretation as if it were a 'dark grove' (***du***). Forms such as those for 1612 and 1703 tend to favour ***tyn¹*** adj. probably with the specific sense 'dense (of growth, etc.)' lenited as *dyn* and apparently confused in 1670 with the def.art. *y* and *dyn* 'man, human being'.

Gellidigen ?SN 1724
Llanboidy
Uncertain
(ht.) *Celli digen, Celli digon* c.1700, *Gellydigen* 1738 (1758)
The first e. is ***celli*** but without further evidence it is too hazardous to attempt to identify the second el. and suggest a meaning though it is unlikely to have any connection with Afon Tigen (q.v.) which was at least two km. away. Its precise location is uncertain but it was certainly in the western part of Llanboidy bordering a stream Nant yr Haidd which rises at SN 181256 near Cilhernin and meets Afon Taf at SN 166235. This should not be confused with Gellidogyn (SN 200211), in Llan-gan p., recorded as *Kelly Dogin* 1772, *Kelydogy* 1828, *Gelly Dogin* or *Killy Dogin* 1874 but the similarity is remarkable and the two pns. may have a common unexplained meaning.

Gelligati SN 291412
Cenarth
'grove of (woman called) Cati': ***celli***, pers.n. **Cati**
(ho.) *Gelli Gatti* c.1700, (hos.) *Gelli-gati-fawr, Gelli-gati-fach* 1831, *Gelli-gati-fawr* 1888, *Gelligatti* 1907
Cati is a familiar form of Catherine or Catrin probably found also in Tŷ Cati (*Ty Cattey*

1766), in Caerleon, co. Monmouth, and Trecati (*Trecatty* 1875), in Merthyr Tudful, co. Glamorgan (*tŷ*, *tref*). Cati is best known in the n. of the folk hero Twm Siôn Cati who has been identified with Thomas Jones (c.1530-1609) associated with the Tregaron area, co. Cardigan. Historical forms of Gelligati are, however, late and we cannot entirely rule out a familiar form of some other pers.n. such as Cadfael or some unidentified el.

Gelliogof　　　　　　　　　　　　SN 2124
Llanboidy
'grove near a ?cave': *celli*, ?*ogof*
Gelly Ogof 1692 (1758), (ht.) *Celli ogof* c.1700, *Gelly Ogof Hamlett* 1738, *Gellyogol Hamlet* 1844
There seems to be no evidence of any cave in the area. Gellyogol may be through misassociation with *hogl* 'ill-designed, ramshackle; shed, shelter for cattle, etc'.

Gelli-wen　　　　　　　　　　　　SN 275237
Trelech a'r Betws
'(the) fair grove': (*y*), *celli*, *gwyn*, *gwen*
y gelli wen 1581, *Gelliwen* 1745, *Gelli-wen* 1831, (fm.) *Gelliwen* 1875, *Gelli-wen* 1889
A small scattered settlement adjoining the wooded valleys of Afon Cynin and an unnamed tributary.

Glanaman (Glanamman)　　　SN 6613, 6713, 6813
Betws
'bank of (r.) Aman': *glan*, rn. **Aman**
Glanamon c.1500, (mess. and tmt.) *Tyr ynglann Aman* 1601, *Y Tiddhin yn Glan Amman* 1629, (mess.) *Glanamman alias Tyr pen y bont* (in *Llandilovawr*) 1714
The village developed near coalworkings from the early 19th cent and particularly after the arrival of the railway. Earlier refs. apply to hos. in Betws and Llandeilo Fawr.

Glanbrân　　　　　　　　　　　　SN 7938
mansion Llanfair-ar-y-bryn
'bank of (r.) Brân': *glan*, rn. **Brân**[1]
glann brann c.1560, *Glann braen* 1592, *Glanbrane* 1618, c.1735, *Glanbraen* 1704, *Glanbrân* 1764

Forms such as *braen*, -*brane*, *Brane* are attempts at representing the long â [ɑ:]. Identical in meaning to Glanbrân (Llangadog/Myddfai) recorded as *Glanbraen* 1704 and *Glanbrane* 1760 and Glanbrân (in Llansamlet) recorded as *Glanbrane* 1633, *llan bran* 1712, *Glanbrane* 1775, *Glan Brân* 1777. The rn. is found elsewhere with an example in co. Brecon (PNBrecon 43). For the mansion see HCH 74.

Glanbrydan　　　　　　　　　　　SN 668265
Llandeilo Fawr
'bank of (r.) Brydan': *glan*, rn. *Brydan*
Glan Brydan 1704, *Llanbrydan* 1721, *Glanbrydan* 1831, *Glanbrydan Park* 1868
The n. applies to a large ho. Glanbrydan Park (HCH 74) and fm. and park delineated by the A40, a tree-lined road (now a foot-path) and the road leading alongside the stream to Glanbrydan House. Glanbrydan is 'bank of (r.) Brydan': *glan*, rn. *Brydan*. The rn. (which enters Tywi at SN 675257) may contain the el. *bryd*- as in *brwd* adj. 'hot, warm', with suffix -*an* possibly in the sense 'gentle, pleasant' or 'foaming (like hot water)' (EANC 44). The suffix also occurs in Nant Beuddan and Glanbeuddan less than 1 km to the south.

Glan-duar　　　　　　　　　　　　SN 527437
Llanybydder
'bank of (r.) Duar': *glan* rn. **Duar**
Glandyar 1831, *Glanduar* 1839, *Glanduar Mill* 1844
The rn. is also recorded in Aberduar (SN 527441) where Duar reaches Teifi recorded as *Aberdyar* late 15th cent, 1636, *Aberdyer* 1691, *Abarduar* 1711, *Aberduar* 1759. Cf. Nant Duar (Llanelli) recorded in *Aberduar* 1839 and *Nant Dyar* 1891 (*aber*, *nant*).

Glandy Cross see **Groesffordd**

Glangwili　　　　　　　　　　　　SN 430221
Abergwili
'bank of (r.) Gwili': *glan*, rn. **Gwili**
Glangwilie 1645, *Langwilly* 1729, *Glangwilly* 1765
Identical ns. occur in Llanllawddog as *Plas glann gwelie* late 16th cent, *Glangwilly* 1700,

Llanpumpsaint as (tmt.) *Glan Gwilly* 1676 and Llannewydd/Newchurch as (lands called) *Glangwilly* 1705.

Glanmôr, Seaside   SN 501992
Llanelli
'(place) by the sea': **glan**, **môr**[1]; E **sea**, **side**, *Seaside* 1831, *Sea Side* 1851, *Glanmore* 1851
From its location. The area developed after the opening of what became known as Carmarthenshire Dock and the n. initially applied to hos. north of former copperworks, later the site of leadworks and the Cambrian and Pemberton tinplate works (opened 1911). The n. probably shifted to the present location east of the works in the late 19th cent but drops off OS maps until after World War II. It may have survived in informal use since OS maps in 1880 and 1907 show *Glanmôr Foundry (Iron)*, a little to the north beyond St David's Lane. The area developed south of Marine Terrace and Marine Street.

Glansefin   SN 730286
mansion Llangadog
?'bank of Sefin': ?**glan**, rn. **Sefin**
Llan Sevin 1596, *Tir Glansevin* 1628, *Glansevin* 1671, *Glansevyn* 1680, *Lansevin* 1743, *Llansefin* 1783, *Glansefin, ~ -isaf* 1831, *Glansefin Issa* 1831
The apparent absence of early sps. means that it is difficult to be certain if the first el. is **glan** or **llan** which share the len. *lan* after a prep. such as *i* and *o*. Most forms favour the suggested meaning implying that Sefin is a rn. The OS map 1831 applies *Sefin River* to Afon Brân[2] which prompted the historian Phillimore to dismiss this as an inference from the ho.n. and to interpret the first el. of this as **llan**. It is not impossible that Sefin simply applied to a stretch of Brân, cf. Dyfri for part of Bawddwr (see Llanymddyfri/Llandovery), or that Sefin is a lost n. for the small stream which passes through Glansefin and Glansefin mill (SN 733288) to Afon Brân. R.J. Thomas (EANC 212-3) suggests that Sefin may also be a pers.n. possibly derived from Latin *Sabīnus* but it is more likely to be related to *Sumina* (the r. Somme, France), Seven (Yorkshire), Sem (Wiltshire) and Syfynwy, co. Pembroke (see PNPemb 20). Glansefin has been identified with a place *Lan Semin* mid 11th cent c.1200 and *Lansemyn* 13th cent (16th cent) in pedigrees relating to the semi-fictional Brychan Brycheiniog but this is better sought in the vicinity of the rn. Sefin, co. Brecon. For the mansion, see HCH 78.

Glan-tren   SN 521430
fm. Llanybydder
'bank of (r.) Tren: **glan**, rn. **Tren**
(*o*) *Lantren* c.1600, *Lantren* 1729, *Llantren* 1789, *Llantrev Vach* 1763, *Lantren-fach, ~ -fawr, ~isaf* 1831, *Glantren* 1874
The n. of three hos. distinguished as 'little' (**bach**[1]), 'great' (**mawr**) and 'lower' (**isaf**) and a former mansion formerly known as Blaen-tren (HCH 11-12) probably *Blaen Tren* 1647 and note *Tir blaen tren issa* 1600 and *Tyr Blaen Trenn vawr* 1654 (**blaen**). The ho. is also likely to be Blaen-tren in the poetry of Hywel Dafi (floruit c.1450-80) (GHD 1: 38n.). R.J. Thomas (EANC 125) states that Tren is a r. which rises north of Banc Du, flowing to Rhyd ar Dren (SN 520433) and Afon Duar but he appears to have confused it with Nant Einon (otherwise known as *Aberddwr Brook* 1729) about a quarter of a mile east of the hos. Tren is actually a very small stream which enters Teifi directly (SN 519435). Tren is found elsewhere as a rn. in Anglesey and in anglicised form as Tern, in Shropshire, where it is thought to have **tren** 'strong, powerful'. That seems inappropriate here unless it is used facetiously.

Glantywi   SN 4721, 4722
Abergwili
'bank of (r.) Tywi', **glan**, rn. **Tywi**
Glan tywi c.1550, (*o*) *Lan Towy* c.1600, *Glantowy* 1768, 1779, *Llantewy* 1777, *Glann Tywi* 1811, *Glan Towy, ~ -fach, ~ -fawr* 1831
From its location. Glantywi-fach is now Glynmyrddin (SN 458208). There are other examples of Glantywi recorded as *Glantowy* 1795 (SN 706295, in Llangadog) and *Glantowy* 1723 (SN 723312, in Llanwrda and Myddfai).

Glanyfferi, Ferryside SN 365104
Llanismel/St Ishmael
'(place at) side of the ferry': *glan, y, fferi*; E *ferry, side*
Boatside (Llanstephan Ferry) 1740, Ferryside, Glanyferry 1811, Ferry-side or Lan-y-ferry 1831, Glanyfferi 1862, 1867
The ferry connected Llansteffan to Ferryside and the road is recorded as *Ferye weye* 1583 running south-eastwards to Cydweli where it is *Port Way* 1831 (E *port, way*). The ferry is *passagium de Tewy* c.1185, *Passagio in Landesteffan* 1316, *la Ferye* 1326, *la Verie* 1462, *The Ferry* 1675 (AFr *la*).

Glanyrannell SN 650370
mansion Talyllychau/Talley
'bank of the (r.) Annell', *glan, yr,* **Annell**
(tmt.) *Tir Glan yr Annell* 1603, (mess.) *tir glan yr Annell* alias *gweyn yr ystrad* 1614, (tmts) *Tir Glan yr Annell ycha* and *Tir Glan yr Annell yssa* 1663, *Glan yr Annell* 1714, *Glan Rannell* 1715, *Glanyrannell* 1760, *Glanrannell, ~ isaf* 1831
The name is borne by a hotel (SN 650370), a farm (SN 670390) near Caeo, and Glanyrannell-isaf (SN 647365) (*isaf*). Annell rises above Caeo in marshland (SN 709427) between Cefn y Bryn and Esgair Ferchon. It is likely that the rn. was earlier Arannell and that initial Ar- has become confused with the def.art. *yr*, `r` seen in forms such as *Glan yr Annell* and *Glanrannell*. For the mansion, see HCH 79-80.

Glasallt Glasallt-fawr SN 731301
Myddfai
'green wooded-slope': *glas, allt*
Glase Allt 1612, (place) *glassalt* 1620, *Glasallt* 1680, *Glasallt* 1702, (ho.) *Glaesallt* c.1735, *Glas-allt-fawr* 1831
The n. survives as that of a ho. Glasallt-fawr, 'greater Glasallt' (*mawr*) (see HCH 80). Glasallt-fach (SN 729299) is recorded as *Glasallt vach* 1800 and *Glas-allt-fach* 1831, 'lesser Glasallt' (*bach¹*). The n. changed on OS maps to Green Grove some time between 1964 and 1978.

Glwydeth, Afon SN 4737
rn. SN 500361 to Talog SN 465379
Uncertain
Nant Glwydeth 1887-8, *Afon Glwydeth* 1905, 1964, (r.) *Afon Gilwydeth* 1979
The n. also appears in the ho.n. *Aberglwydeth* 1851, *Aber-Glwydeth* 1887-8 (*aber*). Historic evidence is far too late for confident interpretation but it is notable that the rn. has both initial G- and final -th which appears to rule out any comparison with, for example, *clwydedd* 'door, gate'. There is in any case just one literary instance of this in GPC. Initial G- also rules out any comparison with Afon Clwyd, cos. Denbigh and Flint. Clwyd has been explained as 'hurdle river' (*clwyd*) (DPNW) or more likely 'river with numerous bends' or 'powerful river' ('afon â nifer o droeon neu un rymus sy'n cludo': Guto Rhys, StC 52: 179-82).

Glyn¹ SN 4908
ht. Llanelli
'(the) valley': (*y*), *glyn*
(place) *y Glyn* 1634, *Glyn* 1787, (ht.) *Glynn* 1804
The ht. seems to be centred on a ho. Glyn (SN 494082), in the northern part of Llanelli p., and this is confirmed by its occurrence in 1787 with Gelli-oer (formerly SN 478083), Cynheidre, Cynheidre-fach, Cwmgelwr (SN 484081), Ffou (SN 480082) and Tir Christopher (SN 496076). A second ho. is implied by *Glinn icha* 1571 (*uchaf*). The valley is possibly that of the small, unnamed stream which flows southwards to Afon Lliedi or that immediately to the west flowing westwards to Gwendraeth Fawr. This may be the location of *the fforreste adioyninge to the Glynne* 1609.

Glyn² SN 5819
ht. Llanfihangel Aberbythych
'valley': *glyn*
Glyn 1767, *Glynn* 1792, 1811, (ho.) *Glyn* 1887-9
The ht. was evidently centred on the ho. now marked as Tiryglyn Farm (SN 589150) (*tir, y*) but shown on OS maps before 1964 as plain *Glyn*. The valley could be that one of two small

streams on the east of Blaen-y-glyn (*blaen*, *y*) running south from near Glyn-yr-henllan (SN 594153) (q.v.) to Afon Lash.

Glyn³ SN 4407, 4408
Llangyndeyrn
'(the) valley': (*y*), *glyn*
Glin 1586, *Glyn* 1587-1804, *y glyn* 1598, *the glyn* 1603, (place) *the Glynn* 1628, *Glynn* c.1735
The southernmost part of Llangyndeyrn p. The valley may be that of an unnamed stream running south from Glascoed (SN 447088 to Pont Newydd (SN 447073) or that which flows southwards through Glyn-fach (SN 452083) (*bach¹*) to Afon Gwendraeth Fawr. The n. survives in that of a ho. (SN 447079) which took on the n. Glyn Abbey in the 1790s (though there is no evidence of an abbey here, see HCH 81-82) described by the topographer Carlisle in 1811.

Glyn⁴ SN 5408
Llan-non
'(the) valley': (*y*), *glyn*
(place) *y Glynn* 1590, *Glyn* c.1700, 1804, (ht.) *Glynn* 1811
The valley of a tributary of Afon Morlais.

Glyn⁵ SN 6035
Llansawel
'(the) valley': (*y*), *glyn*
Glyn Hamlet 1851
The n. is not current but survives in the ho.ns. Banc-y-glyn (SN 596360) and Rhyd-y-glyn (SN 609358) recorded as *Banc Farm* and *Rhyd-y-glyn* 1831. These are probably best understood as meaning 'hill in Y Glyn' (*banc¹*) and 'ford in Y Glyn' (*rhyd*). The valley is likely to be that of Afon Melinddwr.

Glynaman SN 6613, 6813
Llandeilo Fawr
'valley of (r.) Aman': *glyn*, rn. **Aman**
(*terris vocatis*) *Glyn Amon* 1609, (place) *Glynne Aman* 1611, (ht.) *Glin-Aman* 1657, (mess.) *Glynamman* 1695, (ht.) *Glin Amman* 1718, (ht.) *Glynn Aman* 1811

Glynaman has been displaced by Cwmaman (q.v.) which applied to a ht. in Betws p. and Cwm Aman for the valley.

Glyncothi SN 6236, 5833
Llansawel, Llanybydder
'valley of (r.) Cothi', *glyn*, rn. **Cothi**
(*bosco de*) *Glyncothy* 1300, (forest of) *Glyncothi* 1302, 1353, (king's forest) *Glencothi* 1312, *y forest lynn kothi* c.1500, (mills, vill) *Glyn Cothey* c.1550, (forest of) *Glincothy* 1598, (manor) *Clyn Cothy* 1718
A precise location cannot be given as the forest extended from Llanfihangel Rhos-y-corn and Llanybydder into Llansawel to include the valley of Cothi below Abergorlech.

Glyn-hir SN 640151
mansion Llandybïe
'long valley': *glyn*, *hir*
Glynheer 1725, *Glynhir* c.1735, 1786, *Glynhire* 1746, (ho.) *Glynhir* 1815
The narrow valley of the upper reaches of Afon Llwchwr and the location of the ho. Glynhir Mansion (SN 63921511) dated to the 18th cent. For the mansion, see HCH 82-83.

Glyn-tai SN 617130
Llandybïe
'valley or meadow occupied by houses': *glyn* or *clun²*, *tŷ* pl. *tai*
Glyn Tay c.1700, *Glynn Tay* 1811, *Clun-tai*, ~ *fach* 1831, *Glyn-tai* 1878-1953
clun² appears to be a re-interpretation but earlier forms are required for certainty. It acquires the qualifier *Fawr* 'greater' (*mawr*) before 1964. Glyn-tai Fach, 'lesser' (*bach¹*) is not named until 1964-5.

Glynyrhenllan SN 594153
Llandybïe
'valley of the old church': *glyn*, *yr*, *hen*, *llan*
Glynhenllan 1813, 1851, *Glyn-yr-hên-llan* 1831, (ho.) *Glyn-yr-henllan* 1878, *Glyn* 1843, *Glyn-yr-henllan*, *Glynyrhenllan* 1898
Recorded as a ch. or chp. in Llandybïe in 1836 but it is not shown on the tithe apportionment

1843 and OS maps and the supposition may be based on nothing more than a dubious interpretation of the pn. There certainly seems to be no other evidence for a ch. here. Historic forms are late and it is possible that *glyn* has displaced *clun*²; cf. Glyn-tai. The historian G.M. Roberts gives the n. as Glyn or Clun Henllan (HPLland 13).

Glyn y Swisdir, Swiss Valley   SN 524028
Llanelli
'valley likened to Switzerland': *glyn, y, Swistir*;
E *Swiss, valley*
Swiss Valley 1847, 1919
The W form (properly Glyn y Swistir) applies to a modern street. Swiss Valley seems to be first mentioned in a fictional story 'Mary de Clifford' published in 1847 but the n. may not have been formally adopted until some time after 1891 when it is mentioned by the local historian Arthur Mee (CNotes III, 62). The n was applied to hos. in what is properly Cwm Lliedi (*cwm*). The location of waterworks opened in 1903 and public park and later housing developed on hill-slopes east of the valley of Afon Lliedi around Cribyn Lane.

Goetre   SN 5544, 5603
Llan-non
'(the) settlement in woodland': (*y*), *coetref*
The Goytre 1590, *Y Goytre* 1616, *y Goedref* c.1700, *Goytrey* 1700, *Coytre* ~ , *Goytre Isah* 1782, *Goytre* 1804, *Goytre-wen, Goytre-isaf, Goytre-fach* 1830
The southern part of Llan-non p., characterised by small scattered woods, and centred on Hen Goetre (SN 558043) which is *Old Goitre* 1878-1953 and *Hen Goitre* from 1964 (**hen**). This is also likely to be *Sythyn y goedtre* 1615 (**syddyn**). The other hos. are distinguished as 'white, fair' (*gwyn* fem. **gwen**), 'lower' (*isaf*) and 'little' (*bach¹*).

Golden Grove see **Gelli-aur**

Gorsaf Llwyfan Cerrig, Llwyfan Cerrig Station   SN 406258
Llanpumsaint
'station platform composed of stones': *gorsaf, llwyfan², carreg* pl. *cerrig*, E **station**

A new n. for a station opened in 1988 on the Gwili Steam Railway, the successor of the Carmarthen & Cardigan branch railway. Current OS maps do not show the W form.

Gorlech, Afon   SN 5636
SN 541385 to Cothi SN 585336 Llansawel, Llanybydder
?'very swampy or muddy': *gor-, gwor-, llwch²*
(r.) *Gorluch* 1411, *Gorlech* 1584, *River Gorlech* 1887
The rn. may describe the area where it rises or the r. itself, i.e. one which is characterised by pools or mud'. Sps. with *-lech* reflect association with **llech**. The rn. is also found in Abergorlech (q.v.) and Blaengorlech recorded as *Blaen Gorlech* 1584 and in ho.ns. *Blaengorlech* and *Blaengorlech-fach* 1831 (**blaen**).

Gors-goch   SN 572133
area Llanarthne
'(the) red-brown marsh': *cors, coch*
(mess. etc) *Gorse Goch* 1677, *Gors-goch* 1831, *y Gorsgoch* 1867
The n. applied in 1880 to two hos. on the road running from Capel Hendre to Gors-las but was transferred to Gors-goch Colliery before 1842. *cors* is also found in ho.ns. *Gors-fâch* and *Gors-ddu* and nearby Gors-las (q.v.). Gors-goch is now also applied to hos.. The redness refers to the colour of the vegetation; cf. the raised peat bog Gors Goch (SN 365188) near Llan-llwch.

Gors-las   SN 570138
Llanarthne
'(the) green marsh': (*y*), *cors, glas*
Gors-las 1831, *Gorse Laes* 1844, *Gors-lâs* 1880-9, *Gorslas* 1953
Located in a marshy area of Mynydd Mawr which was common land extending from Llyn Llechowen (SN 569151) to Saron (SN 601125). Developed from the mid 19th cent near several small coal pits and levels and the larger California (SN 579137) and Gors-goch (SN 572129) (q.v.) pits. Formed into an ecclesiastical p. in 1879 from parts of Llanarthne and Llangynnwr.

Gothylon SN 5425
Llanfynydd
Uncertain
(ht.) *Gothylon* 1851, 1863
Possibly from a rn. formed from *gwythol* adj. 'fierce, angry' and noun suffix *-on¹*, *-ion¹*. The conjectured variation *gwo-/go-* may be compared with *go¹*, *go-*. The n. is not current and a precise location cannot be given but the 1851 census includes it with Cornorion. The two were centred on the suggested grid ref. in the most southerly part of Llanfynydd.

Graig SN 445017
Pen-bre
'cliff': *craig*
Craig 1813, *Graig Capel* 1839-1954, *Craig-Capel* 1880, 1921, *Graig* 1972-77
Earlier 'chapel cliff' (*capel*). This may also be *the Graige* alias *Alth y Nant y Graige* 1725 and *Graig* alias *Altythunant y Graige* 1813. The chp. is presumably Jerusalem W Independent (founded 1812 CYB 448). The ht. has developed against a steep hill marked by quarries.

Gransh, Grange SN 3938, 3935, 4037
ht. Llangeler
'grange, farm belonging to a religious house': *gransh*, E *grange*
Trayan y Grange 1698, *Grange* 1841, *Hamlet of Grange* 1847, *Grange Hamlet* 1875, *y Gransh* 1899
The gr. is that of Maenor Forion (q.v.) probably centred on Court Farm (SN 394388) (*cwrt¹*) held by Hendy-gwyn/Whitland abbey before the Dissolution. Gransh/Grange made up the easterly part or administrative 'third' (*traean*) of Llangeler (as distinct from Gwlad q.v.) extending southwards from Afon Teifi to the hill above Triolmaengwyn (SN 392346); see also Pentre-cwrt.

Green Castle see **Castell Moel**

Groesffordd, Glandy Cross SN 143266
Cilymaenllwyd, Llandysilio (East), Llanglydwen

'(the) crossroads' (*y*), *croesffordd*, 'cross near Glandy', pn. *Glandy*, E *cross*,
Glandy Cross 1872, *Groesffordd*, (ho.) *Cross Inn* 1889, *Cross Inn* 1907, *Glandy Cross, Cross Inn (PH)* 1977
Glandy Cross first appears as a n. on OS maps after 1964 but it was clearly in use at an earlier date for what is otherwise known as Groesffordd at the meeting of five roads near Meini Gŵyr prehistoric circle. Glandy (in Llandysilio) refers to two hos. recorded as *Glandy Fawr* and *Glandy Fach* 1888, *Glandy-bâch* and *Glandy-mawr* (SN 144277) 1889 (*bach¹*, *mawr*). The second is plain *Glandy* 1632-1819, a n. surviving also in the rn. Afon Glandy which marks the historic border of cos. Carmarthen and Pembroke recorded as *Glandy brook* 1840. The n. appears to be drawn from the ho.ns. meaning 'clean, fair house': *glân* 'clean; clear of sin; holy; fair' and *tŷ*. The earlier n. of the r. may have been Bre recorded as (rill) *Bray* 1603 (PNPemb 3-4) and as *Bray* by the antiquarian Richard Fenton. It is otherwise described as Afon Gorsfach in its upper reaches and Afon yr Ynys. J. Lloyd James suggests in 'The Welshman' in 1909 that this has *bre¹* 'mountain' ('it being a mountain brook') but Charles notes *brau* adj. 'brittle, weak; free' and *breu* as in *breuer, breufer* 'loud, roaring', possibly the root in Breuan, co. Pembroke (PNPemb 4-5).

Gronw, Afon SN 2221
SN 223272 to Taf SN 202161
(?) (r. derived from a ho.) Gronw
Afon Gronw 1887
Gronw appears to derive from a ho.n. but the reason for its substitution is unclear. The occurrence of a ho.n. *Grwnw Issha* (unlocated) 1659, in Llangan, (*isaf*) implies a second ho. Gronw-uchaf (*uchaf*), both apparently containing the pers.n. *Goronwy, Gronw*. The r. was earlier known as Marlais and is recorded as *Marlas flu:* 1578, *the Marlais* 1587 and *Marlais Oley* c.1700. The latter apparently has *golau* 'bright' and is uncorroborated but the historic form is taken from the testimony of a native of Llanboidy (AC 1975: 107). The qualifier distinguishes the r. from another Afon

Marlais (SN 1515), also a tributary of Afon Taf, west of Hendy-gwyn/Whitland; cf. Marlais[1-3]. The identification is confirmed by the ho.n. Dyffryn Marlais (SN 218217) itself transferred from another ho.n. recorded as *Dyffryn Marles otherwise Maes Marles* 1688 and *Dyffryn Marlas otherwise Place Marlas* 1690 (**dyffryn, maes, plas**).

Gwastade SN 6330
Talyllychau/Talley
'flat places, valley bottoms': **gwastad** pl. *gwastadau*
Gwastadae 1520, *Gwasetide* c.1530, *Custoda* c.1538, (gr.) *Custa* 1559, *Custoda alias Gwastoda* 1577, (place) *Costoday* 1578, *Custa, otherwise Custoda* 1591, *Gwastade* 1626
Earlier forms have not been found and it is difficult to explain the form *Custoda* unless it has developed from *gwstodau* < *gwostodau* as a variant of *gwastadau* with the variation *gwo-, go-*. The n. is no longer current but probably applied to the area around Cwm-du and specifically to the flattish ground to the east (Melville Richards in CStudies 110-1).

Gwempa SN 439119
Llangyndeyrn
?'fair place': **gwymp** fem. *gwemp*, ?-*fa*, *ma*[1]
Gwempa 1590-1831, *Wempa* c.1600, (place) *Gweinpaye* 1609
This is very uncertain because historic forms have no instances of *-fa* but no better solution seems to be available. GPC has only one instance of '*Gwempa, i.e. Gwempfa*, a fair place' in 1753. The form *Gweinpaye* is a clerical error for *Gwempaye*.

Gwenffrwd SN 7447
rn. SN 727495 to Tywi SN 765451
'white (through foaming) stream': **gwyn** fem. *gwen*, *ffrwd*
Gwenfrwd 1820, *Gwenffrwd* 1834, 1888
A short but powerful stream in a narrow gorge. Recorded in *Kwm Gwenffrwd* 1545, *Cwm Gwenfrwd* c.1762 (**cwm**).

Gwenlais, Afon SN 7441
rn. SN 726428 to Tywi SN 760388
'white (through foaming) stream': **gwyn** fem. *gwen*, *glais*
(br. of) *Gwenlys, Gwenlays* 1555, *Gwnlas River* c.1700, *River Gwenlais* 1888
The form *gwen* is a little unexpected since **glais** is a nm. but the r. itself may have been regarded as an adj.f. to accord with **afon**. Two nearby hos. are recorded as *Cwm Gwenlas* and *Glan Gwenlas* 1831. There is another Gwenlais (to Marlais SN 619160) in Llandybïe as *Gwenlais* 1561, *Gwenlaish* c.1700, *Nant Gwenlais* 1878 (**nant**), and see Pentregwenlais.

Gwernogle SN 5334
Llanfihangel Rhos-y-corn p.
'marshes, alder-tree marshes': **gwernog** pl. *gwernogau*
Gwaunogan 1757, *Gwernogle* 1826, *Gwern-ogau* 1831, *Gwernoge* 1834, *Wernogle* 1849
The modern form Gwernogle with **lle** len. *le* 'a place' is by popular etymology. The location of an Independent chp. (HER 11765) described as *Gwernogge Chappel* 1851 (RC I, 505) established in 1749 which has been confused with Capel Sant Silin.

Gwestfa
The customary payment known as **gwestfa** was paid in place of food rents (PNCrm I, 223-4; WTLC I, 284-91) by tenants within a particular administrative area such as a maenor, p., cmt. or cantref or by a particular kindred group. Melville Richards (WATU) lists twenty-five examples of what he classifies as pns. but many of these cannot be fixed to a specific location and historic refs. to most of them are scarce. Four '*gwestfa* districts' (Blaen-twrch, Maesllangelynen, Cwm-twrch and Cwmcothi) in Maenor Rhwng Twrch a Chothi can be located on the basis of their constituent farmsteads (StC 28, 89) but none of these seems to have had the formal prefix 'Gwestfa'. Richards (in AMR) provides a grid ref. only for Gwestfa Llandeilo at SN 6726 which properly refers to Maenordeilo (q.v.). The following examples of Gwestfa are cross-referred to their appropriate division. Most are qualified by another pn. (identifiable pns. employed as

locators are excluded from the Glossary) or a pers.n. (included in a list of personal names appended to the Glossary).

Gwestfa Blaen-twrch   SN 6749
Caeo
'(area paying) food-rent in Blaen-twrch: *gwestfa*, pn. *Blaen-twrch*
Blaen-twrch (SN 681495) is 'headwaters of (r.) Twrch': *blaen*, rn. **Twrch**, recorded as *Blaen Torgh* 1574, (tmt.) *Blaen Twrch*, (tmt.) *Tir Blaen Twrch* 1603 and (2 messuages etc) *Tyr Blaen Twrch* 1694 (*tir*) refer to Blaen-twrch-isaf and Blaen-twrch-uchaf (SN 6749) (*isaf, uchaf*) in Caeo. See also Cwm-twrch.

Gwestfa Bleddyn
Mabudrud
'food-rent of (kindred of man called) Bleddyn': *gwestfa*, pers.n. **Bleddyn**
Westua Blethyn 1303
Bleddyn is unidentified.

Gwestfa Cadwgan ab Einion
Mabudrud
'food-rent of (kindred of) Cadwgan son of Einion': *gwestfa*, pers.ns. **Cadwgan, Einion, mab** len. *fab, ab, ap*
Westua Cadugan ab Eynon 1303
Cadwgan ab Einion is unidentified. Melville Richards (WATU) misrecords this as Gwestfa Cadwgan ap Cynon.

Gwestfa Cadwgan ap Tegwared
Maenordeilo
'food-rent of (kindred of) Cadwgan son of Tegwared': *gwestfa*, pers.ns. **Cadwgan, mab** len. *fab, ab, ap,* **Tegwared**
westua Cadugan ab Tegwared 1303
Cadwgan ap Tegwared is unidentified.

Gwestfa Cenarth   SN 2641
Cenarth
'(area paying) food-rent in Cenarth': *gwestfa*, **Cenarth**
Located in Cenarth.

Gwestfa Cilfargen   SN 5724
Llangathen
'(area paying) food-rent in Cilfargen': *gwestfa*, **Cilfargen**
westua Kilvargeyn Kilsaen et Rigilleyd 1303
Recorded with Gwestfa Cilsân and Gwestfa Rhingylliaid.

Gwestfa Cilsân   SN 5922
Llangathen
'(area paying) food-rent in Cilsân': *gwestfa*, **Cilsân**
westua Kilvargeyn Kilsaen et Rigilleyd 1303
The ref. also applies to Gwestfa Rhingylliaid (q.v.).

Gwestfa Cwmblewog   see   **Cwmliog** (Cwmblewog)

Gwestfa Cwmcothi   SN 6846, 6944, 6946
Caeo
'(area paying) food-rent in Cwmcothi': *gwestfa*, **Cwmcothi**
Probably co-terminous with Cwmcothi (q.v.).

Gwestfa Cwm-twrch   SN 6541, 6644, 6745, 6845
Caeo
'(area paying) food-rent in Cwm-twrch': *gwestfa*, **Cwm-twrch**
Probably co-terminous with Cwm-twrch (q.v.).

Gwestfa Cynwyl Elfed   SN 3727
Cynwyl Elfed
'(area paying) food-rent in Cynwyl Elfed': *gwestfa*, **Cynwyl Elfed**
Probably co-terminous with Cynwyl Elfed (q.v.).

Gwestfa Dinefwr   SN 6121
Maenor Deilo
'(area paying) food-rent in Cynwyl Elfed': *gwestfa*, **Dinefwr**
westwa de Dynewr 1303
No other refs. have been found but the *gwestfa* must have been centred on Dinefwr.

PLACE-NAMES OF CARMARTHENSHIRE

Gwestfa Gruffudd ab Elidir
Catheiniog cmt.
'food-rent of (kindred of man called) Gruffudd son of Elidir: ***gwestfa***, pers.ns. ***Gruffudd***, ***Elidir***, ***mab***, len. *fab, ab, ap*
westua Gruffit ab Elidyr 1303
Gruffudd ab Elidir is unidentified.

Gwestfa Gwion Sais a Maredudd ap Heilyn
Mabudrud cmt.
'food-rent of (kindred of men called) Gwion Sais and Maredudd ap Heilyn': ***gwestfa***, pers.ns. ***Gwion***, ***Maredudd***, ***Heilyn***, ***Sais*** 'English, English-speaker', ***mab***, len. *fab, ab, ap*
Gwestua Wyann Seys et Mared' ab Heylin 1303
Gwion Sais and Maredudd ap Heilyn are unidentified. The tenants recorded here in 1303 as *Howel ab Gwyann* and *Meybon Gruffit ab Wyann* may be his successors.

Gwestfa Hengoed SN 5003, 5103
Hengoed
'(area paying) food-rent in Hengoed': ***gwestfa***, **Hengoed**
Hengoed Westva 1609
Probably coterminous with or a constituent part of Maenor Hengoed (q.v.).

Gwestfa Llangathen SN 5822
Llangathen
'(area paying) food-rent in Llangathen': ***gwestfa***, **Llangathen**
Gwestua Lan Gathen 1303
Centred on the p.ch. of Llangathen.

Gwestfa Llanhirnin SN 5421
Llanegwad
'(area paying) food-rent in Llanhirnin': ***gwestfa***, **Llanhirnin**
Gwestua Lan Hehernyn 1303
Probably co-terminous with Llanhirnin.

Gwestfa Llannewydd SN 3824
Llannewydd/Newchurch
'(area paying) food-rent in Llannewydd': ***gwestfa***, **Llannewydd**
Westva llanewidd 1581
Probably co-terminous with Llannewydd.

Gwestfa Llanwrda SN 7131
Llanwrda
'(area paying) food-rent in Llanwrda': ***gwestfa***, **Llanwrda**
Westua de Llanurdaf, ~ Lanurda 1303
Probably co-terminous with Llanwrda.

Gwestfa Llywelyn Gwynnau
Mabelfyw cmt.
'food-rent of (kindred of man called) Llywelyn (?) Gwynnau': ***gwestfa***, pers.n. ***Llywelyn***, (?) epithet or pers.n. (?) ***Gwynnau***
Gwestua Lewel' Wynve 1303
Unlocated and omitted by Richards (AMR and WATU). The same source refers to decayed rents of *heredibus Lewel' Wyneu* ('heirs of Llywelyn *Wyneu*'). Without further historical evidence we cannot be sure whether these forms stand for *Wynue, Wyueu, Wyune, Wyneu* or *Wynen* since -*u*-, -*v*- and -*n*- were often confused in medieval script. These might conceivably be *Gwynfe, Gwyfeu* or *Gwyneu* or some similar form in ModW. An epithet or pers.n. based on ***gwynnau***[1], *gwyniau* adj. 'spirited, lively' offers fewest problems and may be be used as a working form until more evidence comes to light but we cannot rule out a pers.n.

Gwestfa Llywelyn ap Heilyn
Mabudrud cmt.
'food-rent of (kindred of man called) Llywelyn son of Heilyn: ***gwestfa***, pers.n. ***Llywelyn***, ***mab***, len. *fab, ab, ap*, pers.n. ***Heilyn***,
Westua Lewel' ab Heylin et Morithic 1303
Unlocated. Richards (AMR and WATU) supposes that the n. is qualifed by the n. of a single person Llywelyn ap Heilyn ap Moreiddig but it appears to refer to two individuals. Neither has been identified.

Gwestfa Maesllangelynyn SN 6542, 6643
Caeo
'(area paying) food-rent in Maesllangelyn': ***gwestfa***, ***Maesllangelynyn***
Maesllangelynyn is recorded as (tmt.) *Tir Maes Llanglynyn* 1603, *Mays Llangelynyn* 1574-1609, *tyr maes llanglynyn* 1613 apparently

80

'open-field at Llangelynyn' (***maes***). This no longer appears on maps but there is a ho. Cwm-gelynen (SN 658427) and *tyr maes llanglynyn* 1613 (***cwm***, ***tir***) is mentioned with other hos. including Bwlchygilwen (SN 669435). It is possible that any lost church or enclosure was at Henllan (SN 653432) near the r. Twrch.

Gwestfa Maestreuddyn   SN 6341
Caeo
'(area paying) food-rent in Maestreuddyn': ***gwestfa***, **Maestreuddyn**
In Y Faenor Isaf. Probably co-terminous with Maestreuddyn (q.v.).

Gwestfa Merthyr ac Aber-nant   SN 3323, 3520
Aber-nant and Merthyr
'(area paying) food-rent in Merthyr and Aber-nant': ***gwestfa***, **Merthyr**, ***a***, *ac* 'and', **Aber-nant**.
Probably co-terminous with ps. of Aber-nant and Merthyr.

Gwestfa Moreddig
Mabudrud cmt.
'food-rent of (kindred of man called) Moreiddig': ***gwestfa***, pers.n. ***Moreiddig***
See **Gwestfa Llywelyn ap Heilin**

Gwestfa Owain ap Rhydderch
Mabelfyw cmt.
'food-rent of (kindred of man called) Owain son of Rhydderch': ***gwestfa***, pers.ns. ***Owain***, ***Rhydderch***, ***mab***, len. *fab*, *ab*, *ap*
Westua Oweyn ab Ryther 1303
Owain ap Rhydderch is unidentified.

Gwestfa Penarth   SN 6440
Caeo
'(area paying) food-rent in Penarth': ***gwestfa***, *Penarth*
terra Pennarth 1303, *tir Pennarth* 1596, *Pennarth* 1603, *Penarth* 1746
Glanville R. Jones (StC 28: 89-90) describes Penarth as a ***gwestfa*** in Faenor Isaf ht. (q.v.) but instances of prefixed Gwestfa seem to be lacking. Penarth is 'top of the headland': ***pen¹***, ***garth¹***, recorded in two hos. Penarth-isaf and Penarth-uchaf. The fms. are located on the slopes of a hill overlooking Pumsaint. The 1303 ref. places it in Maenor Meibion Seisyll (q.v.).

Gwestfa Perth-lwyd   ?SN 6129
Catheiniog cmt.
'(area paying) food-rent in Perth-lwyd': ***gwestfa***, pn. *Perth-lwyd*
Gwestua Perth Luyd 1303
Perth-lwyd may refer to Berth-lwyd (q.v.) in Llangathen.

Gwestfa Pont Rhyd-coll
Maenordeilo cmt.
'(area paying) food-rent at Pont Rhyd-coll': ***gwestfa***, pn. *Pont Rhyd-coll*
westua Pont Ryd Coll 1303
Unlocated and no other refs. have been found. Pont Rhyd-coll translates as 'bridge at Rhyd-coll' (***pont***) and Rhyd-coll is apparently 'ford near hazel-trees' but the historic form is irregular since ***coll²*** would normally lenite after ***rhyd*** a nf.

Gwestfa Rhingylliaid
Catheiniog cmt.
'(area paying) food-rent of the ringilds': ***gwestfa***, ***rhingyll*** nm. *rhingylliaid*
westua Kilvargeyn Kilsaen et Rigilleyd; *terra Rigilleyd* 1303
Probably a food rent-paid to the ringilds (court officials) rather than a geographical location.

Gwestfa Tre Cynwyl Gaeo   SN 6739
'(area paying) food-rent in township of Cynwyl Gaeo': ***gwestfa***, ***tref***, **Cynwyl Gaeo**
For Cynwyl Gaeo see Caeo.

Gwestfa Trefgynnull
Maenordeilo
'(area paying) food-rent of Trefgynnull': ***gwestfa***, pn. *Trefgynnull*
westua Tref Gynhull 1303
Trefgynnull means 'township of collection', ie. ***tref*** and ***cynnull²***, i.e. where harvest or

rents might be collected; cf. Cynnull-bach and Cynull-mawr. No historic forms of this have been identified, however, and it is possible that it refers not to a specific place but to a particular type of food-rent of uncertain definition.

Gwestfa Tre-lech a'r Betws　　　　SN 3026
Tre-lech a'r Betws
'(area paying) food-rent in Tre-lech a'r Betws': *gwestfa*, **Tre-lech a'r Betws**
Gwestva Trelech ar Bettus 1581
Centred on Tre-lech a'r Betws.

Gwestfa Wyrion Idnerth
Catheiniog cmt.
'(area paying) food-rent of kindred of Idnerth: *gwestfa, gŵr* pl. *gwyrion*, pers.n. **Idnerth**
Gwestua Vyron Ithnerth 1303
The n. probably refers to a food-rent paid by a kindred group rather than a specific location.

Gwestfa Wyrion Ieuan ac Wyrion Seisyll
Maenordeilo
'food-rent of kindred of Ieuan and Seisyll': *gwestfa, gŵr* pl. *gwyrion*, pers.ns. **Ieuan, Seisyll**
westua Vyron Jeuan et Vyron Seysill 1303
Ieuan and Seisyll are unidentified. The n. probably refers to a food-rent paid by a kindred group rather than a specific location.

Gwestfa y Faenor Isaf　SN 6736, 6838, 6939
Caeo
'(area paying) food-rent in Y Faenor Isaf': *gwestfa*, pn. **Faenor Isaf**
No historic forms have been identified.

Gwestfa Ysgolheigion　　　　　SN 6121
Llandyfeisant
'settlement of the clerks': *gwestfa, ysgolhaig* pl. *ysgolheigion*
villa de Scleygon 1280, *westua de Sclogans* 1283, *westua sglogyon* 1303, *Trefscoleygyon* 1318
See Dinefwr for its administrative significance.

Gwestfa Ystradmynys　　　SN 7333, 7433
Mallaen cmt.; ?Llanwrda
'(area paying) food-rent in Ystradmynys': *gwestfa*, **Ystradmynys**
westua Estravynvs 1303
Ystradmynys otherwise recorded as (place called) *Ystradbynis* 1650, is 'valley of (r.) Mynys', *ystrad*, rn. **Mynys**, and probably survives in Ystrad (SN 745329) recorded as *Ystrad isaf* 1831 (*isaf* 'lower'). *Ystrad uchaf* (*uchaf* 'upper') refers to the house Pantyllwyfen (*pant*, def.art. *y, llwyfen*). The r. Mynys marks the western edge of Llanwrda p. with Llansadwrn.

Gwidigada
Cantref Mawr
uncertain
(land) *[Widi]gada* 1220, *Wydigada* c.1250, (land) *Gwydygada* 1281, (cmt.) *Widigada* 1275-6, 1328, *Ewydugada* 1284, (cmt.) *Wytigada* 1302, *Commoto Whitigada, Witigada* 1304, *Comm. Withigada* 1536-9, *Gwy digoda* 1600-7
The scholar Melville Richards (ETG 50) was unable to suggest any meaning or identify any of its els. Possibly a pers.n. **Gwidigada* but further research is clearly needed. A cmt. covering the ps. of Abergwili, Llanllawddog and Llanpumsaint in Cantref Mawr.

Gwili[1]　　　　　　　　　SN 4223, 4428
rn. SN 503276 to Tywi SN 432205
aqua de Wyly 1326, *Wyli* 1345-80, *Nant Gwyly* 1400, (*drwy*) *Wili* 1455-85, (r.) *Guily* 1536-9, (r.) *Gwilly* 1612
R.J. Thomas (EANC 147-8) suggests it may have *gŵyl*[1] nf. in the specific sense 'watch, guard' or *gŵyl*[2] adj. 'modest, gentle', with suffix *-i*. See Abergwili and Cwmgwili. Applied on OS maps to the r. flowing northwards from about SN 489272 towards Dolgafros.

Gwili[2] see **Cwmgwili** (Llanedi)

Gwlad　　　　　　　SN 3639, 3739, 3735
Llangeler
'country, open country': *gwlad*
Gwlad (& *Grange*) c.1700, *Gwlad* 1841, *Hamlet of Wlad* 1847, (*i*) *Faesdref y Wlad* (*Gwlad Hamlet*) 1899
The precise sense of *gwlad* here is uncertain but it included the p. ch. and may be employed to mean 'country under the direct administration

of the parish' as distinct from Grange (q.v.). The boundary between Gwlad and Grange is shown on the tithe plan. *gwlad* is apparently recorded in ho.ns. in Caeo, viz. Gwlad-eithaf (SN 655421) recorded as *Gwladitha* 1831, with *eithaf* 'furthermost'.

Gwydderig, Afon　　　　　　SN 8734, 7734
rn. SN 864343 to Brân SN 771342
?'wild (r.)': *gŵydd³*, suffixes *-gar*, *-ig*
(mill of) *Gwytherig* 1396-7, *Gwitherik* c.1537, *Gwetherik flu:* 1578, *the Gutherijc* 1587, *Gwethrik* 1612, *Gwyderig* 1729, 1754, *Gwtherig* 1783, *River Gwtherig* 1785, *Gwdderig* 1832
Rises as Nant Gwydderig (*nant*) with a main source, in Llywel, co. Brecon. R.J. Thomas (EANC 190) noted the similarity of Gwydderig with the pers.n. *Gŵydd(i)ar* (< OW *Guidcar*) with the same suffix *-gar* found in other pers. ns., rns. and pns. though he was uncertain whether the first el. was *gŵydd³* or *gŵydd⁴* nm. 'presence; sight' or *gŵydd³* adj. 'wild, untamed' (v. Gwyddgrug). The latter makes far better sense describing the nature of a forceful and uncontrolled torrent.

Gwyddgrug　　　　　　　　SN 464356
Llanfihangel-ar-arth
'prominent mound', *gŵydd¹*, *crug*
Castellum Guidgruc late 13th cent, *Gudgruc* 12th cent (1332), *Gowithgrege* 1535, *Gwythegrege* c.1550, *Goyddgrug* 1633, *Gwyrgrige* 1649, *Gwyderig alias Goithgrug* 1691, *Grange of Gwyddgrig otherwise Gwyrgrig* 1759
The mound probably refers to the small motte-and-bailey castle (SN 47693563) (HER 1792) near Castell fm. Identical in meaning to Yr Wyddgrug (= Mold), co. Flint (discussed in DPNW 326 and PNF 203). The first el. was identified as *gŵydd⁴* 'memorial, cairn' by Sir Ifor Williams (ELl 14) and Melville Richards (CStudies 119; ETG 32), now discounted. A second Gwyddgrug appears to have been near Crug (SN 626231) near Llandeilo and is recorded as (mess.) *Woythgrygge* 1559 and *Gwythgrig* 1609 but this may be a natural mound.

Gwyddyl　　　　　　　　　SN 4238
Llanfihangel-ar-arth
(area near stream called) Gwyddyl: rn. *Gwyddyl* (manor) *Gwyddyl* 1693, *Gwyddil* 1811
Recalled in ho.ns. Aber-gwyddil (SN 420440) and Felin-gwyddil (SN 421383) recorded as *Aber-gwiddel* and *Melin-gwiddel* 1831 (*aber*, *melin*), the former Blaen-gwyddil (SN 415343) which is *Blaen-Gwyddel* 1889 (*blaen*) and Tir Dyffryn Gwyddyl which is *Tyr Dyffryn gwythill* 1564 and *Tir ystelis Gwyddel alias Tir diffrin Gwyddel* 1609 (*tir*, *dyffryn*, *ystlys*). All are ultimately named from a rn. Nant Gwyddyl (*nant*) recorded as *Gwyddell* 1542, *Gwydhil* c.1700, *Gwyddel* 1889, possibly 'river in a thicket *or* brambles' with *gwyddel* as a variant of *gwyddwal* nmf. 'thicket, bush, brambles'.

Gwydre　　　　　　　　　SN 789275
Llanddeusant
?'prominent settlement hill': ?*gwydd¹*, ?*tref*
Gwydre 1763, 1886-7, *Goytre* 1831
Goytre is through false association with *coetref* typically found as (Y) Goetre in pns. Historic forms are too late for certainty. The ho. is located on hill-slopes in an exposed and prominent position. Gwydre is recorded as a pers.n. in the tale of Culhwch and Olwen (CO³ nos. 258, 1116) but I think we can rule this out here. Pers.ns. nearly always act as qualifiers and historic forms give no hint of any lost generic.

Gwynfe see **Capel Gwynfe**

Halfpenny Furze to Horeb

Hendy-gwyn/Whitland. Llan-gan Road looking towards Market Street c.1915.

Halfpenny Furze SN 271132
Llanddowror
?'area of furze paying halfpenny rent': E *halfpenny, furze*
Halfpenny Furze 1746, 1888, *Halfpenny-furze* 1849
A *Haulffpenie meadow* is recorded in 1598, in Llanblethian, co. Glamorgan, and there are similar fns. in England, such as Halfpenny Croft and Halfpenny Hay which may allude to rent or to a payment for temporary grazing (NDEFN 189). Furze was used as fodder and for burning. Alternatively a mocking n. for poor land (CStudies 124).

Halfway[1] SN 525004
Llanelli
Half-way 1830, *Halfway* 1880
Named from *Halfway House (P.H.)* 1880 on the Llanelli-Llangennech road. A second Halfway (*Half way* 1830) located near a former lodge to Stradey Park disappears from maps before 1880. This was roughly halfway between Llanelli and Pwll.

Halfway[2] SN 829329
Myddfai, Llywel (Traean-mawr)
'(place) near Halfway Inn'

Halfway 1832, *(tafarndy a elwir) Half Way* 1839, *Halfway,* (inn) *Half-way* 1875
The inn (**tafarn, tŷ**), former smithy and ht. are roughly half way between Trecastell, co. Brecon, and Llanymyddyfri/Llandovery. Earlier known as Tir Ymryson, 'disputed land' (**tir, ymryson**[1]) (PNDH 10.68, 69, 141), and recorded as (*Old house with garden called*) *Tir yr Umrusson or Tir ym russon* 1786. The village and its immediate area straddle the border of cos. Brecon and Carmarthen.

Halfway[3] SN 648307
Talyllychau
(PH) *the Halfway House* 1828, *Halfway House Inn* 1849, (ho.) *Halfway, Halfway House (P.H.)* 1888, *Old Halfway,* (P.H.) *Halfway, ~ Forge* 1906, *Halfway* 1947
Shown as *Glan-yr-afonddu-isaf* 1831 near *Glan-yr-afonddu-ganol* (SN 643309) and *~ -uchaf* (SN 643315) named from Afon Ddu which flows north to Talyllychau/Talley (**isaf, canol, uchaf**) recorded undistinguished as *Glanyrafonddu* 1742 and *Glanyr-afon-ddu* 1756.

Harddfan SN 549007
Llangennech
'beautiful place': **hardd, man**[1]
Harddfan c.1960 to present
A modern n. for hos. constructed on and near Heol y Mynydd from the late 1950s possibly derived from a very common house n.

Hebron[1] SN 417356
Llanfihangel-ar-arth
'(place near chapel called) Hebron: **Hebron**
The n. derives from Hebron Baptist chp. recorded on OS maps from 1889 but has achieved prominence only on recent editions as plain Hebron applying to a few hos. extending eastwards to Nant-y-gwair (*Nant-y-gwair* 1831) (**nant, y, gwair**). Hebron is the name of a Palestinian city south of Jerusalem in the Judaean mountains. The choice of name for the chp. would have been made on the basis of its biblical associations with Abraham or perhaps its location among hills.

Hebron[2] SN 181277
Llanglydwen
'(place near chapel called) Hebron: **Hebron**
Hebron 1843
Named from an Independent chp. erected in 1805 (RC I, 362).

Hen Briordy SN 473412
Llanllwni
'old priory': **hen, priordy** len. *briordy*
Llanllownye priory 1576, *an old Priory near the Church* c.1700
The scholar Edward Lhuyd c.1700 says that there were ruins of a priory near the ch. of Llanllwni but nothing is shown on OS maps and despite the refs. there seems to be no good evidence that one ever existed here. The supposition may have arisen because it is known that a messuage in Llanllwni was granted to Caerfyrddin/Carmarthen priory some time before 1330 when the priory acquired an additional two acres of land and a moiety of the advowson of the p. ch. from William Wynter, archdeacon of Caerfyrddin/Carmarthen.

Hendy SN 582038
Llanedi
'(the) old house': (**yr**), **hen, tŷ**
the Hendy 1627, *yr Hendy* c.1677, (lands) *Hendy and Penybenallt* 1701, (mess. and lands) *Hendy* alias *Hendy bennalt* 1789, *Hêndy* 1833
The ho. is also likely to be (tmt.) *Sythyn yr hendy* 1564 (**syddyn**) and is occasionally qualified as *Penybenallt* and *Benallt* 'hill with a wooded slope' (**pen**[1], **allt**). The n. referred to a former fm. Benallt (SN 58250453), recorded as *Pen'allt* 1830, *Pen y rallt* 1842, and *Pen-allt* 1879, located on the northern side of a small hill in the ht. of Fforest. Hendy may have the particular sense of 'home farm', a lowland dwelling occupied in winter months in contrast to Benallt on higher land occupied in summer to manage stock; cf. the use of **hendref** and **hafod**. The br. over Gwili is *Hendy Bridge* 1792, *Pont-hendy* 1833 (**pont**). The ht. developed in the 19th cent south of Hendy ho. along the Llanelli-Pontarddulais

road near Hendy tin works which was built on part of Hendy farm in 1866.

Hendy-gwyn, Whitland SN 1916, 1917
Llan-gain
'old white-house': **hen**, **tŷ**, **gwyn**; 'white land': E **white**, **land**
(de) *Alba domo*, *Albæ landæ monasterio* c.1191, *Ecclesia beate Marie de Alba Landa* 1214, *Alba Landa* 1222 (1239), (monastery) *Alba Launda al(ia)s Whitlaunde* 1535, (abbey) *Blauncheland* 1277, *le Blaunchelande* 1318, *Wytelond* c.1291, *Whitland* 1309, (manor) *Olde Whiteland in Wales* 1314, (ch. of) *St. Mary, Whitland* 1352, *the chapple of Whiteland* 1613, (mills) *Hendigwyn*, *Hentigwen* 1535, *Hentegwyn* 1590, *y Ty Gwyn* c.1400, (abbey) *Teguin ar Taue* c.1538, *the Ty Gwynne* 1559, (o'r) *Ty gwyn ar Daf* c.1600, *Tŷ-Gwyn-ar-Daf* 1831
The earliest Latin forms show that the E form Whitland replaced 'white house' with **alba** and **domus** current over the period c.1191-late 13th cent and this form is reflected in *y Ty Gwyn*, *Teguin ar Taue* etc, the last standing for Tŷ-gwyn ar Daf, 'white house on (r.) Taf' (**tŷ**, **gwyn**, **ar** 'on', rn. **Taf**). The 'white house' may have referred specifically to the first location of the Cistercian monastery and 'white land' (Alba Landa) adopted when the monastery moved one mile up the valley of Afon Gronw in 1151-2. Historic forms of Whitland seem to first occur in c.1291 but the n. is clearly older as Latin *Alba Landa* and Fr *Blauncheland* 1277 (**blanche**) show. The modern form Hendy-gwyn is first recorded in 1535 and seems to have applied initially and specifically to the water-mills on Afon Taf while Tŷ-gwyn – or its fuller form Tŷ-gwyn ar Daf – applied to the town down to the 19th cent. The purpose of the qualifier **hen** 'old' is uncertain but the likeliest explanation is that Hendy-gwyn referred specifically to the mills (note the refs. for 1535) located close to the site of the present town. The latter is thought to have been an earlier and *older* location of the monastery (WCist 14). Hendy-gwyn was simply transferred at a very late date from the mills to the adjoining town and displacing (Y) Tŷ-gwyn.

Hengil SN 4423, 4525
Abergwili
?'old hazel-trees: **hen**, ?**cyll**
Hengyll 1716, *Hengill* 1725, *Hengil* 1801, 1845
The n. of a former ht. also found in that of the hos. Hengil-isaf (SN 454246) and Hengil-uchaf (SN 454251) recorded as *Hengil Ucha* 1786 and *Hengil Issa* 1800, *Hengil-Issa* 1837, and a *Hengil-fach* 1838 (**uchaf**, **isaf**, **bach**[1]). Francis Jones also records a *Diffrin Henegill* (**dyffryn**) here in 1595 (HCH 93-94). It is possible that *Hengil* is a dialect variant with -gil standing for -gyl (-gyll) comparable with *sticyl* for **sticil**, *sticill*, an anglicisation (*l* for *ll*) or has been influenced by association with *cil*[1] 'nook' and **hen** reinterpreted as 'old nook'. It is difficult, however, relating such a meaning to the topography. All three hos. are on south-facing slopes above Nant Crychiau.

Hengoed SN 5003, 5103
Llanelli
'old wood, old trees': **hen**, **coed**
(manor) *Hengoed* c.1500, 1609, *Hengoide* 1562, *Hencoed* 1775, *Hengod* 1800
The n. is also recalled in that of two hos. Hengoed-fach (SN 501035) and Hengoed-fawr (SN 509034) recorded as *Hengoed vach* 1717, *Hengoed-fâch* 1878-80, *Hengod vaure* 1740 and *Hengoed-fawr* 1878-80 (**bach**[1], **mawr**). See also Capel Dyddgu and Gwestfa Hengoed.

Henllan[1]
Abergwili
'old church': **hen**, **llan**
Henllan 1836
A reputed chp. but not otherwise recorded unless it was near Cefn Henllan (SN 458241) recorded as *Kevenhenllan or Tuymawr Coed y Kestill* 1747, (2 mess. etc) *Keven henllan* (*Foesykay* and *Coed y Kastill*) 1800, *Cefn'r-hen-llan* 1831, *Cefn Henllan* 1841 (**cefn**).

Henllan[2] SN 653432
Caeo
'old church': **hen**, **llan**
Henllan 1800, 1824, *Henllan or Bryneglwys* 1836
There seems to be no other evidence of a ch. or at Bryneglwys roughly 2km from

Henllan. Bryneglwys is *Bryn Egluys* 1303 and *tir bryn eglwys* 1614, 'hill with a church' (***bryn***, ***eglwys***, ***tir***) referring to the hill between the hos. Bryneglwys-fawr (SN 644420) and Bryneglwys-fach (***mawr***, ***bach***).

Henllan Amgoed SN 180200
Llanfallteg
'old church in Amgoed': ***hen***, ***llan***, **Amgoed**
Henllan Amgoid c.1291, *Henllan Amgoyd* 1404, *Henllan Amgoid* 1535, *ll.ðewi o henllan* c.1566, *Henllan-Amgoed* 1641, *Henllan Amgoed* 1668, 1843
Amgoed (q.v.) is the n. of the former cmt., usually coupled with Peuliniog (q.v.), which distinguishes it from Henllan (SN 356410), co. Cardigan, (DPNW 192) – Henllan Deifi in full – also dedicated to Dewi/David. The focus of Henllan Amgoed has shifted from the p. ch. (SN 185207) to the area around the Independent chp. established in 1797.

Henllanfallteg see **Llanfallteg**

Heol-ddu SN 462037
Pen-bre
'dark road': ***heol***, ***du***
Heol-ddu 1969-70
The area was developed around a ho. Brynmelyn (***bryn***, ***melyn***) after 1964 and is now shown as part of Penymynydd (q.v.).

Heolgaled see **Salem**

Hermon[1] SN 363308
Cynwyl Elfed
Hermon 1831
The settlement developed along the B4333 (Cynwyl Elfed to Aberteifi/Cardigan) road near a smithy and Hermon Independent chp. recorded in 1799. The chp. is located on a hill and is named from biblical Hermon which was located on a mountain. The present building was apparently built in 1884.

Hermon[2] SN 671281
Llandeilo Fawr
Hermon Chapel (Indt) 1887

Plain Hermon on current OS maps. The n. is taken from an Independent chp. built in 1813 located appropriately on a hill; cf. Hermon[1]. The cause began at a ho. Tŷ-isaf (SN 661283) in 1810 (HEAC III, 574-5).

Hernin SN 5421
Llanegwad
?pers.n. ***Hiernin***
(place) *maynor Jernin* 1608, *Hermin* 1668, (ht.) *Heirnyn* 1710, *Heirnin Hamlet* 1724, (ht. Egwad and) *Hiernin* 1804, *Hîrnin* 1811, *Ernyn, otherwise Hernin* 1845, (ht.) *Hernin* 1863
See also Llanhernin. David Jones (Yr Haul 23: 258) calls this Maenor Hernin, a form supported by *Meynor Hiernyn* 1609 (***maenor***), and from the places he mentions – Tŷ-pica (*Typica*) (SN 541207), Nantergwynlliw (SN 54682180) and Twŷn (SN 536220) – it is clear that it made up the easterly part of Llanegwad p. reaching to Afon Dulas which marks the border with Llangathen. He describes *Cefn-hernin* 1888-9 (SN 53902161) on OS 1:2,500 as *Cefn Saint Hernin* though earlier evidence calls it plain Cefn (***cefn***). References to *Tir hiernin* and *agrum hiernin* c.1145 (LL 150) (***tir***, ***ager***) cannot be identified with Hernin without further evidence.

Hiraeth SN 172210
Llanfallteg
?'(place or r. where there is) long gorse': ***hir***, ?***aith***
Hiraeth 1819, *Hyreth* 1851, (area north of crossroads) *Hiraeth* 1889-90
There are doubts with the suggested meaning because historic forms are late. There are some doubts too with regard to the origin of ***aith*** though it may be a back-formation of ***eithin*** (GPC). The name is applied on the current OS Landranger to the area around the crossroads (SN 17172210) but earlier OS maps assign it to an area (SN 172213) a little to the north. Hiraeth may also have been the name of the now unnamed stream flowing north-westwards into Afon Taf rising by *Blaenhiraeth* 1761, *Blane hireth* 1779, *Blaen-Hiraeth* 1907

(*blaen*). Cf. Mynydd Hiraethog, 'long gorse-land mountain' (*mynydd*) (DPNW 333).

Hirfryn
'long hill': ***hir, bryn***
(cmt.) *Hyriurn* 1210 (1323), *Hyrthrin* 1257, *Hyrefrin* 1281, (lands of) *Hyrefryn, Hirefrin* 1292, *Irefryn, Irfryn* 1299, (cmt.) *Hyrvrin, Hirvryn* 1317, *yc6m6t Hirvryn* 15th cent, *Hirfrynn* before 1564, *Hevrin* 1589
The cmt. covered the ps. of Llandingad, Llanfair-ar-y-bryn and Llandovery but gradually drops out of historical records after its merger with Perfedd (q.v.) in the 16th cent. The hill is either Cefn Hirfryn recalled in the ho.n. Cefnhirfryn (SN 797411) (HCrm I, 9) (***cefn***) or that recalled in Llanfair-ar-y-bryn.

Honeycorse SN 282091
Llansadyrnin
?'sweet or sticky marsh': ME ***honnye***, E dial. ***cors, -e***
Haynygtors 1307, *Honeycorse farm* 1673, *Honey Corse* 1831, *Honey-cors* 1889
This may also be (pasture in marsh of Laugharne called) *Menecors* c.1280 as a scribal error for *Henecors* or some similar form. OE ***hūnig*** (> ***honnye***) generally means 'sticky' in fns. and might describe sticky soil or a marsh notable for sticky mud. That seems most appropriate here but it is also thought to have been used in the complimentary sense 'sweet' or a place where bees were kept and honey gathered (NDEFN 212-3). Cf. Honeyborough (for 'sweet') and for ***cors, -e***, see Corston, both in co. co. Pembroke (PNPemb 618-9, 690 and glossary).

Hopkinstown see **Trehopcyn**

Horeb[1] SN 514282
Brechfa
Horeb Chapel 1889, *Capel Horeb* 1906
Named from an Independent chp. (SN 514276) built in 1830. The n. has shifted northwards to include the area around crossroads on the B4310. The chp. is located on the side of a hill. Horeb was a Biblical mountain.

Horeb[2] SN 498057
Llanelli
Eglwys Horeb 1839, *Horeb* 1880
Named from a Particular Baptist chp. erected 1832 in succession to an older meeting-ho. at Five Roads. The latter is apparently that named as *Horeb* in a bequest to deacons of Baptist congregations in 1793 and seems to have been established c.1759.

Iddole to Is-morlais

Iddole. The former Capel Seion.

Iddole SN 423157
Llandyfaelog
?pers.n. *Iddole*
(*Vacca*) *Ithole* 14th cent, *Ithole* c.1501, 1565, *y ddolee* 1590, *Iddole* 1591, 1858, *Idole* 1624, 1888, *Iddol* 1811
It is difficult to find comparable ns. Iddole was evidently the centre of a **maenor** recorded as *Maynor y ddole* 1619 seemingly for Maenor y Ddôl (the preferred form in WATU) as if it has Y Ddôl 'the water-meadow' (*y, dôl*). This is very unlikely in view of other historic forms and because Iddole is generally pronounced locally as 'Idole' [i`dole] with final -e for standard pl. *dolau*. The n. may actually have a pers.n. containing the el. **udd** 'lord, chief' found in pers.ns. such as Ithel, Idnerth and Maredudd. Iddole evidently paid a manorial customary due described as **vacca**, Latin for 'a cow', probably with the meaning 'area paying a render of one cow'. Development seems to have begun with a school established in a private ho. after 1888 and Capel Seion-Idole Baptist chp.

PLACE-NAMES OF CARMARTHENSHIRE

Iet-y-bwlch SN 162284
Llanglydwen
'gate at the entrance': *iet¹*, *y*, *bwlch*
Yat y Bulcheu 1729, *Iet-y-bwlch* 1889
The n. could be used in a broader sense for a route giving access to commons and it is interesting that the first historic form has the pl. *bylchau* which would indicate several gaps or entrances. A gate marked the entrance from the A478 leading northwestwards to the south slopes of Foel Dyrch.

Is Cennen
Cantref Bychan
'(land) below (r.) Cennen': *is*, rn. **Cennen**
(*terre*) *Iskennyn* 1203, *Hyskennen* 1265, *Iskennen* 1284-1613, (cmt.) *Yskennen*, (*homines de*) *Yskennen* 1292, (cmt.) *Iskennyn* 1361, 1441, *Comm. Hiskennen* 1536-9, *Commotte de Iskennen* (*in dominio Kydwelly*) 1609
The n. of a cmt. and hd. in Cantref Bychan (q.v.) largely on the south side of Afon Cennen. The other two cmts. of Cantref Bychan, viz. Perfedd and Hirfryn, were on the north side of the r.

Iscoed¹ SN 208182
Llanboidy, Llan-gan
'(area) below the trees': *is*, *coed*
(gr.) *Iscoyde* 1574, 1590, *Iscoide grange* 1577, *Iscoyde Grange* 1617, *the Grainge of yscoed* 1670, *Ishcoed* 1760
The n. is no longer current. David H. Williams (WCist 313) identifies it as the final site of the Cistercian abbey of Hendy-gwyn/Whitland. The property associated with it extended over southerly parts of the ps. of Llan-gan and Llanboidy including Fforest Gaerdydd (SN 224166). Woodland is found on the eastern side of Afon Gronw and extends on both banks upriver.

Iscoed² SN 3812
Llandyfaelog
'(area) below the trees': *is*, *coed*
Yskoyed 1590, *Iskodd* 1591, *Iskoed* 1692, *Is Coed* 1811, *Iscoed* 1844
The n. survives in Iscoed-uchaf (SN 384122) recorded thus in 1831; 'upper' (*uchaf*) in contrast to Iscoed, in Llanismel/St Ishmael.

Iscoed³ SN 5603, 5703
Llanedi
'(area) below the trees': *is*, *coed*
Iscoed Hamlet 1792, (ht.) *Ishwed* 1804; *Is Coed* 1811
The southernmost part of Llanedi. Most woodland is near Fforest (q.v.). The n. survives in Iscoed Road, Hendy, shown as Hendy Road on OS maps.

Iscoed⁴ (Iscoed Morris) SN 3811
Llanismel/St Ishmael
'(area) below the wood associated with (man called) Morris': *is*, *coed*, pers.n. **Morris**
Iskoyt and the quarter of *Moriz* 1361, *Iscoid Morrice* 1552, *Iscoyd Morris* 1609, (cmt.) *Iskoed Moris* 1616, (*Uchcoid Morris and*) *Iscoyd Morris* 1619, *Ishcoed* 1760
Centred on Iscoed Farm (SN 382119) recorded as *Iscoed Farm* 1831 in the northerly part of Llanismel/St Ishmael in contrast to Uwchcoed Morris (q.v.).The fm. or the ho. Iscoed is *Iskoed* 1692. There are small woods in the adjoining valley. Probably the same person as in Treforis (q.v.).

Is-morlais SN 5310, 5311
Llan-non
'(area) below (r.) Morlais': *is*, rn. **Morlais**
(ht.) *Is morlais* c.1700, (ht.) *Ishmorlais* 1804, (ht.) *Ismorlais* 1811
The n. is not current but seems to have applied to the general area of Llan-non village and Pont Morlais (SN 537073) and may have the particular sense of 'lower down (r.) Morlais'.

Johnstown

Johnstown/Tre Ioan. Looking west towards the Toll House with Tafarn y Cyfeillion/The Friends Arms behind, 2021. (© Sionk)

Johnstown see **Tre Ioan**

Kidwelly to Kingsland

Kingsland hamlet, in Llanboidy. Taken from Thomas Kitchin's 'Accurate Map of Carmarthen Shire' c.1762.

Kidwelly see **Cydweli**

Kingsland SN 1924
Llanboidy
'land belonging to the king': E **king**, **land**
Kingsland 1692 (1758), *Kings Land* 1738, *Kingsland Hamlet* 1844
Kingsland appears to have been formed some time before 1692 (taken from a later transcript) as an amalgamation of Ffynnon-oer and Bronysgawen (q.v.). Nothing appears to be known with regard to the significance of its origin. Hendy-gwyn/Whitland abbey possessed a gr. here, probably known as Dyffryn-tawel recorded as *Dyffryn tawel* 1692 (1758), near Cwrtau-bach (SN 201251) or at Cwrtau-mawr (AC 1975: 107-8) recorded as *Cwrte bach* and *Cwrte mawr* 1738 (1758). It is possible that the n. was coined after the acquisition of monastic property by the Crown at the Dissolution.

Lacques to Loughor

Llanelli. Station Road looking north towards the Town Hall c.1910.

Lacques, The   SN 296107
Talacharn²/Laugharne
'streams': ME *lake²* pl. *lakes*
(mess.) *the Lakes, Tythyn y laackes* 1616, *Lakes* 1678, *ye Lakes* 1730, *the Laques* 1791
The n. is probably taken from the small stream flowing through Laugharne into Afon Taf. Cf. the nearby stream Macrels Lake flowing past Roche Castle into Talacharn²/Laugharne and Laques (SN 336103) in Llansteffan recorded as *the Lakes* 1610, *y Lackes* 1614, *Laques* 1695, *Lakes-vach* or *Laques* 1700, *Llaques* 1734 and nearby Laques-fawr recorded as *Laques fawr* 1759 and *Laques-fawr* 1831 (***mawr***). *Lacques* may have developed through supposition that the pn. is Fr (cf. *laque* 'sealing wax' and E *lacquer*). The second historic ref. has ***tyddyn***.

Lash, Afon   SN 5914
rn. SN 580150 to Afon Llwchwr SN 622132
'(the) stream': (***y***), ***glais***
Y Lash springs at Pylheu Gloiwon ... and falls into Lhychwr vawr at Aber laish c.1700, *River Lash* 1878
Earlier plain Y Lash with len. after the def. art. (later dropped) and aspirate -sh [ʃ] for -s commonly found after or before -i-. Aber-lash (SN 6214) is also recorded as *Aber Lash* 1679, *Aberlash* c.1735-1815 (***aber***). Cf. Lan-lash (SN 570220), in Llangathen, located next to an

unnamed stream, is recorded as (*o*) *Lan Lais* 15th cent, *Lanlash* 1778, *Llanlais* 1792 (***glan***).

Laugharne see **Talacharn²**

Llan¹ SN 5919
Llanfihangel Aberbythych
'(ht. called) church': ***llan***
(ht.) *Llan* 1811, 1851
Llan (and Tre'r-llan) is typically a t. or ht. containing the p. ch. with other examples in Llanboidy, Llanelli, Llanfallteg, Llansteffan and Pen-bre, and elsewhere in Wales.

Llan² SN 5822
Llangathen
'(ht. called) church': ***llan***
(ht.) *Llan* 1851, 1863
Llan made up most of the south-east part of the p. centred on the p.ch.

Llan³ SN 4514, 4515
Llangyndeyrn
'(ht. called) church': ***llan***
(hts. of *Blayne* and) *Llan* 1650, (ht.) *Lhan* c.1700, (ht.) *Llan* 1804
One of the eight hts. or ts. of Llangyndeyrn.

Llanarthne (Llanarthney) SN 533202
'church of ?Arthneu or ?Arddneu', ***llan***, ?pers.n. **Arthneu** or **Arddneu**
(*ecclesia*) *lan hardneu* c.1170, *Lanarthneu* 1281, *Lannarth'neu* 1326, *Llann Arthnev* c.1400, *Llanartheney* 1436, *Llanarthney* 1535, 1831, *ll.arthne* c.1566
See DPNW 217. Arddneu or Arthneu need not be a saint (none is recorded) and the ch. is in any case dedicated to Dewi/David. Llanarthne has been linked to *Llann adnau* in the poetry of Gwynfardd Brycheiniog late 12th cent. It is very difficult to accept that this refers to Llanarthne. *Llann adnau* is thought to contain ***llan*** 'church' and ***adnau¹*** with the particular meaning of 'deposit, thing stored for safekeeping'. Partly on these grounds, the scholar Wade-Evans (LStD 60) thought that *Llann adnau* should be identified with 'the Monastery of the Deposit' (*Depositi Monasterium*) located at Llanfeugan,

co. Brecon, in the Life of St David. The significance of 'the Deposit' is unclear though it has been taken to refer to a deed kept for safety at Llanfeugan (ETG 133; HGC 46, 190) but it could as easily refer to the location of a burial or refuge (see further Ann Parry Owen in DewiGB, n.66; GLlF 468-9) perhaps of a saint. The ch. of Llanarthne is traditionally said to have stood 400 yds north from the original p. ch. 'which was carried away by an overflowing of the river *Tywi:* the Site whereof is to this day called *Hên Llan*, i.e., *The Old Church*' (CarlisleTD) (***hen***, ***llan***), presumably in the vicinity of Parc Henllan (SN 536200) (***parc***).

Llanbedr see **Caerfyrddin/Carmarthen**

Llanboidy SN 216232
'stream near a cow-house': ***nant*** displaced by ***llan***, ***beudy***
Lanbedeu 1175-6, *Nambeude* 1264, *Nanthendu* c.1291, *Nantbeudy* 1361, *Lanbedu* 1377, *Nantbeudi* 1432, *Llanbeyde* 1501, *llanbeydy* 1595, *Llanbeydey* 1603, *Llanboidy* 1740, 1831, *Llanbeudŷ* 1801
See DPNW 221. *Llanboidy* in late sources reflects dialect and is comparable with 'coy' [kɔi] for *cau*, 'coynant' for ***ceunant***, and 'moydir' [mɔidr] for ***meidr***. Substitution of ***llan*** for ***nant*** is witnessed in Llancarfan, co. Glamorgan, Llaneglwys, co. Brecon, Llanthony, co. Monmouth, and Llangwnnadl, co. Caernarfon (PNGlamorgan 118). The antiquity of the site of St Brynach's church is indicated by two late 6th/early 7th cent inscribed stones (one now lost) recorded here (Corpus II, 227).

Llan-crwys see **Llan-y-crwys**

Llan-dawg, Llandawke SN 282111
'church of ?Tyddawg': ***llan***, pers.n. **Tyddawg**
(rector of) *Llandethauk* 1353, *Landauk* 1377, *Landethauke* 1465-7, *Llandauke* 1513, *llandawk* 1564, *Llandawke* 1535, 1763, *ll.dawc* c.1566
Two recorded forms suggest a pers.n. **Tyddawg** containing honorific **ty-** and perhaps **dog** (earlier orthography *dawc*) 'to take, snatch' (DPNW 223) as in the pers.n. Dogmael (see

PNPemb 178-9). An earlier Llandyddawg would typically develop into Llandyddog but evidence for this is lacking. It is possible that Llandyddawg was stressed on the final syllable (cf. Llandyfân). That might account for *dyddawg* > *-ddawg*, *-dawc* with loss of unstressed *-dy-* or *-ydd-*. The form *Llandawk*, *-e* typical in E-language sources from the 16th cent to the present is now generally pronounced as [lan`dɔ:k] (rhyming with 'hawk') and is drawn from the reduced W form. This is represented in historic sources as *Llandawc*, *Llan Dawc* and *Llandawk* 1687 for [łan'dauk]. The antiquity of the site of the ch. is indicated by a 5th/6th cent inscribed stone (Corpus II, 232-3) but the original dedication of the ch. seems to be unrecorded. It was re-dedicated to St Margaret of Scotland in the 14th cent but it is now dedicated to Oudoceus/Euddogwy probably an antiquarian reinterpretation of Llan-dawg/Llandawke.

Llanddarog SN 503166
'church of Darog': ***llan***, pers.n. ***Darog***
Landarauk 1284, *Landarok* 1280, *Llandarok* 1333, 1500, *Llantharog* 1513, 1752, *Llantharock* 1555, *Llanddarog* 1599-1831
This is probably also the location of *Mainaur de Lantarauch* 1222, standing for Maenor Llanddarog (***maenor***), an otherwise unattested n. of an administrative division of Is Cennen. There seems to be no evidence of any saint called Darog (DPNW 223) and he may be a secular person. The ch. is now mistakenly dedicated to Twrog.

Llanddeusant SN 777245
'church dedicated to two saints': ***llan***, (***y***), ***dau***, ***deu***, ***sant***
Thlanadusant in Dehubert c.1282, *Landusant* 1284, *Thlandeusand*, *-seynt* 1295, *Llannedoysant* 1506, *ll.yðey saint* c.1566, *Llanthewyesant* 1578, *Llanthoysant* 1576, *llanddeysaint* 1600, *Llanddoisant* 1609, *Llanddwysant* c.1700, *Llanddeusant* 1831
Historic forms with *-thoy-* and *-ddoi-* indicate a dialect pron. 'Llanddousant' [łan`ðɔisant]. The saints are generally said to be Simon and Jude

(CarlisleTD; ETG 149) or Potolius and Notolius in Wrmonoc's Life of St Paul (or Peulin) 884. Little is known of these reputed saints but there is a *Petheloc* in *Lambezellec*, Brittany, and an Ir *sancti Natali* in the Martyrology of Tallaght (LWS 152, n.11) 9th cent. The first historic form locates it in Deheubarth, the 'south part' (***deau***, ***parth***) distinguishing it from Llanddeusant in Anglesey (DPNW 224).

Llanddowror SN 255145
'ch. of the water-men, or water-drinkers', ***llan***, ***dŵr***, ***dyfr***, ***gŵr*** pl. ***gwyr***
territorii ecclesię aquilensium; lannteliau ... Lanndÿfrguÿr; Lannteliau lanndibrguir mainaur c.1140, *Landubrguir* c.1100 (c.1200), *Landoverour* 1339, *Llandeuerour* 1398, *ll.ddyfrwr* 1590, *llandowror*, *llandoverour* 1608, *Llandoveror* 1682, (*o*) *Landdowrwr* 1778, *Llandowror* 1831
Early refs. show that the ch. was dedicated to St Teilo and one legend credits him with saving seven sons whose father was attempting to drown them; another legend associates the place with fishermen (AC 1915: 401). The pn. may allude to the practice of abstinence and fasting (ETG 140; DPNW 226-7). Llanddowror may also be *Taui urbis* 'city of (r.) Taf' (***urbs*** gen. *urbis*), held c.680 by a certain Sadwrn (*saturn princeps*). The identification is based partly on Llanddowror's location next to the r. Taf and partly on its pre-eminence among other churches dedicated to Teilo in the land of Dyfed.

Llandeilo (Llandeilo Fawr) SN 629224
'(principal) church of (St) Teilo': ***llan***, pers.n. **Teilo**
lann teliau maur c.1145, *ecclesia Sancti Teleauci* c.1191, (*ch.*) *Lanteilawmawr'* 1222, *Lanteylavaur* 1281, *Thlanthilogh Vaur* 1295, *Landeilau Vaur* late 13th cent, *Llann Deilaw* c.1400, *ll.deilo fawr* c.1566, *Llandilo* 1612, *Llandilo Iskennen* 1609; *Llandeilo-fawr* 1831
Llandeilo Fawr 'the greater' ***mawr*** to distinguish it from other places known as Llandeilo (DPNW 228). Llandeilo was the centre of the cult of St Teilo whose grave (*tumulum sancti Thilawi*) is recorded in 1294-5 (SGStudies 205-8 and refs.,

245-9). The town of Llandeilo lay partly in the cmt. of Is Cennen and partly in Maenordeilo (q.v.).

Llandeilo Abercywyn SN 310131
'church of (St) Teilo at Abercywyn': **llan**, pers.n. **Teilo**, pn. *Abercywyn*
Lann teliau aper Couin, ~ aper coguin c.1145, *St. Teliau of Abercowen* 1250, *Lantelo* 1292, *Nanteliow* 1339, *ll.deilo* c.1566, *llandiloabercowyn* 1597, *Llandilo Vach, Llandilo (Aber cowin)* 1740, *Llandilo-fach* 1778, *Llandilo fechan, Llandeilo Abercywyn* 1831
On the east side of the estuary (**aber**) of the r. Cywyn (q.v.). This may also be *Lanteliau bechan* in the Life of St Teilo c.1145. Occasionally described as Llandeilo Fach/Fechan 'small, lesser' (**bach**[1], **bechan**).

Llandeilo'r-ynys (Llandeilo Rwnws) SN 4920
Llanegwad
'church of (St) Teilo in Maenor Frwnws': **llan**, pers.n. **Teilo**, **Maenor Frwnws**
lann teliau mainaur brunus c.1140, *lanteliau mainaur brunuis* 13th cent, *Lanteilau Brunus* 1324, *Landewrennes* 1353, *Llandilowronus* 1559, *Llan Deilo rwnws* 1710, *Llandilo yr Unys* 1808, *Llandilo-rhwnws* 1887-88, *Llandilo-yr-ynys* 1906
A former chp. of Llanegwad and located in Maenor Frwnws (q.v.). David Jones (Yr Haul 23 (1879), 98) says it was 300 yards from the railway br. (SN 50122072) over Afon Cothi and that the stones were re-used in building at Llandeilo'r-ynys fm. (SN 49462036). This would place it north-east of the fm. where there is a standing stone (HER 11063; Ludlow in EMESP) and west of the fm. where there are traces of a circular earthwork bank. A presumed medieval window was also identified at the fm. (HER 7557). The br. is *Pont Llandiloyrynys* 1812 and *Pont Rhwnws* 1887-88 (**pont**). The qualifer Rwnws is a reduced form of Maenor Frwnws though it has been misinterpreted as *yr ynys*, 'of the island, of the water-meadow' (**yr, ynys**) (CStudies 119). Brwnws is also found in Ystrad Brwnws where Owain ap Cadwgan, a prince of Powys, was killed by Flemings 1116. This appears to have been the location of Ynysdeilo, a gr. of Talyllychau/Talley abbey, recorded as *Ynysteliau* 13th cent (1332), *ynisdilo* 1540-1, *Enysdilo* 1537-8, (gr.) *Ennys Dillow* 1559, meaning 'river-meadow of Teilo' with the elliptical sense of 'river-meadow in Llandeilo'. Ffynnon Deilo (SN 496217) (**ffynnon**) on the Caerfyrddin/Carmarthen road is *Funnon Dilo* 1828 and *Ffynondilo* 1831.

Llandeulyddog SN 4120
Caerfyrddin/Carmarthen
'church of (St) Teulyddog': **llan**, pers.n. **Teulyddog**
Lann Toulidauc icair c.1145, *lann toulidauc ig cair mirdin* c.1170
The pn. pre-dates the Norman conquest of Caerfyrddin/Carmarthen in 1093 and the cited refs. confirm that the ch. lay within the Roman fort of Caerfyrddin/Carmarthen, almost certainly to be identified with a ch. dedicated to St Theodore the Martyr granted by Henry I to Battle abbey in 1100. Theodore is thought to be a confused form of Teulyddog who was reputedly a follower of St Teilo (dedicated at Llandeilo and others places in co. Carmarthen). It has also been argued by J. Wyn Evans (SGStudies 249-251) that he is in actuality the same person as Teilo (**Teilo**) partly on the basis that both Teilo and Teulyddog are familiar or hypocoristic pers.ns. composed of an honorary prefix **ty-** with pers.n. *Eliud* and suffixes *-o* and *-og* (earlier *-auc*) respectively. Eliud was Teilo's formal n. in marginalia added to the Lichfield Gospels ('Book of St Chad') 8th cent and in the Life of St David c.1200. Taken as a whole, the argument that Teulyddog is the same person as Teilo is doubtful. Llandeulyddog seems to be last recorded in the 13th cent.

Llandingad (Llandingat) SN 7633403
'church of (St) Dingad': **llan**, pers.n. **Dingad**
(St.) *Dingat apud Llandeuery* after 1130 (c.1200), *ecclesiam Dyngad* 1282, *Llandyngat* 1389, *Llandyngatte* 1535, *Llandyngad* late 16th cent, *Llandingad* 1622, *Llandingatt* 1676
Dingad is thought to be one of the numerous offspring of Brychan Brycheiniog recorded in

medieval pedigrees. The expected form would be *Llanddingad* with len. *d > dd* but this form is very rare; contrast Llanddingad which is the W form for Dingestow (PNGwent 87); absence of len. of -d- after n- is also found in Llandinam, co. Montgomery (DPNW 229) and Llandovery (q.v.).

Llandovery see Llanymddyfri

Llandre SN 4320
Llangynnwr
?'church township': ?*llandref*
Llandre 1796-7, 1845
Melville Richards describes this as a t. of Llangynnwr but it is poorly evidenced and is not current. It may have applied to the area around the p. ch. and village corresponding with *Llangunnor District* 1792. The pn. has been connected with *Laundry Ishulle* and *Laundreishulle* 1308-9 (E *hill*) held by John Laundry in 1308 and which may be identified with lands in the bailiwick of Caerfyrddin/ Carmarthen for which William *Landri* or *Laundry*' paid homage in 1251. This is very unlikely, particularly in view of the paucity of recorded forms. The first members of the family connected with co. Carmarthen are Ernulf son of *Landrici* 1130 and Walter son of *Landri* c.1170, and a Philip Laundry is recorded in connection with *Kyldewolthe Vaur* in 1312. If this is Llandyfaelog Fawr, then *Laundreishulle* is better sought further south in the Cydweli area, perhaps (close called) *Bryn Llandery* 1726 near Llandyry (q.v.).

Llandre Egremont see Egrmwnt, Egremont

Llandybïe SN 619155
'church of (St) Tybïe': *llan*, pers.n. *Tybïe*
Landebyeu 1284, *Landebyeu* 1348, *Llandebie* 1495-1831, *Llandebye* 1535, *ll.dybie* c.1566, *Llandebea* 1609, *llandybie* late 16th cent, *Llandybie* 1831
Tybïau is recorded as *Tibyei filia Brachan* 11th cent (c.1200), one of the numerous reputed daughters of the semi-legendary Brychan Brycheiniog: *Tebie* 13th cent (16th cent) (EWGT 18; VSBG 317, 319). Her feast was three days before that of the Virgin Mary according to Edward Lhuyd c.1700 and she possessed a well Ffynnon Dybïe (*ffynnon*) (HPLland 27-28; HWW 164).

Llandyfaelog SN 417118
'church of (St) Tyfaelog': *llan*, pers.n. *Tyfaelog*
lann diuailauc c.1170, *landivayloc* c.1155, *Landevayloc* 1283, *Landevailoc* 1495, *Llandevayloge* 1560, *ll.dy fayloc* c.1566, *Llandevailogge, als Llandevallocke* 1618, *Llandeveylogg* 1763
The pers.n. is composed of the honorific prefix *ty-* and pers.n. *Maelog*. The saint is generally identified with Maelog in the Life of St Gildas where he is connected with Llowes, in Elfael, co. Radnor (PNRad 73-74), and with Meilig in the tale of Culhwch ac Olwen (CO³ no. 209; LBS III, 401-6). Tyfaelog is otherwise recalled in Llandyfaelog Fach and Llandyfaelog Tre'r-graig, co. Brecon (DPNW 233; PNBrec 95-6).

Llandyfân SN 642171
'church of Tyfân': *llan*, pers.n. *Tyfân*
Lh.dyvân c.1700, *The Welch Bath call'd Llanduvaen* 1729, *Llandyfaen Chapel* 1752, *Llandyfaen* 1789, *Landefane* 1800
Llandyfân possessed a small chp., used by Methodists and Nonconformists, recorded as the location of a medicinal well and bath in 1754 and a holy well alternatively known as Ffynnon Gwyddfaen (from the rn.) and Ffynnon Llandyfaen (HWW 167) (*ffynnon*). Welsh Bath is also applied to the small settlement around the chp. in 1811 and 1831. The chp. was returned to the Anglicans in 1838 and replaced by a chp.-of-ease on a different site in 1865. Later OS maps describe it as *St. Dyfan's Church* but historic forms favour Tyfân stressed on the final syllable. Forms with *-faen* are probably hypercorrections based on the supposition that *-â-* is a colloquial form of *-ae-*. There seems to be no record of any saint bearing this n. (DPNW 235). The pers.n. may survive in that of the nearby hill Carreg Dwfn (SN 655174) (*carreg*) which is recorded as *Careg y Dwfan* 1879 and in a ho.n. recorded

as *Carregydwfan* 1831 and *Carreg-Dwyfan* 1879 now known as Pen-y-garn (SN 654175).

Llandyfeisant SN 622222
'church of St Tyfái': ***llan***, pers.n. **Tyfái, sant**
Lantesvassan after 1271 (1331), *Landevaysen* c.1291, *Landeuaysan'* 1326, *Llandevayson* 1535, *ll.dyfaysan* c.1566, *Llandevaisen* 1593, *Llandefoysaint* 1710, *Llandefeison* 1785, *Llandyfeisant* 1831
Tyfái was reputedly a nephew of St Teilo (StoryCarm I, 62) but the Life of St Oudoceus (Euddogwy) describes *Tÿfei* as one of the sons of a certain *Anaumed* (with Isfael or *Ismael*). Tyfái is described in the Book of Llandaf (LL 130) as 'a martyr lying in Penalun (Penally)', co. Pembroke c.1145. Tyfái is also dedicated at Llandyfái otherwise Lamphey, co. Pembroke.

Llandyry SN 434049
Pen-bre
'church of (?)Tydyri': ***llan***, pers.n. (?)*Tydyri
Lanthedury 1358, *Llandedury* 1520, *Llandydyry* 1609, *Llandidery* 1623, *Cappell Dery* 1658, *Lhanderi, Lhan Dyri, Lhandyrri Chapel* c.1700, *Llandery* 1709, *Landyrry Hamlet* 1792, *Capel Llandurry* 1833, (vicar of Pen-bre/Pembrey cum) *Llandyry* 1848
A former chp.-of-ease (***capel***) to Pen-bre. The pers.n. is not on record and there seems to be no evidence for the dedication of the chp. here. The modern form appears to be a contraction through dropping the second unstressed *-dy-* (syncope) in Llandydyry.

Llandysilio Pemb SN 121215
'church of (St) Tysilio': ***llan***, **Tysilio**
Landesilion c.1291, *Landesilian'* 1326 (16th cent), *Llandesilio* 1389, *Llandissilio* 1535, *ll.dyseilio yn yfed; p. dysilio ynyfed* c.1566, *Llansillio Llawhadden* 1585, *Llandissilio Neved* 1613, *llandissilio, llansillio* 1623, *Llandissilio, St. Tyssilio P.* 1763, *Llandysilio* 1843
Historic forms with *-on* and *-an* are scribal errors for *-ou* and *-au*. Occasionally recorded as Llandysilio yn Nyfed, distinguishing it from other places called Llandysilio (DPNW 235). Dyfed (q.v.) was the n. of the historic

gwlad, revived between 1974 and 1996 for a county covering the three historic cos. Cardigan, Carmarthen and Pembroke. The cp. of Llandysilio West (SN 1124) lay in co. Carmarthen, Llandysilio East in co. Pembroke till 2003. Associated with Llawhaden in 1585 for which see PNPemb 420.

Llanedi SN 588066
'church of (St?) Edi', ***llan***, pers.n. **Edi**
Lanedy 1407, *Lannedy* 1423, *Llanedi* c.1500, 1742, *Llanedy* 1535, 1671, *Lanedi* 1550, *ll. edi* c.1566
There seems to be no reliable evidence of any St Edi though M.H. Jones (AC 1915: 329) refers to a 'St. Edi, Confessor'. Some popular association with a saint Edi must account for Ogof Gŵyl Edi, 'cave of the feast of Edi' (***ogof, gŵyl***[1]), otherwise known as Ogo'r Cawr and Ogof Edi'r Cawr, 'cave of Edi the giant' (***cawr***) referring to a cave in rocks here. Melville Richards rejects the traditional dedication of the ch. to Edith (ETG 154) given by Ecton and others. The village now encompasses an area extending northwards to the B4297.

Llanegwad SN 5121, 5221
'church of (St) Egwad', ***llan***, pers.n. **Egwad**
Lanegwade Vaur c.1220, *Thlanhegodvaour* 1287, *Lanogwat Vaur* c.1291, *Lanegwat Vaur (cum capellis)* after 1271 (1332), *Llanegwat* 1389, *LL.egwad vawr* c.1500, *ll. egwad fawr* c.1566, *Llanegwad* 1594, 1671, *Llanegwad Vawre* 1655, *Llanegwad in Elvett* 1710
The fuller form Llanegwad Fawr with ***mawr*** 'the greater' distinguishes the ch. from that dedicated to Egwad at Llanfynydd (q.v.) (DPNW 237). Llanegwad also has Eisteddfa Egwad (SN 535240), 'seat of (St) Egwad'; cf. Llangynnwr (q.v.) for the use of ***eisteddfa***. Dydd Egwad, St Egwad's Day, is mentioned by the poet Lewys Glyn Cothi in a cywydd of praise to William Siôn of Llanegwad but the actual date of the feast is apparently lost. The 1710 sp. refers to the location of Llanegwad in the hd. of Elfed. Egwad occurs as the n. of Cwm Egwad, the valley of a short stream

running from SN 464261 into Nant Crychiau (SN 461249).

Llanelli SN 507006
'church of (St) Elli': *llan*, pers.n. *Elli*
ecclesiam Sancti Eclini c.1165, *lann elli* c.1170, *monasterium Elli* c.1100 (c.1200), (ch. of) *St. eslini of Carnewallon* c.1235, *Lanetly* c.1291, *Lanelthy* 1346, *Llanelly* 1446-1785, *Llanhelli* 1542, *Llannellthi* 1564, *ll. elli* c.1566, *Llanelly V. alias Llan-Elli, St. Elliew* 1763
Elli (*Ellinus*) was a reputed disciple of St Cadog according to the latter's 'Life' c.1100 (VSB 56, 58, et seq.). He is also mentioned in the poem 'Y Seintiau' by Huw Cae Llwyd (fl. 1455-1505) (GHCL 108) and possessed a well recorded as *ffynnon elli* 1604 (DPNW 238; HWW 164) (*ffynnon*). *Llanelly* is the usual OS sp. 1880-c.1970 before the pn. was standardised to Llanelli. The ch. dedication to St Ellyw is a misinterpretation of Elli possibly under the influence of Capel Gwynllyw² (q.v.).

Llanfair-ar-y-bryn SN 770351
'church of (St) Mary on the hill': *llan*, *Mair*, *ar*, *y*, *bryn*
Lanveyr 1317, *Llanvaire Arbryn*, ~ *Arbrynne* 1559, *ll. fair y bryn* c.1566, *Llanvair ar y Brynn* 1567, *llan vayer ar y bryn* 1629, *Llanvair ar y bryn* 1672, *Llanfair-ar-y-bryn* 1831
The paucity of early refs. may be attributed to its former status as a chp. to Llandingad (Ecton 1763: 474). The ch. is on a low hill and within the banks of a Roman auxiliary fort identifiable with *Alabum* 8th cent possibly a latinised form of Brit *alabon 'hill, crest' (PNRB 242). The fort was once identified with Loventium, properly Luentinum, which refers more likely to the Roman mining settlement at Pumsaint (PNRB 400).

Llanfallteg SN 147193
'church dedicated to (St) Mallteg (or Ballteg)': *llan*, pers.n. **Mallteg** or **Ballteg**
Landenayth'teg' 1326, *Llanvalteg* 1499-1798, *LLam Valdec* c.1500, *Llanbaltege* 1535, *ll. fallde* c.1566, *Llanvaltegg p(ar)ish* 1672, *Llanfallteg* 1819

The first instance probably stands for ModW Llandyfallteg with the honorific form *Tyfallteg*. The ch. is dedicated to Mallteg but it stands on a mound 'locally called Rhiw Bylltig' (AC 1915: 295) (*rhiw*) and, taken in conjunction with the 1535 form, it is possible that the saint is properly Ballteg. The ch. lies in the former cp. of Llanfallteg West in historic Pembrokeshire. The remainder of the p. is in the former cp. of Llanfallteg East in historic Carmarthenshire. Llanfallteg is now more prominently applied on OS maps to hos. which developed around the former railway station (SN 156200).

Llanfihangel see **Capel Mihangel** (Cydweli)

Llanfihangel (Llanfihangel Cynros) SN 63263128
Talyllychau/Talley
'church of (St) Michael': *llan*, **Mihangel**
Lanvihangell Kynros (in cmt. Manordeilo) 1506, (p.) *Lanvihengl Kynros* 1507, (ruinated chp.) *Llanvihangel* 1742, '*Mynwent Cappel Llanfihangel*, i.e., The Churchyard of St. Michael's Chapel' 1811, (ht.) *Llanfihangel* 1811, *Mynwent Capel Llanfihangel (Site of)*, (ho.) *Banc-Llanfihangel* 1888, 1964
A former chp. of Cefn-blaidd gr. belonging to Talyllychau/Talley abbey. Otherwise known as Capel Mihangel (*capel*), evidently with its own graveyard (*mynwent*). Located near Tŷ Hywel a mile south of the abbey (RCAHM V, 264) on a spur of Mynydd Cynros (SN 6132) (*mynydd*) but there were no physical traces apart from some shaped stones in 2011 (HER 1900). Cynros or Cefnros is 'moorland ridge' (*cefn*, *rhos²*); see also Maenordeilo.

Llanfihangel Aberbythych SN 590198
'church of (St) Michael at Aberbythych': *llan*, pers.n. **Mihangel**, pn. *Aberbythych*
(ch.) *Aberbetthek* c.1291, *ecclesia Sancti Michaelis de Aberbythych after 1271* (1331), *LL. V'el aber bythych vwch ryd arwer* c.1550, *Llanvyhangell Aberbuthick* 1555, *Llanvihangell Aberbuthicke* 1590, *Llanvyhangell abberbithige* 1615,

llanvihangel Aberbythych 1616, *Llanfihangel Aberbythych* 1887

Aberbythych is 'mouth of (stream called) Bythych': ***aber*** and a lost rn. *Bythych* presumably once applying to the small stream running northwards to Tywi (SN 580206). Earlier forms suggest that Bythych was substituted for Bythig (with suffix *-ig*) but the precise meaning is unclear; ***byth*** 'ever, always' might describe a r. whose flow was continuous throughout the year but it is difficult to find comparable examples. The 1550 ref. to *vwch ryd arwer* is probably 'uwch Rhydhirwen', 'above Rhydhirwen' (at SN 559194). The area is generally known locally as Gelli-aur/Golden Grove (q.v.).

Llanfihangel Abercywyn SN 303134
'church of (St) Michael at Abercywyn': ***llan***, pers.n. ***Mihangel***, pn. *Abercywyn*
(ch.) *St. Michael, Abercownen* 1351, (chp.) *Sancti Michaelis de Abercowyn* 1535, *ll. V'el aber kowyn* c.1566, *Llanhangell Abercowyn* 1574, *Llanfihangel Aber-cowyn* 1612, *Llanfihangel Abercywyn* 1831
At the confluence (***aber***) of Afon Cywyn (v. **Cywyn**) and Afon Taf. The old ch., possibly 12th or early 13th cent in date (VCW 113), is located opposite to Llandeilo Abercywyn (q.v.) and is in ruins. It was replaced by the present one (SN 300170) on the A40 in 1848 (AC 1915: 398). The old ch. was familiarly known as the Pilgrim's Church since it possessed graves attributed at least as early as 1837 to three poor pilgrims, possibly on the basis that cockleshells had been found in one of the graves when it was opened. Bruce Coplestone-Crow (CAS 46: 8-9) suggests these may actually have belonged to members of the Norman family Torte or Courtmain.

Llanfihangel-ar-arth SN 455399
'church dedicated to (St) Michael at great hill': ***llan, Mihangel, gor-, garth***
Lanvyhangel Orarth c.1291, *Lanyhangel Orarth* 1342, *Llanvihangelorarth* 1401, *Llanvihangel Yerrorth* 1514, (rectory) *St. Michael of Orroth* 1551, *Llanvihangell ararth* 1563, *ll. V'el Joroth* c.1566, *Llanvihangell Yeroth* 1580, *Llanihangle Yerworthe* 1580, *Llanvihangelyerwarth* 1690, *Llanvihangel Ararth* 1751, *Llanfihangel ar Arth* 1796
The unfamiliarity of *gorarth* in its lenited form *Orarth* prompted association with the pers.n. ***Iorath*** (variants Ieroth, Iorwerth) recorded in historic forms 1516-1831. The presumed pers.n. was then regarded as the n. of the founder of the ch. The modern OS form is also a re-interpretation as 'on *or* over against a hill' (***ar, garth¹*** lenited *arth*) which dominates historic forms from the late 18th cent to the present. The village gradually developed southwards in the 19th cent to crossroads on the B4336 at Cross Inn. A prominent hill extends westwards from the village reaching its summit at SN 441393.

Llanfihangel Cilfargen SN 573241
Llangathen
'church of (St) Michael at Cilfargen: ***llan, Mihangel***, pn. **Cilfargen**
(rector of) *Kylvargyn* 1430, *Llanvihangell Gilbergen* 1535, *ll. V'el* c.1566, *Llanvihangell Kilyvargen* 1628, *Llanvihangel Kylvargen* 1638, *Llanvihangel-vach-cilvargen* 1763, *Llanfihangel-fach-cilfargen* 1831, *Llanfihangel-Cilfargen* 1887
Cilfargen was the location of a gr. of Hendy-gwyn/Whitland abbey – possibly coterminous with the p. (CAS 28: 51) – granted by Lord Rhys (d. 1197) and it has been suggested that the gr. chp. may have preceded the p. ch. (WCist 182, 198, 202). The 1430 ref., however, shows that the p. ch. existed before the Dissolution probably serving also as chp. to the gr.

Llanfihangel Rhos-y-corn SN 550348
'church of (St) Michael at Rhos-y-corn': ***llan, Mihangel***, pn. *Rhos-y-corn*
Llan Viangel Roscornew 1388, *Llanvyangel Roscornowe* 1395, *Llanvihangel Roscorne* c.1545, *Llanvyhangell Rose y corne* 1552, (chp. of) *St. Michaell Roscorny* otherwise *Llanvyhangell Rhoscorny* 1632, *Llanvihangell Ros y Corn* 1689, *Llanfihangel-rhôs-y-corn* 1831
A former chp. of Llanllwni. Earlier forms show that the qualifier Rhos-y-corn is properly 'moor or promontory in Cornyw or Cornwy',

composed of **rhos**² and perhaps a territorial n. which was later re-interpreted as *cornau* as a supposed pl. of **corn** 'horn, point' with the extended sense of 'hill'. From the late 17th cent *cornau* was itself displaced by the sg. form (DPNW 248).

Llanfihangel-uwch-Gwili SN 490229
'church of (St) Michael above (r.) Gwili', **llan**, pers.n. **Mihangel**, **uwch**, rn. **Gwili**¹
capella Sancti Michaelis de Lechmeilir 1331, *Lanvihangell Llez Veiler* 1395, *Llanuihangle ugwely* 1578, *llanvihangle ywch gwilly* 1585, *Capel Llan Uchwilly* 1715, *Lan-Fihangel-uchwily, St. Michael, Chapel to Abergwilly* 1763, *Llanfihangel uch Gwilly* 1792, *Capel Llanfihangel* 1813
'Uwch Gwili' may be used in the broad sense of 'area above Gwili' or 'upper part of Abergwili' since Llanfihangel is four miles east of the confluence of Gwili and Tywi. Llanfihangel was a former chp. in Abergwili p., possibly granted to Talyllychau/Talley abbey before c.1200 (HER 49238) and is described as 'decayed' in 1710. The earlier qualifier Llechfeilir meaning 'Meilir's slab (of rock)', **llech**¹, pers.n. **Meilir**, has not been fully explained. The n. could be an older n. for the ch., possibly taken from a natural feature or may refer to a lost stone monument.

Llanfynydd SN 559275
'ch. at the mountain': **llan**, **mynydd**
Laneguadveniz 1281, *Lanvenyth* 1276 (late 13th cent), *Llannenyth*' 1326, (prebend) *Llanvenyth* (in Abergwili) 1391, *Llannvynydd* c.1500, *ll. fyny* c.1566, *Llanvyneth* 1566, *Llanvynyth* 1671, *Llanvynith V.* alias *Llanfynnydd, St. Egwad* 1763, *Llanfynydd* 1804
Earlier Llanegwad Fynydd from the ch. dedication, now St John; cf. Llanegwad. The ch. stands on a spur though the village has extended down into the valley and southwards up a second hill to Spite/Sbeit (**sbeit**).

Llangadog¹ SN 706283
'church of (St) Cadog': **llan**, pers.n. **Cadog**
lanncadauc c.1170, *Lankadoc* 1284, *Llann Gadawc* early 14th cent, *Llangadok* 1535, *ll. gadog fawr* c.1566, *Llangadog* late 16th cent, 1831, *Lhangadogvawr* 1627
Occasionally qualified in later sources with *Fawr* 'the greater', *mawr*, perhaps to distinguish it from Llangadog (Cydweli). Cadog is an alternative form of Cadfael (**cad**¹ 'battle' and **mael**² 'prince') (ETG 174). Cadog is almost certainly Cadog ap Gwynllyw (VSB 24-141), usually associated with the south of the county and Glamorgan.

Llangadog² ?SN 422079
Cydweli
'church of (St) Cadog': **llan**, pers.n. **Cadog**
ecclesiam sancti Cadoci c.1160, *Llangadog* 1777, 1880, *Langadoc* 1830
The paucity of early refs. to the chp. is difficult to explain. It appears to have been a chp. to Cydweli/Kidwelly belonging to Sherborne abbey (HER 49238) and a ch. is mentioned in 1835 but this is in an unreliable source (WSS 329). There are few clues to its location unless it was at Sanctuary Bank. Gruffydd Evans found 'not a vestige of the church' in 1915 (Cymm 25: 148) and it is likely that it had disappeared before the 18th cent. The hill to the south is *Cadok is More* 1458, *Cadox More* 1593, and *Waun gadoc* 1830 and a mill here is *Cadokysmyll* 1529 and *Cadockes Mylle* 1609 (E **moor**, **gwaun**, E **mill**) probably referring to Upper Mill 1889 (SN 418073). Llangadog has been identified with an unnamed chp. mentioned in 1720 (CAS XIV, 9-10) on Llechdwni estate but this is doubtful.

Llan-gain SN 383160
'church of (St) Cain': **llan**, pers.n. **Cain**
St. Keyn c.1175, *Eglwyskeyn* c.1182, *Langau* c.1291, *Manorgayne, otherwise Llangayn* 1551, *ll. gain* c.1566, *Llangaynge* 1597, *Llangain* 1831
Cain was a reputed daughter of the semi-legendary Brychan Brycheiniog recorded in 12th cent pedigrees and is probably identical to Ceinwen and Ceinwyr. Eglwys Cain (last recorded 1247) (**eglwys**) and Llan-gain refer specifically to the church (closed in 2017). Llan-gain alternates from 1288 down to the early 17th cent with Maenor Gain (q.v.) which probably originated as a secular n. Llan-gain

is also the W n. of Kentchurch, co. Hereford (PNHer 123).

Llan-gan SN 177187
'church of (St) Can *or* Cann': *llan*, pers.n. *Can(n)*
Llangan late 12th cent-1819, *Llan Gan* 1710, *Llan-Gan, St. Gan P.* 1763, *Llangan West, Llangan East* 1890
The ch. is attributed to St Canna on OS maps 1890 and 1891 and there is a reputed prehistoric stone Cadair St Canna otherwise St. Canna's Chair (*cadair¹* 'chair, rock with the appearance of a chair') inscribed CANV on the north-east side of the churchyard with a well Ffynnon Ganna (*ffynnon*). The supposed chair is now thought to be a fake (Corpus II, 531). Air photography and a geophysical survey nonetheless confirm the antiquity of the ch. and its site within a complex of enclosures. Llan-gan has been compared as a pn. with Llan-gan otherwise Llanganna, co. Glamorgan (PNGlamorgan 123). Historic forms, however, consistently favour Can(n) as Charles says (PNPemb 413). Can(n) is not independently recorded but it may well contain the el. *can¹* 'white' found in *Canna* and *Cannou* 12th cent. The former cp. Llan-gan West, which includes the ch., was in co. Pembroke and Llan-gan East in co. Carmarthen covering an area centred on Cwmfelin-boeth (q.v.), extending eastwards to Afon Gronw.

Llangathen SN 582222
'church of (St) Cathen', *llan*, pers.n. *Cathen* or *Cathan*
Langattkek c.1291, *Langathen* 1318, 1535, *Langathan* 1323, *Lankaththen in Cantremaure* 1325, *Langatheyn* 1348, *Llangathen* 1555, 1831, *ll. gathen* c.1566, *Llangathan* 1763
Cathen or Cathan also occurs in Catheiniog (q.v.). R.J. Thomas regards this as an Ir pers.n. composed of *cath* corresponding to W *cad¹* (EANC 51: Cathan). There was also a Breton saint *Cazen* identified in Langazen (Finisterre). Jones notes a ref. to a field *kae ffynnon gathen* 1675 on Allt y gaer farm (SN 572210) without trace of a well (HWW 165). The field is *Fynnon Gathen* 1839 (1211) (*ffynnon*). Melville Richards (BBCS 23, 325) notes the variation *gathen/gathan* and suggested that the pers.n. may actually be Cathan as in Myhathan (q.v.) in the adjoining p. of Llanarthne.

Llangeler SN 374394
'church of (St) Celer': *llan*, pers.n. *Celer*
Martir Keler' c.1291, *Merthyr Keler* 1303, *Merthirkeler* 1326, 1418, *Llangeler in Emblyn* 1532, *Llangeler* 1535, 1831, *ll.geler* c.1566
Earlier Merthyr Celer, 'graveyard consecrated with the bones of Celer' with *merthyr²* but this was confused with *merthyr¹* leading to the supposition that Celer was 'a martyr' with the further suggestion that he was associated with Beddgelert, co. Caernarfon (as HPLlangeler 12, 85). The p. also has Rhosgeler (*rhos²*), Plasgeler (*plas*) and a healing well Ffynnon Geler (AC 1915: 330) (*ffynnon*). The last is recorded as *St. Celert's Spring* 1811 (CarlisleTD) which probably lay at the foot of Allt Geler hill near Plas Geler (HWW 164). Neither the pn. nor the well has anything to do with Celert (DPNW 26).

Llangennech SN 561019
'church of (St) Cennych': *llan*, pers.n. *Cennych*
L. gennyz ynglann LLychwr c.1500, *Llangenech* 1545, *ll. genych* c.1566, *Llangennech* 1609, (village) *llangenneche* 1620, *Llangennech, St. Gwynnoc* 1763
Cennych is likely to be the saint Cennych mentioned in a poem in praise of Tomas ap Syr Rhosier Fychan late 15th cent (GyN 12). Cennych, later Cennech, has been identified with Cennech, abbot and confessor, and a reputed disciple of St Cadog at Llancarfan, co. Glamorgan (AC 1915: 332). The dedication to Gwynnog is probably through misassociation with the saint dedicated at Llanwnnog, co. Montgomery (see DPNW 286). Llangennech may be the location of the chp. of *Stratkenny* recorded in 1286, the n. surviving in *Estrakennies myll* 1643, *Estra kennis* 1665 and *Ystrad Kemisse mill* 1686 in ref. to Llangennech Mill (SN 557025). The first el. refers to its location in the 'valley' (*ystrad*) of Afon Morlais but what -kenny, kenni(e)s are intended to represent is

unclear. They are unlikely to contain some form of **Cennych** because later examples co-exist with *Llangennech*. The paucity of early refs. to Llangennech may be attributed to its former status as a chp. to Llanelli. Earlier evidence suggests that the eastern part of the p. was in the p. of Llandilo Tal-y-bont. Little of the earlier ch. survives. Located near r. Llwchwr, hence the sp. c.1500 (*glan*, rn. **Llwchwr**).

Llanglydwen SN 181268
'church of Clydwen': *llan*, pers.n. **Clydwen**
Llangledewen c.1291, 1355, *ll. glydwen ar daf* 1590-1, *Llangludwen* 1637, 1672, *Llanglwydwen* 1679, 1843, *Lhanglwydwen* c.1700, *Llangloydwen* 1757
Melville Richards (ETG 179) prefers the pers.n. *Clydwyn* but the pn. is almost invariably spelt with -wen. This also cancels out the suggestion by J.E. Lloyd (HCrm I, 118) that he is identifiable with Clydwyn, one of the sons of the legendary Brychan Brycheiniog mentioned in early pedigrees and associated with Deheubarth. The ch. was dedicated to All Saints c.1700 but later reverted to Clydwen. Several prehistoric or early Christian stones are recorded near the ch. The 1590-1 ref. seems to be unmatched but refers to the location of Llanglydwen near the r. Taf (*ar*, **Taf**).

Llangyndeyrn SN 456140
'church of (St) Cyndeyrn': *llan*, pers.n. **Cyndeyrn**
Lankederne 1358, (*o*) *L. gynn daúyrn* c.1500, *Llangindern* 1545, *Llangyndeyrn* c.1550, *ll. gyndeyrn* c.1566, *Llangendeirne* 1579, *Llangendeyrne* 1580, *llangyndeirn* 1600-1888
The pers.n. is very uncommon and it is tempting to identify Cyndeyrn with St Cyndeyrn or Kentigern, otherwise known as St Mungo, who is believed to have lived in the late sixth century. Evidence, however, is lacking and the only known dedication to Cyndeyrn in Wales is recorded in 1657 at Llanelwy/St Asaph (DPNW 430). St Cyndeyrn/Kentigern is widely dedicated in Scotland and the north of England and is thought to be identical to Cyndeyrn Garthwys, described as 'Chief of Bishops', in the medieval Welsh triads (TYP 1 and 322n.). The paucity of early sps. probably reflects Llangyndeyrn's former status as a chp. of Llandyfaelog. The irregular form *Llangendeirne* is the usual form on OS maps from 1906 down to the 1990s.

Llangynfab see **Capel Cynfab**

Llangynheiddon SN 430151
Llandyfaelog
'church of (St) Cynheiddon': *llan*, pers.n. **Cynheiddon**
(chp.) *Lankeneythou* 1358, *Lankneython* 1361, *llan gan heiddon* c.1550 *Llangenhython* 1552, *ll. gynheiðon* c.1566, *Capelllangellhithon* 1578, *chaple of llangenhythen* late 16th cent, *Llangenithon* 1608, *Capel Llangenheiddon* 1811, *Capel-llanganheiddon* 1831, *Llangenheiddon* 1851
A former chp.-of-ease to Llandyfaelog (RCAHM V, 106), the site taken over by a Calvinistic Methodist chp. (*capel*) (HER 49271). Cynheiddon was one of the numerous daughters of the legendary Brychan Brycheiniog (ETG 16; MWG) recorded as a saint in medieval pedigrees, *Keneython filia Brachan* and associated with Mynydd Cyfor (SN 4416) more than a mile away, apparently in error. Llandyfaelog also possessed a well Ffynnon Gynheiddon recorded as *fynon ganhython* 1636 and *ffynnon gyn haython* 1716 (**ffynnon**). Llangynheiddon may be the location of *Bettus* (**betws**) recorded with Pen-bre in 1613. Displaced as a pn. by Bancycapel (q.v.).

Llangynin SN 251197
'church of (St) Cynin': *llan*, pers.n. **Cynin**
Lankenyn 1325, (chp.) *St. Kennyns* 1446, *Llangynyn* 1570, *Llangynin* 1599, *llangynnyn* 1602, *Llangining* 1624, *Llangynnin* 1755, *Llanginning Chap. to St. Clear* 1763, *Llanginning* 1804
The pers.n. Cynin may be derived from OW *Cunein* borrowed from Ir *Co(i)nin* < **Cunegna(s)* (nominative of *Cunigni*). A St Cynin is mentioned in late medieval poetry and it is possible that was the father of Avittorix recorded on a 5th/early 6th cent inscribed

stone at Eglwys Gymyn (q.v.) (Corpus II, 214-7). Cynin has also been identified with a son of Brychan Brycheiniog (HCrm I, 118) apparently on the basis of late copies of the pedigree known as 'Plant Brychan' which refer to *Kynin ap brychan y sydd yn sant yngwlad Ddyvet yn y lle a elwir Llann gynin* (EWGT 82). There are doubts with this supposition, however, because earlier manuscripts show that the form of the pers.n. is Cynon and that the association with Llangynin may be an error (EWGT 15, 18, 43, 147). Many late historic forms of Llangynin have inorganic -*g*. Llangynin ch. was formerly a chp. attached to Sanclêr/St Clears priory.

Llangynnwr (Llangunnor)   SN 430201
'church of (St) Cynnwr': *llan*, pers.n. **Cynnwr**
Lanconor alias *Langonefor* 1283, *Lanconur* 1361, *Llangonnor* 1445, 1609, *ll. gwnwr* c.1566, *Llangwnnwr* 1600, *Llangunnor* 1671-1993, *Llangynnwr* 1741, *Llan-Gunnor V. St. Gwinnour* 1763, *Llangynnor* 1831
The pers.n. was earlier *Cynfor, Cynfwr* (ETG 184) as the 1283 form suggests though OS maps mistakenly attribute the ch. dedication to St Cynyr. See also Capel Cynnor (Pen-bre). The same p. has the ho.ns. Pistyllgynnwr (**pistyll**) (now Penbryn Cottage SN 419188) and Eisteddfa Gynnwr (now Y Garth SN 452184) with **eisteddfa** 'seat, throne' used topographically for hills or rocks resembling a chair and sometimes associated with the patron saint of a particular p. as in the case of Eisteddfa Egwad in Llanegwad (q.v.) and Eisteddfa Gurig in Llangurig, co. Montgomery.

Llangynog   SN 339163
'church of (St) Cynog': *llan*, pers.n. **Cynog**
Llangonoke 1552, *Llangenocke* 1558, *ll. gynog* c.1566, *Llangynock, Langynock* 1594-5, *Llangonnock* 1598, *llangwnnocke* 1610, *Llangynock, St. Cynog, Chap. to Llanstephan* 1763, *Llangynog* 1831
Cynog may be one the numerous sons of Brychan Brycheiniog dedicated at Llangynog and Merthyr Cynog in co. Brecon, or perhaps Cynog, reputed successor of St Padarn in Llanbadarn, cos. Cardigan and Radnor.

Identical pns. also occur in cos. Montgomery and Monmouth (DPNW 266) but it is not certain that these have the same saint. The apparent absence of early forms may be because Llangynog was formerly a chp. of Llansteffan.

Llanhernin   SN 5421
Llanegwad
'church of Hiernin': *llan*, ?pers.n. **Hiernin**
(chp.) *Lanehernyn* after 1271 (1332), (chp.) *Lanyhervyn* [?*Lanyhernyn*] 1342, (chp. to Llangathen) *Lanyharnyn* 1373, (chp.) *Lannyhernyn* 1400, *Llanherenyn* 1552, *Llanyhayrnin* 1645, *chappell of Llanyhernin* 1703, *Llanhiernyn* 1708, *Llanhirnin* 1836
A lost chp. granted to Talyllychau/Talley abbey c.1200 probably located in Hernin (q.v.) (SN 5421) in the eastern part of Llanegwad p. One source (1373 above) places the chp. in the adjoining p. of Llangathen but later sources favour Llanegwad. Hernin or Hiernin may derive from Heyernin which is thought to be a dim. of Haearn (ETG 186) containing **haearn** 'iron' found in other pers.ns. such as Cynhaearn and Trahaearn (TYP 396 and cited sources). There was a saint and confessor Hoiernin, Hernin in Brittany (LBS III, 281-2; EANC 207). The great variation in historic forms must be due to the unfamiliarity of the pers.n. There are few reliable clues to the precise location of the chp. David Jones (Yr Haul 23: 258) in 1879 states that bones – possibly human bones – were found during building work on a new cowshed at Tŷ-pica fm. (SN 541207) which he identifies as **Capel Gwynllyw**[1]. He adds that St Gwynllyw lived in *Cefn Sant Hernin*, ie. Cefnhernin (SN 53902161) and built a chp. in a valley between Cefnhernin and Twyn (SN 53672195). Unfortunately, he provides no evidence to support this statement. Archaeological proof is also lacking and there are no clues in fn. evidence apart from *Llandraw* (no. 81) on Cefn farm in the tithe apportionment. Neil Ludlow (EMESP) was sceptical of this and suggests that a cropmark (SN 54082138) south-east of Cefn is probably an Iron Age enclosure.

Llanismel, St Ishmael SN 362080
'church of (St) Ismael': *llan*, E *saint*, pers.n. *Ismael*
(ch.) *Sancti Ismaelis* 1141, (ch. *de*) *Sancto Ysmaele*, (ch.) *Sancti Ysmaelis* c.1240, (p. ch.) *Saint Ysmael* 1308, *St. Ishmaels* 1364, 1671, *St. Ismaell* 1565, *Saynt Tismaell* 1566, *ll. ismel* c.1566, *Llanyesmaell* 1573, *Lanysmell* 1597, *Llanishmell* 1686, *Llan Ishmael* 1831

Ismael, or Ishmael in local dialect, is probably derived from *Ysfael* (< OW *Osmail*) rather than the biblical Ishmael (ETG 188). This might have been expected to be represented in W historical sources as 'Llanysfael' with -f- [v] rather than Llanismel with -m- but W evidence before the 16th cent is lacking and it is likely that the pn. has been influenced by *Ishmael*. Nearly all the medieval evidence is found in Latin texts which typically represent the n. of the ch. and p. as *Sancti Ismaelis*, ~ *Ysmaelis* with *-m-* and it seems likely these forms have influenced the pn. Ysfael is a reputed nephew of St Teilo (EWGT 28).

Llanllawddog SN 458294
'church of ?Llawddog or ?Llawyddog': *llan*, pers.n. **Llawddog** or **Llawyddog**
Llanllawothog 1395, *Llanllawddok* 1535, *ll. llowðoc* c.1566, *LL. llowddog* c.1560, *Llanllwythoge* 1608, *Llanllowthog* 1613, *Llanllawddog, St. Lawdog, Chapel to Abergwilly* 1763, *Llanllawddog* 1806

llan and *Llawddog* would regularly produce the form *Llanlloddog*. *Llanllawddog* is more likely if the pn. has lost an unstressed vowel, ie. *Llanllawoddog* (-awodd- > -awdd-) or less likely *Llanllwyddog*, (-wydd- > -wdd-) (note the forms for 1395 and 1608) as Melville Richards suggests (AMR). Llawddog has been identified as the patron saint of Cenarth and Penboyr, and Cilgerran, co. Pembroke (WSS 329, 331, 347). This is unlikely since latinised forms (ch. of) *Sancti Ludoci* c.1191 and *Sancti Ludoci* 1221 x 1229 tend to favour a W form Llwydog (DPNW 269).

Llanlluan see **Capel Llanlluan**

Llan-llwch SN 385188
'church at a pool (or marsh)', *llan*, *llwch*²
Landlothe (*extra Muros Villæ de Carmarthen*) 1230 (16th cent), *Lanlok* c.1299, *Lanlogh* 1302, (chp.) *Llanllogh* 1395, *ll. llwch* c.1566, *Capel Llanlloch* 1578, *Llanllwch* 1710-1831

A former chp. to Caerfyrddin/Carmarthen in 1763 (Ecton 1763: 471) located near the eastern end of a marsh Gors-goch, described in 1811 as 'formerly a large Lake' (CarlisleTD).

Llanllwni SN 487394
'church of (St) Llewenni *or* Llywynni': *llan*, pers.n. **Llewenni* or **Llywenni*
Lanthleweny 1329, *Llanllewony* 1395, *Llanlony* 1490, *Llanllony* 1535, *Llanlloyny* 1560, *ll. llwni* c.1566, *Llanlloony* 1654, *Llan-llwyni* 1831, *Llanllwni* 1889

Late sources attribute the dedication to St Llwni (Ecton 1763: 473) but the only support is in later sps. The proper form of the pers.n. is uncertain (ETG 191). *Llewenni* is possible composed of another pers.n. **Llawen** and suffix *-i* and may be compared with Llewenni, co. Denbigh, with *-i* as a territorial suffix (DPNW 270). The alternative is a pers.n. **Llywenni** containing the pers.n. Llywen (recorded in a 13th cent pedigree as one of the saints of Bardsey/Ynys Enlli). Shift of stress to *-llyw-* before 1490, might account for Llanllywénni > Llanllỳwenni > Llanllwni. The evidence shows that the pn. cannot be Llonio as some late sources suggest, though a Llonio Llawhir of Brittany is recorded 13th cent. The pn. shifted from the vicinity of Maesycrugiau c.1970 to hos. extending along the A485 centred on Talardd Arms PH and the school. Llanllwni is recorded in 1576 as the location of a priory and Edward Lhuyd identified ruins near the church c.1700. There seem to be no early refs. to it, however, and the confusion may have arisen because Caerfyrddin/Carmarthen priory obtained a grant of a ho. near the ch. in 1291 and the advowson of the ch. and land in 1330 (HCrm I, 357, 361).

Llanllyddgen see **Capel Dyddgen**

Llanmilo SN 260087
Llanddowror
Uncertain
Llanmyllo c.1580, *llanmyloe* 1583, *Llan: melon* 1675, *Lanmilo* 1723, *Llanmilo* 1739-1853, *Llanmiloe* 1745, *Llanmiloe, ~ Pentre* 1777
The n. is applied to two hos. Llanmiloe (originally at SN 250088) and Llanmiloebach (SN 245086) (***bach***[1]) and was transferred to the modern ht. largely developed since 1945. There is no evidence of a ch. here which probably rules out ***llan*** as the first el. This would be expected to cause len. in the following *-milo(e)* though its absence could be attributed to E influence, cf. Llanmihangel, co. Glamorgan (PNGlamorgan 126-7). Association of ***llan*** with other els. such as ***glan***, ***llwyn*** and ***nant*** (cf. Llanboidy) is possible but without more evidence it is too hazardous to suggest anything. Richard Fenton identified Castell Llwyd as the old n. of Llanmilo (OPemb II, 306) but there is no conclusive proof. There is nothing to substantiate the suggestion that the pn. was associated with any AN lord of Milo (as ALaugharne 261).

Llannerch SN 513011
Llanelli
'clearing, pasture': ***llannerch***
(place) *Llan arche* 1550, *Llanerch* 1650-1880, *Llanerth* 1709, *Llannerch* 1745, (mess.) *Lenarch* alias *Llanerch* 1779, *Llannerch* 1952-3
The n. of a ho. transferred to a row of hos. Llannerch Cottages on the same site (SN 51340119) before 1880 and a former colliery. Developed for housing after World War II.

Llannewydd, Newchurch SN 384243
'new church': ***llan***, ***newydd***; E ***new***, ***church***,
Eglusnewit c.1140 (17th cent), *Eglusnewith* c.1165, *Egluysnewith* c.1182, (chp.) *Neuchirch* 1395, *Llanneuyth* 1443, *Nove Eccl(es)ie* 1535, *Newchurch* 1551, *yr eglwys newyδ* c.1566, *Llannewydd* 1754
Earlier forms show that ***llan*** has replaced ***eglwys*** but the two forms run concurrently in 15th-16th cent sources. The E form probably derives directly from Eglwys Newydd. Richards (ETG 193) believed that Llannewydd/Newchurch was alternatively known as Llanfihangel or Llanfihangel Croesfeini but see Capel y Groesfeini.

Llan-non SN 539084
'church dedicated to Non', ***llan***, pers.n. ***Non***
LLann Onn c.1500, (manor) *Llannon* 1529, *LL. nonn* c.1550, *Llanon* 1556-1831, *ll. onn* c.1566, *Llannon* 1561-1831
Our written historical evidence for Llan-non is late which may be attributed to its earlier status as a chapelry to Llanelli. Parts of the ch. are medieval but we do not know when it was first attributed to St Non. Some earlier sps. appear to suggest that the second el. could be ***onn*** 'ash-trees' although the combining of ***llan*** with plant-ns. is rare. All we can be certain is that popular association of Llan-non with St Non had taken root by c.1700 and that the saint's feast-day was kept on 2 March. Non is said to be mother of St Dewi/David in a medieval pedigree 'Bonedd y Saint', the earliest manuscript of which dates c.1320 (EWGT 54-5). In the Latin life of St David by Rhygyfarch (d. 1099) her name is given in latinised form as Nonnita which may be an early variable form of Nonna, either hypocoristic or diminutive (NJacobs). This seems likelier than the argument that *Non* is a short form of a male pers.n. latinised as *Nonnita* (CAS 49: 5-15). Non was otherwise dedicated at the former chp. of Llan-non, co. Cardigan (PNCrd 729-730).

Llanpumsaint SN 418290
'church of the five saints': ***llan***, **(*y*)**, ***pump***, ***sant*** pl. ***saint***
Llanypynsaynt 1535, *Llampemsaunt* 1547, *ll. y pymsaint* c.1566, *llanipimsaint in elvet* 1608, *Llanypymsaint* 1683, *Llan y Pimp Sant* 1765, *Llanpumsaint* 1782, *Llanpumsaint* 1831
A former chp. to Abergwili and located in the hd. of Elfed (q.v.). Traditionally dedicated to five brothers Gwyn, Gwynno, Gwynoro, Celynnin and Ceitho who are said to have been abducted by a magician to caves to sleep and wait until the return of the legendary Arthur or until a saintly bishop should come to the bishopric

(OPemb IV, p.410; ETG, p.194). Cf. Pumsaint. In 1710 it was said that Llanpumsaint had five wells or pools in which the individual saints were said to have bathed. In summer time local people visited the wells to wash in them in order 'to cure aches'.

Llansadwrn　　　　　　　　　　SN 695314
'church of (St) Sadwrn': *llan*, pers.n. **Sadwrn**
Laisadurn 1229, (chp.) *Lansadurn* 1331, *Lensadorn* c.1285, (*o*) *lan ssadwrn* early 14th cent, *Lansadorn* 1308, *Lansadourn* 1335, *ll. sadwrn* c.1566, *Llansadorn* 1671, *Llansadwrn* 1760, 1831
Cf. Llansadwrn, Anglesey (DPNW 275-6). The dedication of the ch. to Sadwrn is recorded in 1763 (Ecton 1763: 473); he was also known as Sadwrn Farchog (ETG 196) (***marchog***[1]). The ch. was granted to Talyllychau/Talley abbey in 1196 by Rhys ap Gruffudd (the Lord Rhys) of Deheubarth.

Llansadyrnin (Llansadurnen)　　SN 281102
'church of (?St) Sadyrnin': ***llan***, pers.n. *Sadyrnin*
Lasedurny 1307, *Lansadornen* 1513, *Lansadurnon* 1531, *Llansadernen* 1535, *Llanvaduren* late 16th cent, *Llansadurnen* 1763, *Llansadurnen* 1889
It is also possible that this is *Launcedurni* 1328 and *Lansadorney* 1401. Nothing is known of any saint by this n. which may account for the ch. dedication being attributed to Sadwrn (Ecton 1763: 470); see Llansadwrn. Sadyrnin is ultimately derived from Latin Saturninus (ETG 186). A 6th cent inscribed stone, recorded by the scholar Edward Lhuyd c.1698, built into the churchyard wall (Corpus II, 261-3) is testimony to the antiquity of the ch. site.

Llan-saint　　　　　　　　　　SN 385081
'church dedicated to (All) Saints': ***llan***, (*y*), ***sant*** pl. *saint*
Ecclesiam Omnium Sanctorum 12th cent, *Halwencherche* 1280, *Halthenchirche* 1319, *llann y saint yng hyd weli* c.1500, *Allhalon church* 1525, *Halgynchirch* 1529, *ll. y saint* c.1547, *Haukinge Churche* 1609, *Cap: Llansant* 1675, *Llan-saint* 1831

The suggested meaning is confirmed by the first ref. to the 'church of All Saints'(in translation) and *the chappell of All Saints* 1609 reflecting its former status as a chp. in Llanismel/St Ismaels (HCrm I, 355-6). The E forms recorded from 1280 down to 1706 (*Haulkin Church*) are less easy to explain. B.G. Charles (LMS I, pt.2) interpreted the first form *Halwencherce* as OE *hāligan*, *hālgan* adj.sg and pl. (acc., gen. and dat.) of **hālig**, pl. *hālge*, 'holy', in the sense of 'saints', and *cirice*. *Hāl(i)gancirice* might be expected to have produced ME *Halwencherche* 1280 but not *Halthenchirche* (probably a copyist's error for *Halchenchirche*) 1319, *Halgenchirche* 1505, *Hawlkyng Churche* 1552 etc. The discovery of two early medieval inscribed stones built into the outer wall of the ch. (Corpus II, 277-80) and a tradition of a third stone in the churchyard has prompted the suggestion that the ch. was formerly dedicated to local saints but this cannot be proven (DPNW 276). Llan-saint appears twice on Saxton's county map 1578 as *Llansant* and *Hawton*. The latter is plagiarised on many later maps and was thought to be a distinctive pn. but it is probably an erratic form of *Halgyn-*, *Haukinge*, *Hawlkyng*, etc.

Llansawel　　　　　　　　　　SN 620362
'church dedicated to (St) Sawyl': ***llan***, pers.n. *Sawyl*
Lansawel 1302, *Lansawyl* 1303, 1473, *Llanysawell* 1535, *Llansawell* 1550, *LL. sawyl* c.1555, *Llannisawell* 1732
Formerly a chp. to Caeo. Many historic forms and current sp. *Sawel* as a variant of *Sawyl* derive from earlier *Sawl* and ultimately Latin *Samuel* (ETG 198). Cf. Llansawel, the W n. of Briton Ferry, co. Glamorgan (PNGlamorgan 21). Local pron. includes 'Llansewyl' though this is not represented in written evidence (DPNW 278). A well dedicated to Sawyl is recorded as *Pistillsawil* after 1271 has been identified with Pistyll Sawyl (***pistyll***) at Pen-y-garn fm. (AC 1915: 332; HWW 27, 166).

Llansteffan　　　　　　　　　　SN 350108
'church (St) Steffan': ***llan***, pers.n. **Steffan** or *Ystyffan*

(*ecclesiam Sancti Stephani de*) *Landestephan* c.1185, *Landestephan* c.1191-1308, *Lanstephan* 1276, (*castellum*) *Sancti Stephani* 1214 (late 13th cent), *Lantstephann* 1347, *Llann Ystyphan* c.1400, *LL. ystyffan* c.1500, *Llanstephan* 1546, 1831, *Lhanystyffant* 1559

The ch. is generally thought to be dedicated to Steffan, variant Ystyffan, a reputed native of Powys, traditionally a friend of St Teilo, dedicated at Llandeilo Graban, co. Radnor (ETG 198) rather than the biblical saint Stephen/Steffan. The -d- found in some historic forms – notable in AN sources – may be intrusive before the prosthetic vowel of Ystyffan or less likely a lenited form of -t- in the honorific prefix *ty-* in unattested Tysteffan (DPNW 280). The area fell into the hands of the Anglo-Normans who constructed the castle and established the borough in the 12th cent.

Llanwinio SN 261264
'church of (St) Gwinio': ***llan, Gwinio***
(chs.) *de Sancto Wynnoco (et de Kylkemara)* 1260, (ch.) *Lanwynnean* c.1291, *Lanwynnyan* 1321, *ll. wnio* c.1566, *Llanwnio* 1600-1674, *Llanwonio* 1592, *Llanwynio* 1619, *Llanwinio* 1653-1831

Gwinio is a dialect form of Gwynio, a familiar form of Gwyn according to Melville Richards (ETG 201), but it is more likely to be a later form of MW Gwyniaw derived from Brit *Uinniau, probably a familiar form of *Uinn(o)barr(os). Historic forms with -*an* are misreadings or miscopyings for -*au* standing for archaic -*aw*; cf. *Lanweneaw* 1378. The first ref. has been misidentified with Llanwenog, co. Cardigan, Llanwinio is mentioned in conjunction with Cilgryman (q.v.). Two early medieval stones found in the churchyard are testimony to the antiquity of the site of the ch. (Corpus II, 264-7).

Llanwrda SN 714315
'church of Gwrdaf': ***llan*, pers.n. *Gwrdaf***
Launwrdaf 1282, *Llanwrdaf* 1303, *Lanurdam* 1331, (chp.) *Llanurda* c.1538, *ll. wrda* c.1566, *Llanworda* 1715, *Llanwrda* 1763

Gwrdaf is unattested as a saint (DPNW 286) and may be a secular person. The false form Cawrdaf appears on OS maps. Described c.1538-1763 as a chp. to Llansadwrn. There is also mention of a mess. *Tythin y Velindre or Cappell Gwrda* 1739 (***tyddyn, capel***). The pers.n. is composed of ***gŵr*** and -***taf*** ('great hero') as in Gwyndaf (***gwyn***) (ETG 201), not ***gwrda*** 'nobleman, lord; good man'.

Llan-y-bri SN 336126
'church of ?Morbri': ***llan*, pers.n. *Morbri***
Morabri 1292, 1369, *Moraburia* 1419, *ll. fair y byri* c.1566, *Morebichurch, otherwise Morbulchurch* 1567, *Marblechurche* 1598, (town) *Morbillchurche* 1611, *Llanbrea* 1675, *Llan-y-bree* otherwise *Marblechurch* 1718, *Llanybri* or *Llanvair y bri* 1733, *Llan y Bri* 1753, *Llan-y-bre* 1889

The suggested pers.n. seems to be otherwise unattested, prompting the suggestion that it is a compound of ***môr*[1] and ***bre*** but that would produce *Morfre, Morfry* with len. (b > f). The development of Llan-y-bri is especially difficult to understand owing to the paucity of early evidence. Melville Richards (WATU) supposed derivation from unrecorded *Llanforbri and that remains the best explanation but it has to be added that the collective evidence before the 16th cent suggests that church (***cirice***) and Llan- are later accretions: *Morabri* and *Moraburia* also have a middle vowel -*a*- although these may be explained as products of latinisation. Perhaps too much should not be made of such flimsy evidence. What is plausible is that hypothetical Llanforbri was re-interpreted by W-speakers as Llanfair-y-bri. Note the forms for c.1566 and 1733 influenced by the more familiar pn. Llanfair with the presumption that the ch. was dedicated to Mair/Mary. Morbri, Morabri appears to have been regarded as a saint by many since the p. possessed a Ffynnon Morbri recorded in a fn. *Parke fonon Morbre* 1672 (***parc, ffynnon***) otherwise known as Ffynnon Olbri (HWW 171). *Marblechurch* and similar forms, frequent from 1534 down to the 18th cent, are re-interpretations of *Morbri Church, Morbrichurch* later influenced by association with E ***marble***. Richards also cites an historic form Llanddewi Forbri but this has not been identified and there is no evidence that Llan-y-bri was ever dedicated to Dewi/David. Llan-y-bri was a chp. (SN 33701255) to Llansteffan. Ruinous in the 17th century, it was replaced by a Nonconformist chp. before 1790. This closed in 1962. Anglican services resumed in Holy Trinity (built 1851).

Llanybydder SN 5243-5244
'church dedicated to the deaf ones' or 'church of the bird-of-prey *or* bittern': ***llan***, ***byddair*** or ***buddair***
Lanbytheyr 1303, *Thlaunebadeyr* 1318, *Lannabedeir* 1319, *Lanybeddeir* 1395, *Llanybyddeyr* 1401, *Llanybydder* 1535, 1611, *ll. y byðar* c.1566, *Llanybyther* 1612-1964
Traditionally interpreted as 'church of the deaf ones' with ***byddar*** nm., *byddair*, though this has been questioned because there is no tradition of any 'deaf saints' here (DPNW 288). There are instances of holy wells recorded elsewhere in Britain and Ireland which were reputed to cure deafness notably at St Winefrides' Well/ Ffynnon Gwenffrewi, Holywell, mentioned by the poet Tudur Aled (c.1470-1525), but no comparable ch. dedications. On these grounds we have to admit the possibility of the rare el. ***buddair***, *byd(d)air* 'bird of prey' and 'bittern' which is probably found in two fm.ns. *Cae byddar isa* [no. 1878], *Cae byddar ucha* [no. 1879] (***cae***) at Rhydafallen-isaf (***rhyd***, ***afallen***, ***isaf***) in the p. of Pencarreg and in a number of topographical ns. This may seem unlikely in combination with ***llan*** but it is possible that Llanybydder derives from earlier and unrecorded Nantybydder with the meaning 'valley of the bird-of-prey *or* bittern'. We can rule out ***nant*** in its more common sense of 'stream' because no r. Bydder has been identified here though ***buddair*** does occur in all likelihood as the main el. in the rn. Bytherig, co. Montgomery, and perhaps *Bydderi the conjectured n. of a stream flowing through Llanbydderi Moor at Llanbethery (Llanbydderi), co. Glamorgan (PNGlamorgan 116). Replacement of ***nant*** by ***llan*** would find a parallel in pns. such as Llancarfan (< Nant Carfan), Llanthony (< Nant Hodni) and Llanboidy (q.v.). Additional support for this interpretation may be found in an older n. for the farm Gwar-graig (SN 512429), close to the village, which is recorded as *Gwar y Graig* or *Prisk y Bythar* in 1718 and *Priskybyddar* in 1722. This is also likely to be *Tir prysk y byddeir ycha* or *Tir prysk y byddeir issa* in 1599 (***tir***, ***prysg***, ***uchaf***, ***isaf***). This has been taken to mean 'copse in a quiet place' but that seems unlikely because the fm. is next to Afon Teifi, a r. not notable for silence and tranquillity. The ch. dedication to Pedr/Peter is a false re-interpretation of -bydder as -bedr.

Llan-y-crwys (Llancrwys) SN 645453
'church of the cross, church of Llanddewi possessing a cross': ***llan***, ***Dewi***, ***y***, ***crwys***
Lanecros c.1291, *Landewi Crus* 13th cent (1331), *Llandewy Crus* 1495, *Llanycrois* 1520, *Lancroys* c.1550, *ll. y krwys* c.1566, *Llanycrwys* 1576-to present, *Rodechurche* 1535
Llan-y-crwys has been identified with *llann de6i y cr6ys* in the poem 'Canu y Dewi' composed by Gwynfardd Brycheiniog c.1176 (early 14th cent). Doubt has been cast on this, however, because it is named among churches located in Brycheiniog, Buallt and Elfael, later in cos. Brecon and Radnor (WSaints DewiGB n.73). That has prompted identification with Llanddewi Fach, co. Radnor, or Llan-ddew, co. Brecon. No early refs. to Llanddewi Fach have been found (the first seems to be *Llandewy* in 1544) probably because it was simply a chapelry of Llywes (also dedicated to Dewi). Llanddewi Fach has also been misidentified with *ecclesia de Landov* in 1402 (StudiaC 46: 122) but this probably refers to the church of Llandrindod, co. Radnor, formerly dedicated to God (***llan***, ***Duw*** earlier ***Dwyw***). The arguments in favour of Llan-ddew, co. Brecon, are largely circumstantial. Historical evidence (*Lando*, *Landu* 1162 (1351), 1242, and *Landou* 1280) shows that the church of Llan-ddew was dedicated to God from c.1191 to 1750 and to the Holy Trinity in 1763. The association with Dewi/David is late and based on a reinterpretation of historic forms such as *Llanthewe* 1595. Whatever the correct

identification of *llann de6i y cr6ys*, Llan-y-crwys certainly bore the fuller n. Llanddewi'r-crwys (in ModW spelling) down to 1495 when it is recorded as *Llandewy Cruys*. The precise significance of the n. is uncertain but Richards (ETG 205) notes that **crwys** was sometimes used to describe someone under a cross such as a corpse with hands crossed on its chest. That may imply that Llan-y-crwys once possessed a notable cross dedicated to Dewi/David. It is also worth noting that a ho. Esgair-crwys (SN 640461), north of the village, is recorded as *Tir Esceir y Krwys* 1633 (**tir**) and *Esgair crwys* 1819 (**esgair**), and located near a short scarp of rock recorded as *Cerrig tair Croes* 1834 (**tair** 'three') and *Careg Crŵys* 1888 (**carreg**). Both Elwyn Davies (GazWPN) and Melville Richards (WATU and CStudies 116-7) prefer the form Llan-crwys with loss of unstressed *y* but most historic forms from c.1530 to the present favour Llan-y-crwys. *Rodechurche* 1535 containing E **rood** 'cross, crucifix' and **church** seems to be unmatched and is probably a learned translation. Llan-y-crwys was a relatively small p. probably carved out of the territory of Trefwyddog which belonged in the early Middle Ages to a monastery at Llandeilo Fawr.

Llanymddyfri, Llandovery SN 767343
'church near (stream called) Dyfri': **llan, am,** rn. *Dyfri*
Lanamdeveri 1194, *Lanamdebery, Lanamzevery* 1222 (1239), *Lanemdovri* c.1250, *Lanamdebery, Lanamzevery* c.1291, *Thlanadevery* 1299, *Landoveroi* 1312, *Lanymdevery* 1383, *Llanymdyfri* c.1400, *LL. ym ddyfri* c.1500, *Llanddyfri* 1738
The development of the pn. is fairly clear: *Llanamddyfri* > *Llanemddyfri* > *Llanymddyfri* through shift of the second vowel -*a*- to -*y*- under the influence of the third vowel -*y*- and association with **yn, ym** 'in' > *Llanddyfri, Llandyfri* (by syncope). The anglicised form Llandovery probably derives from colloquial *Llandyfri represented by historic forms such as *Landyuery* 1540s, *Llandeveroe* 1579 and *Landefry* 1831. The lack of expected len. of -d- > -dd- is paralleled in Llandingad (q.v.). The precise meaning remains uncertain; -*dyfri* has generally been taken to be an unrecorded pl. of **dŵr**, *dwfr* (ETG 206) describing the watery location of Llanymddyfri near the meeting-points of the rs. Tywi, Brân, Gwydderig and Bawddwr (DPNW 231). A rn. is more likely since the boundaries of the old borough are described (in translation) in 1485 as extending 'from the water of Tewy to the water of Devery and in breadth from the water of Fulbroke to the dike of Krenchey'. The last probably ran parallel with Garden Street north of Broad Street (TMW 163), *Ffulbroke* applied to the lowermost part of Nant Bawddwr near the castle and *Devery* must refer to a part of Nant Bawddwr (TCymm 1911-12: 52-53, 59-62) above the meeting-point of Broad Street and High Street. *Devery* is also likely to be the brook *Eueri, Euery* recorded by the antiquarian John Leland c.1537. *Dyfri may have the meaning 'river which spreads or floods': **dŵr**, *dwfr* and suffix -*i*. Both *Ffulbroke* (E **foul, brook**) and its semantic match Bawddwr (**baw, dŵr**) recorded in (meadow of) *Baudour* 1317 would describe a stream which is particularly muddy or polluted. There is no evidence that Bawddwr or any specific area or location bore the n. Amddyfri or Ymddyfri.

Llechach, Afon SN 8025
rn. SN 810254 to Sawdde SN 757241
'rocky (r.), (r.) in a rocky place or valley": **llech**[1], -**ach**
(brook) *Llechach* 1709, (brook) *Nant Lleechach* 1813, *River Llechach* 1887
Rising near Blaenllechach (**blaen**) which meets Sawdde at Aberllechach (**aber**). R.J. Thomas (EANC 15) notes Ir **leacach** which he translates as 'lle caregog' – better still 'area of flat rocks' – and suggests that Llechach is either a cognate W form or (more likely, in his opinion) a borrowing from Ir. Historic evidence is far too late to confirm either interpretation.

Llechdwni SN 428100
Llandyfaelog
'stony land, stony turf': *llech¹*, ?**tonni* len. *donni*
lleghtonny Vcha, ~ Isa c.1500, (place) *Llechetonney* 1549, *llechdwni* c.1550, *Llechdonny* 1590, *Llechdonney* 1662, *Llechdunny* 1737, *Llechdwnny* 1880
**tonni* could be an unrecorded pl. of *ton²* nmf. 'ley; turf, sward', variant *twn²* found in compounds such as *gwyndwn* 'unploughed land'. This is apparently *Llechdoun* 14th cent but the original source has not been identified.

Llech-fraith SN 5124, 5224
Llanegwad
'speckled stone': *llech¹*, *brith* fem. *braith*
(ht.) *Llech fraith* 1710, *Llêch Fraeth* 1811, *Llechfraith Hamlet* 1851
The n. would appear to suggest the existence of a notable perhaps prehistoric stone but neither stone nor n. appear on OS maps. Cf. Llechgron. The t. known as Maenor Llech-fraith to David Jones (Yr Haul 23: 175-6) is centred on a roundish hill (SN 518244) and a second, steeply-sided hill (SN 519236).

Llech-gron SN 5126, 5225
Llanegwad
'round stone': *llech*, *crwn* fem. *cron*
(*chappell of Llanyhernin and*) *Llechgron* 1703, (ht.) *Llech gron* 1710, *Llechgron* 1811, *Lech-gron Hamlet* 1851
The stone has not been identified. There is a Bronze Age stone (HER 663; at SN 51422832) but this is at the northern end of the t. which is centred on Capel Pont-yr-ynys-wen Calvinistic Methodist chp. (SN 530250) in the east of the p. David Jones (Yr Haul 23: 259) describes it as Maenor Llech-gron (*maenor*).

Llethrgele SN 5024
Llanegwad
?'slope notable for leeches': *llethr*, ?*gelau¹*, *gele* nmf. 'leech
(ht.) *Llether gele* 1710, *Llethr Gele* 1811
Known to David Jones as Maenor Llethrgelau (Yr Haul 23: 259) (*maenor*). Centred on a ho.

Llethrgele (SN 506247) recorded as *Llethergelen*, *Llethrgelly* 1839 in the south-central part of the p. The second el. could otherwise be *gelau²* 'blade, spear' but it is difficult understanding how that might apply to the topography. There is no r. Gele here to prompt comparison with the r. Gele, co. Denbigh, recorded in Abergele (DPNW 7).

Llidiardnenog SN 546374
Llanfihangel Rhos-y-corn, Llanybydder
'gate near (r.) Nenog': *llidiart*, rn. **Nenog**
Llidiart-Nenog 1831, 1881, *Llidiadnenog* 1869, *Llidiardnenog* 1874, *Llidiartnenog* 1904
Located on a ridgeway at the head of Afon Nenog (q.v.). The gate may have marked the uppermost limit of enclosed land. The Independent chp. is known as *Llidiadnenog* 1833 and *Capel Nenog* 1906 (*capel*). The location of *Cross Inn* in 1888.

Lliedi, Afon SN 5104
SN 516085 to Afon Llwchwr/River Loughor SS 501985
?'muddy river': ?*llaid* 'mud, dirt, clay', suffix *-i*
the Aberlheddie water 1587, *llydey* 1602, *Lheidi River* c.1700
*Lleidi > Lliedi with inversion of -ei-. R.J. Thomas (EANC 157) suggests *llied- containing *lli-* as in *lliant* 'flood, flow, sea' and Llanllieni, the W n. of Leominster, co. Hereford (PNHer 17-18, 137), but he was unaware of the first historic form and does not explain -ed- which is presumably the suffix *-ed³* (in addition to *-i*). The ho. Blaenlliedi (SN 509083) recorded as *Blaen-llihedi* 1831 (*blaen*) is about 0.5km below its source on Mynydd Sylen.

Llwchwr, Afon : River Loughor SN 5701, 6007, 6213, 6617
'bright, shining (r.)': Brit **leuco-*, suffix *-ar(a)*
(r.) *Leuca, Leucaro* 1-4th cent (8-12th cent), (r.) *Luchur, lychur* c.1145, *Locher* c.1191, (water) *Lozcharne* 1306
Brit *Leucara* would regularly have developed into **Llugar* rather than Llychwr or Llwchwr but Rivet and Smith (PNRB 388-9) suggest that the normal phonetic process may have

been disturbed by analogy or folk-etymology. It is possible that it has been influenced by **llwch**. The n. has been anglicised as Loughor with current pron. among E-speakers as approximately [ˈləχor]. The r. rises at Llygad Llwchwr (SN 673180) (**llygad** 'eye; source of river') and meets Bae Caerfyrddin/Carmarthen Bay between Porth Tywyn (q.v.) and Whiteford Point (SS 4496), Glamorgan. The rn. is discussed more fully under Loughor in DPNW 302 and PNGlamorgan 134.

Llwyfan Cerrig station see **Gorsaf Llwyfan Cerrig**

Llwynhendy　　　　　　　　　　SS 540997
Llanelli
'grove at Hendy': **llwyn**, ho.n. Hendy
llwyn hendu 1553, (tmt.) *Lloyn hendye* 1588, (place) *Llwyn hendy* 1604, *Llwynhendy* 1709-1839, *Lloynhendy* 1752, *Llwyn-r-hendy* 1833, *Llwyn-hendy* 1879-80
The n. appears to be drawn from a ho. and fm. (SS 537996) but was transferred to the industrial ht. which developed along the Swansea-Caerfyrddin/Carmarthen road and two roads – now known as Parc Gitto and Heol Hendre – which link near the fm. Hendy, 'old house' (**hen**, **tŷ**) has not been identified.

Llwyn-swch　　　　　　　　　　SN 5117
Llanddarog
?'pointed grove, grove shaped like a ploughshare': **llwyn**, **swch**¹
(place) *llwyn swch* (etc. *kynyll mawr*) 1602, (ht.) *Lloynswch* 1703, *Llwynswch* 1804, 1845, (ht.) *Llwynys Uwch* 1811
Centred on Llwyn-swch a former ho. (SN 5171750 recorded as *Llwynswch* 1826 and (homestead) *Llwyn Swch* 1847 – but the site of a quarry by 1888 – and (ho.) *Blaen-swch* 1831 (SN 508172). There is a group of trees (SN 510175) near Blaen-llwyn-swch on OS maps (**blaen**) but this is nearly 0.75 km west of the site of Llwyn-swch. A hill between the two hos. is notably pointed at its western and eastern ends.

Llwyn-teg　　　　　　　　　　　SN 552081
Llan-non
'fair grove': **llwyn**, **teg**
Llwynteg 1851, *Llwyn-teg-Chapel (Independent)* 1878-9, (area) *Llwyn-têg* 1906, 1960, *Capel, Llwyn-têg* 1988
Probably a new n. coined at the time the chp. was erected in 1845 on former open moorland Mynydd Bach, recorded as *Mynydd Bach Llannon* 1813 (**mynydd**), later partly afforested.

Llwynyrebol　　　　　　　　　SN 132261
Llandysilio
'grove of the colt *or* foal': **llwyn**, **yr**, **ebol**
(gr.) *Lloyneryboll*, (mill) *Lloynerebell* 1535, (gr.) *Lloynrevel* 1582, *Lloinereball* 1605, *lloyne yr Eboll Grange* 1617, (the grange of) *Llwyn yr Ebol* 1652, (tmt.) *Llwyn Rebol* 1701, (ho.) *Llwyn-yr-ebol* 1889
The grove may refer to the wooded slopes above r. Cleddau Wen/Eastern Cleddau recorded as *Allt Llwyn-yr-ebol* 1889 (**allt**). A former gr. with a water-mill belonging to Hendy-gwyn/Whitland abbey (WCist 314).

Llyn Brianne　　　　　　　SN 7949, 7950
resr.
'lake at (Nant) Brianne': **llyn**, rn. Nant Brianne
A reservoir authorised by the West Glamorgan Water Board (Llyn Brianne) Order 1968 and officially opened in May 1973. Located partly in Llanfair-ar-y-bryn, co. Carmarthen, Caron, co. Cardigan, and Llanwrtud, co. Brecon. Nant Brianne rises at SN 779503 and meets the reservoir at SN 795496. The n. is recorded as *Nant Brianne* 1887 and later OS maps. The rn. has been supposed to mean 'stream in the hills' on the assumption that it is a misspelling of Nant y Bryniau (**bryn** pl. *bryniau*) but it is more likely **breuant** 'throat' pl. *breuannau* which would describe narrow gorges or defiles (DPNW 296). This also seems better than **breuan** 'quern, millstone', pl. *breuanau*. The development -eu- > -ei- > i- and final -au > -e in either case is colloquial.

Login SN 164233
Cilymaenllwyd, Llanboidy
'dirty (brook)': ***halog**, -yn*
Loggin 1684, *Llogin* 1782, *Login* 1843, *Logyn* 1851, 1889
Loss of *Ha-* is likely to be a result of association with the def.art. *y*; cf. several rns. Login, Logyn (EANC 5 and PNGlamorgan 133) and Lygan, co. Flint (PNF 93). The n. may once have applied to Afon Wenallt which meets Afon Taf (SN 166232) on the south side of the village or the unnamed stream rising above Fro-wen (SN 191243) meeting Taf above the village.

Loughor see **Llwchwr, Afon**

Mabelfyw to Myrtle Hill

Meidrum. Viewed from the south, 1905.

Mabelfyw
Cantref Mawr
'(cmt. of) son of Elfyw': **mab**, pers.n. **Elfyw*
Mabelvew 1257, *Mabelvyw* 1261, (land) *Mabelueu* 1277, (cmt.) *Mab Eluyu* 1292, (ryngildship of *Mabuderud* and) *Mabelveu* 1310, *Mab Elfyw* late 15th cent late 15th cent, *Comm. Mabeluye* 1536-9, *comote of the sonne of Elviw* 1559
Nothing is known of Elfyw and the n. is rare. **mab** may have the sense of 'former prince, leader' (ETG 50). A former cmt. in Cantref Mawr; cf. Mabudrud (q.v.). Lloyd states (HCrm I, 8) that Mabelfyw covered the ps. of Llanybydder and Pencarreg with the ht. of Fforest in Llan-y-crwys.

Mabudrud
Cantref Mawr
'(cmt. of) son of Uderydd': **mab**, pers.n. **Uderydd*
Mabuderith 1257, (cmt.) *Mabuderith* c.1287, *Mabuderud* 1302, (ringildship of) *Mabuderud* (and *Mabelveu*) 1310, *Comm. Mabudrid* 1536-9, *mab vchdrud* before 1564, *the Commut of mabedred* 1604
For pers.ns. containing **mab**, cf. Mabelfyw and perhaps Mebwynnion (PNCrd, 1330-1), co. Cardigan. A former cmt. in Cantref Mawr covering the ps. of Llanfihangel-ar-arth, Llanfihangel Rhos-y-corn and Llanllwni (HCrm I, 8).

Machynys SS 513984
Llanelli
?'plain *or* open-land of (man called) Cynys': *ma-*, pers.n. ?*Cynys*
Machynys c.1145, *Maughenes* 1529, 1536, *Bachannie Iland, Bachannies Ile* 1586, *Bachenis* 1590, *Maughennes, Bach Ynis* 1609, *Machunnis* c.1735, *Machynis Island or Farm* 1785, *Machynys* 1787, 1830
Early historic forms are scarce but favour initial M- rather than B- hence the suggested meaning. Forms with B- account for the suggestion that the n. means 'island shaped like a hook' or 'small island' *bach²* or *bach¹* and *ynys* under the influence of shared len., ie. reinterpreting *ym Machynys* 'in Machynys' as 'in Bachynys'. The variation B-/M- is common; cf. historic forms of Baglan and Machynlleth (PNGlamorgan 10). The association with *ynys* is understandable because Machynys is a former island (LPC viii-x), once nearly separated from the mainland by creeks, later drained by ditches and embanked to prevent tidal flooding. Melville Richards (BBCS 25: 420-1) identified the pers.n. as Cynys which he suggested is also found in Treginnis, co. Pembroke, with *tref*. This was rejected by B.G. Charles (PNPemb 302) who interprets it as 'farm over against an island' with *rhag* and *ynys*.

Macrels SN 294103
Laugharne
?'(place or land belonging to) Maquerel': surname ?*Maquerel*
(stream) *Makerellis* c.1280, *Macrels* 1307, (Thomas *de Rupe*, lord of) *Makerels*, (garden of *William de*) *Makerels* 1330, *Makkereleswallis* 1393, *Roche alias Machrellswalles* 1573
Historic forms suggest that Macrels is possessive favouring Fr Maquerel rather than AFr *makerel* < *maquerel*, 'mackerel' (the marine fish). The n. allegedly refers to a Norman settler Macquerelle. The pn. also appears in the n. of a stream Mackerel Lake (*lacu*) recorded as *Mackwell Lake* 1720, *Macoral Lake* 1758-99, and *Mackerel Lake* 1891. B.G. Charles (NCPN 110; CStudies 127) suggested that forms containing -wallis, -walles may be substitutes for OE *wella* 'a well, spring, stream' but they could as easily represent OE *weall* 'a wall' in ref. to the castle of Roch (q.v.).

Maenor Aberbargoed SN 3440
Llangeler, Penboyr
'administrative unit at Aberbargoed': *maenor*, pn. Aberbargoed
Maenaur Aberbargaud, Maenaur Aber Bargaud 1303
Located in the cmt. Emlyn. The n. must refer to the general area at the confluence (*aber*) where Nant Bargod (q.v.) meets the r. Teifi although no other historic forms have been found.

Maenor Bachsylw SN 165257
Cilymaenllwyd
'administrative unit at Bachsylw': *maenor*, pn. **Bachsylw**
Maynor Baghselowe 1592, *Maynor Bachyselow* 1595, (ht.) *maenor bach y sulw*, (ht.) *Bach y Sulw* c.1700
Cf. Bachynys. Part of Traean Clinton. The ho. is located above a small stream leading down to Afon Taf.

Maenor Bach-y-ffrainc SN 2222
Llanboidy
'administrative unit at Bach-y-ffrainc': *maenor*, pn. **Bach-y-ffrainc**
Maynor bach y frainke 1592, *mainor ~, maynor bach y ffraincke* 1607, *maynor bach y ffrainke* 1637, *the maynor of Bach y ffrainke ycha & bach y ffrainke Issa* 1638, *maynor bach y ffraincke ycha* 1671
Part of Traean March (q.v.).

Maenor Berwig SN 5498
Llanelli
'administrative unit at Berwig': *maenor*, pn. **Berwig**
the Maynor of Burwick (at a place *Penvan*) 1519, *maynor of Burwig yn Llanerch Vorwig* 1533, *maynor de Burwig* 1543, (manor) *Maynor Burwicke; Maner de Burwicke, Mannor of Burwicke* 1609

115

Centred on Berwig (q.v.) and located in lp. of Cydweli 1543. The el. ***maenor*** is applied in some parts of Carmarthenshire to small administrative units equivalent to a township or group of townships. One form has ***llannerch*** 'a clearing, a glade'.

Maenor Betws   SN 6311, 6511, 6811
Betws
'administrative unit in (parish of) Betws': ***maenor***, pn. **Betws**
Stryflond 1323, (manor) *Bettws alias Stryveland,* (p.) *Bethos alias Striveland,* (lands called) *Striveland (contayninge the parishe of Bettws)* 1609
Otherwise known as Stryveland, 'disputed land': E ***strife***, ***land***, although as an E n. this is unlikely to have had any currency beyond legal and administrative records. Maenor Betws appears to have originally formed part of Gower/Gŵyr but the area was disputed between Rhys ap Fychan, lord of Dinefwr, and William de Braose, lord of Gower, in 1252. William is recorded as having built 'a new castle in Gower' (*Novum Castrum de Gower*) which Rhys burnt some time between 1250 and 1271. J. Beverley Smith suggests (Morgannwg 9: 5-10) that this was at Penlle'rcastell (SN 665096) recorded as (place called) *Llercastell* 1609, *Penka'er Castle* 1650, *Penllyr Castell* 1764, *Penclear Cast.* 1777, *Penller Castell* 1785, 'place on a hill with a castle' (***pen¹***, ***lle¹***, ***yr***, ***'r***, ***castell***). The area was probably transferred to Is Cennen in the late 13th cent or perhaps early 14th cent.

Maenor Blaen-cuch   SN 2932
Cilrhedyn
'administrative unit at Blaen-cuch': ***maenor***, pn. *Blaen-cuch*
(*Dotagio de*) *Maenaur Blaen Cuch,* (~) *Maenaurblaenchuch* 1303
In Emlyn Is Cuch. Blaen-cuch (SN 290325) is 'headwaters of (r.) Cuch' with ***blaen*** and rn. Cuch (q.v.) and is recorded as *Blaenkych* 1603, *Blaen-cych* 1831 and *Blaen-cŷch* 1889-90. The ***maenor*** probably extended along the right bank of Afon Cuch towards Cwm-morgan.

Maenor Cadwgan   SN 4715
Llangyndeyrn
'administrative unit of (man called) Cadwgan': ***maenor***, pers.n. ***Cadwgan***
Maynor Cadogan c.1500, (manor) *Maynor Cadogan* 1529, (manor of) *Cadwgan* 1661, (*y glyn'* in manor) *Cadogan* 1598, *Meinor Cadogan, Maynor Cadogan* 1606, *Maynor Cadowgan* 1609
This may also be (quarter of) *Cadugan* 1361 and (manor) *Quartercodogan* 1561 (***cwarter***). Cadwgan is apparently unidentified. The pn. is not current but was probably centred on Cadwgan Fawr (SN 470151) and Blaenau (SN 482151) at the southern end of Llangyndeyrn. Historic forms for Cadwgan Fawr and Cadwgan Fach include *Coed Wogan* 1666, *Coedwgan* 1769 and 1804 and *Coed-Wgan-fâch* 1889 for Cadwgan Fawr (influenced by ***coed*** and pers.n. ***Gwgan***). See also Blaenau³.

Maenor Capel Martin
Traean Clinton
'administrative unit at the chapel': ***maenor***, ***capel***
Maynor cappell m'tyne 1592
Other refs. to this place have not been found. The chp. probably lay in Sanclêr/St Clears near the priory which was a daughter-ho. of St Martin-des-Champs, Paris.

Maenor Castell Draenog   SN 2021, 2121
Llan-gan
'administrative unit at Castell Draenog': ***maenor***, pn. *Castell Draenog*
Maynor Castell drynock 1581, *Maynor castell drynock* 1592, *Maynor Casteldandocke* 1677
The ho. Castell Draenog (SN 208214) is *Castle Draynock* 1680, (capital mess.) *Castell Drinogg* otherwise *Blaen y Dyffrin* 1688 and *Castelldraenog* 1819, with ***castell*** and ***draenog*** either in the sense of 'hedgehog's castle' or perhaps 'castle notable for thorns'. The n. refers to an enclosure, possibly prehistoric (HER 3954) (SN 21292145). Located in Traean Morgan (q.v.). Cf. Castell Draenog in Llanarthne recorded as *Castell y draynog* 1650 (with def. art. ***y***), a n. belonging to a group of pns. combining ***castell***, sometimes in the looser

sense of 'place which has the appearance of a castle, a high mound', with the n. of an animal such as *llygoden* 'mouse' (Mouse Castle alias Castell y Llygod, Llanblethian, co. Glamorgan), *dryw¹* 'wren' (Castell-y-dryw, Llanarthne and Llanddarog), *gwenyn* 'bees' (Castellygwenyn, in Llangyndeyrn) and *gwiwer* (Castell-y-wiwer, in Llanfabon, co. Glamorgan) in a mocking sense for something small or insignificant.

Maenor Castelldwyran   SN 1418
Castelldwyran, Cilymaenllwyd
'administrative unit at Castelldwyran': *maenor*, pn. **Castelldwyran**
Maynor castle dyran 1592, *Maynor Castledyran* 1597
Centred on Castelldwyran and probably coterminous with the chapelry, in the division of Traean March. The n. is not current.

Maenor Castell Madog   ?SN 1223
Llandysilio
'administrative unit at Castell Madog': *maenor*, pn. *Castell Madog*
Maynor castell madock, Maynor Castle Madd' 1592
Castell Madog is 'Madog's castle' with pers.n. **Madog** and **castell**, though he has not been positively identified. The pn. is also recorded in *Escair castell madok* 1581 and *Esker Castle Madocke* 1661 (*esgair* 'ridge') which appears to have been located near Dyffryn (SN 120233) recorded as *Dyffryncaerrhug* 1819, *Dyffryn-cae-rhyg* 1890 (*cae*, *rhyg* 'rye (grain)').

Maenor Cefndaufynydd   SN 3020
Meidrum
'administrative division at Cefndaufynydd'; *maenor*, Cefndaufynydd
Maynor Keven doyvenyth 1592, (2 'maynors' *Trecoz* and) *Keven doy venith* 1602
Probably also *Kevendeneuyth* 1326. Centred on a ho. recorded on OS maps since 1972 as *Cefn Farm* (SN 303200) but earlier (ho.) *Pen-dau-ffynydd* 1831, (fm.) *Cevendoivinidd* 1832, (ho.) *Cefn-dau-fynydd* 1889, 1964, meaning 'ridge at two mountains' (*cefn*, *dau*, *mynydd*). The ho. is located on the western side of a col between two small hills (SN 304203, 307199).

Maenor Cenarth   SN 2641
Cenarth
'administrative unit at Cenarth': *maenor*, pn. **Cenarth**
Maenaur Kenarth 1303
Centred on Cenarth (q.v.) in Emlyn Is Cuch.

Maenor Cilau   SN 1224
Llandysilio
'administrative unit at Cilau'
Maynor Kyllye 1592, *Maynor Killie* 1597
Located in Traean Morgan. Melville Richards (WATU) uses the form Maenor Ciliau but many historic forms favour the pl. form Cilau (*cil¹* pl. *cilau, ciliau*) meaning 'nooks' in ref. to several short valleys above Afon Cleddau. Centred on Ciliau-fach (SN 121241) and Ciliau-fawr (SN 121251) recorded undifferentiated as *Kyle* 1589 and *Kille* 1723 and as *Kile-vach* 1767 and *Killevawr* 1790 (*bach¹*, *mawr*). The n. survives in Cilau (SN 124245) recorded as *Ciliau-uchaf* 1889-90 (*uchaf*) and known earlier as part of *Pentre-cille* 1819 (*pentref*) with Cilau-ganol (*canol*). Cilau-fach (SN 121245) is variously *Cillefach* 1819, *Ciliau-isaf* 1889-90 and *Cilaufach* 1977 (*isaf*). Cilau-fawr (SN 121251) is *Cillefawr* 1819 and *Ciliau-fawr* 1889-90.

Maenor Cilcenawedd   SN 3021
Meidrum
'administrative unit at Cilcenawedd': *maenor*, pn. *Cilcenawedd*
Kilgennawed 1572, *Maynor Kilkenawedd, Maynor Kilkenawed* 1592, *Maynor Kylknawed* (later called *Trecoch*) 1619
Cilcenawedd is probably 'nook of (person called) Cenawedd', *cil¹* and pers.n. **Cenawedd** possibly composed of **cenaw** 'cub, puppy and *gwedd¹* nmf. 'sight, appearance; face', cf. the pers.n. **Blodeuedd** (*blodau* 'flowers, blooms'). If the 1619 description is reliable then it may be a duplicate of Maenor Tre-goch (q.v.).

Maenor Cilellyn SN 1522
Cilymaenllwyd
'administrative unit at Cilellyn: *maenor*, pn. **Cilellyn**
Maynore Kyllellyn, Maynor Kylellyn 1592, *Maynor Killellyn* 1597, (ht.) *maenor Cil Ellyn* c.1700
One of four hts. in Cilymaenllwyd and Traean March, centred on Cilellyn (q.v.).

Maenor Cilgynfyn ?SN 1823
Llanboidy
'administrative unit at Cilgynfyn': *maenor, Cilgynfyn*
Maynor Killgynvyn 1592, (ht.) *Kilgynnydd* c.1700
Cilgynfyn is 'Cynfyn's nook' (*cil¹*, pers.n. *Cynfyn*) referring to Cilgynydd (SN 184235), in Traean March, recorded as *Kilgynnydd* 1692 (1758), *Kilgynnyth* 1738 (1758), *Kilginith* 1741 and *Cilgynidd* 1819. The variation -gynfyn/-gynydd is difficult to explain but it is possible that later forms reflect unfamiliarity with the pers.n. and association with a common pn. el. such as **mynydd** len. *fynydd*. There are close parallels elsewhere. Cilgynfydd, in Llangeler, is *Kilgynvyn* 1563, *Kilgynwidd ycha* 1749 and *Kilgynnydd* 1841 and Cilgynfydd, in Diserth, co. Radnor, is *Kilgynven* 1604, *Kilgyuven* 1626 and *Cilgunfidd-fach* and *Cilgunfidd-fawr* 1833.

Maenor Cilhengroes SN 1921
Llanboidy
'administrative unit at Cilhengroes': *maenor, Cilhengroes*
Maynor Kyll hengroese, Maynor Kilhengroese 1592, *Kilhengroes* c.1700, *Kilhengros Hamlett, Kilhengros* 1738, *Killhengross Hamlet, Cilhengros Uchaf, ~ Isaf* 1844
Centred on hos. Cilhengroes Uchaf (SN 192218) and Cilhengroes Isaf (SN 192216) recorded as *Kilhengros ycha, Kilhengros issa* 1692 (1758), *Cil-hen-rhos-uchaf* and ~ *-isaf* 1889 and 1977. Probably 'nook near the old cross' (*cil¹*, **hen, croes, isaf, uchaf**). Also recorded as *Cilhenroes* 1819 confusing **rhos²** with lenited *groes*. In Traean March.

Maenor Cilhernin SN 1725
Llanboidy
'administrative unit at Cilhernin': *maenor*, **Cilhernin**
Maynor kilhernyn 1592, *Maynor Kylheyrnyn* 1597, (ht.) *Kilhernin* c.1700
Centred on Cilhernin (q.v.) and located in Traean March.

Maenor Cilnawen SN 1822
Llanboidy
'administrative unit at Cilnawen': *maenor*, pn. *Cilnawen*
Maynor kill nawan 1592, (manor or ht.) *Killnawen* 1690, *Kilnawen Hamlett* 1738 (1758), 1738, *Killawen* ~ , *Killhawen Hamlet* 1844
Centred on Cilowen (SN 180226), earlier Cilnawen Uchaf, and the former Cilnawen Isaf (SN 173225) *Kilnawen Issa, Kilnawen ycha* 1692 (1758), *Kilnawen isha, ~ icha* 1735, (tmt.) *Kilnawen* 1717, *Lower Kilnawen* 1800, *Kilnawen genol* 1788, *Cillawenucha, Cillawen-isâ* 1819, *Cilawen Uchaf, ~ Isaf* 1844. Cilnawen has *cil¹* apparently with a rn. recorded as *Nant Nawen* (SN 182221 to Afon Taf SN 165224) probably containing the el. *naw-* found in *dinau* 'to pour, flow; stream out' and *nawes* 'running, flowing' which R.J. Thomas (EANC 32) identified in a rn. Nawe, co. Monmouth, with suffix *-an*. That might describe a r. which pours from its source into its valley misassociated in some sources with **awen¹** nf. 'muse, inclination, desire' and pers.n. **Owen**. In Traean March.

Maenor Clun-tŷ SN 1319
Llandysilio
'administrative unit at Clun-ty': *maenor*, *Clun-tŷ*
Mainor clyntny 1592, *Maynor Clyntuy* 1597
Probably centred on a former ho. Glyn-ty (SN 133191) and nearby Glyn-ty-fach no longer recorded on OS maps, meaning 'house meadow': **clun²** associated in late sources with **glyn** and **tŷ**. Recorded as *Clyn tu* 1556, *Glyn y tuye* 1581, *Klynntw* 1611, *Clyn Ty* 1637-1765, *Clyn y ty* 1686, *Glynty, Glynty-fâch* 1890. Two sources suggest a lost def.art. *y*. In Traean Morgan.

Maenor Cwarter Trysgyrch SN 4512, 4611
Llangyndeyrn
'administrative unit called Cwarter Trysgyrch':
maenor, **Cwarter Trysgyrch**
(manor) *Maynor Quarter Triskirch* 1609, *Quarter triskyrch* 1611; (place) *Maynor dryscyrch* 1648
Cwarter Trysgyrch is 'quarter-part called Trysgyrch' (***cwarter***) perhaps identifiable with unlocated lands held by Hendy-gwyn/Whitland priory recorded as *Tresgrych* 1316, ultimately drawn from the rn. Nant Trysgyrch recorded as *Nant Rhys-crych* 1888 (***nant***) better recorded in the n. of a mill Melin Drysgyrch (SN 460123) at Felindre (***melin, tref***) found on OS maps since 1969 as *Tryscych Mill*. Earlier forms include *Tryscyrch Mille* 1609, *Treskerch Mill* 1644 (1685), mill called) *Truscyrch mill* 1716, *Felin Drysgyrch* 1848, *Drysgeirch Mill (Corn)* 1888. The rn. may be a compound of ***tres***² nm. 'battle, raid' qualified by ***cyrch***¹ adj. 'direct, aggressive' (and nm. 'attack, assault' etc) perhaps describing a stream which was particularly strong, forceful enough to power water-mills.

Maenor Ddwylan Isaf SN 2817
Llanfihangel Abercywyn
'lower administrative unit within two banks':
maenor, dau, ddwy, glan, isaf
Maynor doyllanyssa 1592, (fm.) *Maynor dwlan* 1737, (hos.) *Pen-ar-ddau-lan-isaf* 1831, *Manardd-wylan-isaf* 1889
Both Maenor Ddwylan Isaf (SN 280176) and Maenor Ddwylan Uchaf occupy an area between the rs. Cynin and Dewi Fawr. Maenor Ddwylan Isaf is plain *Manarddwylan* from 1971 on OS maps.

Maenor Ddwylan Uchaf SN 2818
Llanfihangel Abercywyn
'upper administrative unit within two banks':
maenor, dau, ddwy, glan, uchaf
Maynor doyllanycha 1592, (hos.) *Pen-ar-ddau-lan-uchaf* 1831, *Penarddwylan* 1843, *Manardd-wylan-uchaf* 1889
See Maenor Ddwylan Isaf. Centred on Maenor Ddwylan Uchaf (SN 280180). The form *Penarddwylan* may be a false reconstruction of some form such as 'ym Maenor Ddwylan' represented as 'ym Mhenarddwylan. Middle Maenor Ddwylan is and *Manor Farm* 1843.

Maenordeilo SN 677266
Llandeilo Fawr
'administrative unit in Llandeilo': ***maenor***, pers.n. **Teilo**
Meynaur Teylau 1257, *Meynerdelow* 1280, *Maynardeilau* 1288, (cmt.) *Maenaur Teylav* 1303, (rhaglawry) *Meynordelawe* 1337, *Maynordeylo* 1386, *Comm. Maenaur theylū* 1536-9, *Maynor dilo* 1630, *Manordeilo* 1851, *Manordilo* 1947
The n. is elliptical for 'lord's court in Llandeilo' and probably originated as the centre of administration for the cmt. (***cwmwd***) described as *commotum de Lanteilawmawr'* 1222 (1239). It is also described as a cmt. c.1287 and in 1303 as making up that part of Llandeilo Fawr north of the r. Tywi and is described in 1750 and 1790 as divided into two administrative hts. Later sources sometimes confuse the first el. with E ***manor*** hence *Manordeilo* (Landranger 2016). Modern OS maps down to 1967 apply the n. to an area centred on SN 669260.

Maenor Egrmwnt SN 0920
Egrmwnt
'administrative unit in Egrmwnt (Egremont)':
maenor, **Egrmwnt**
Maynor Egermonte 1597
The ***maenor*** probably coincided in area with the p. Other refs. have not been found. Located in Traean Morgan.

Maenor Egwad SN 5121
Llanegwad
'administrative unit of Egwad': ***maenor***, **Egwad**
Maynor egwad 1548
Probably the area centred on the village of Llanegwad and p.ch. dedicated to St Egwad. It is almost certainly identical to Egwad (q.v.).

Maenor Fabon (Manorafon) SN 6523
Llandeilo Fawr, Llangadog
'administrative unit of (man called) Mabon':
maenor, pers.n. **Mabon**

Mainorvabon, Maynor vabon 1317, *y vaenor Vabon* 15th cent, (place) *Maynervabon* 1614, *Mainor-Vabon* 1657, (ht.) *mayner avon* 1665, *Mannor vabon* 1671, *Manorvabon* 1765, (ht.) *Manerfabon* 1804
One of the four subdivisions of the cmt. of Perfedd, in Cantref Bychan (HCrm I, 235), later a ht. or t., centred on Maenorfabon, Llandeilo Fawr. Maenor Fabon is described in later sources as a single manor 'Gwynfe and Fabon' with Maenor Gwynfe (SGStudies 24, 28-29). The pers.n. occurs in early Welsh pedigrees but it is impossible to identify any of them with Maenor Fabon with certainty. Cf. Llanfabon (PNGlamorgan 120).

Maenor Forion SN 394388
Llangeler
'administrative unit of (man called) ?Borion: *maenor*, pers.n. ?**Borion*
Maynornoreen' 1215, *Maynoruorion* 1309, *Maynorvoreon* 1315-6, (gr.) *Man'verian*, (mill) *Man'voirian* 1535, (gr.) *Courtmanarborion* 1562, (gr.) *Manorvorion, Court manor vorion* 1591, mess. etc) *Tîr Courte Manervorion* 1672, *Manaravon* 1831
This probably originated as a pre-conquest *maenor* centred on Llysnewydd and may have been co-terminous with the p. of Llangeler (MWST 59-60). The smaller area known as Cwrt Maenorforion and Grange was acquired by Hendy-gwyn/Whitland abbey (WCist 182, 192, 314) and centred on Pentre-cwrt (q.v.).

Maenor Fouwen ?SN 4916
Llanarthne, Llanddarog
'administrative unit of (man called) Mouwen': *maenor*, ?pers.n. **Mouwen*
Maynor vauwen c.1501, *Manervoyon* 1574, (manor) *Maynor Vouwen, Maner de Fouwen* 1609
Apparently centred on and evidenced in Trefreyan (q.v.) in Llanarthne. Rees (SDL 174) identifies it with an area north of Llanddarog extending towards Llanarthne.

Maenor Frwnws SN 4920
Mainaur brunus c.1145
'administrative unit of (man called) Brwnws': *maenor*, pers.n. **Brwnws**
The area around Llandeilo'r-ynys (q.v.). The pn. first loses **maenor** and **Brwnws** survives only as a qualifier and is confused with **yr** and **ynys** in very late sources.

Maenor Gain SN 3816
Llan-gain
'administrative unit of (man called) Cain: *maenor*, pers.n. **Cain**
Mainergein 1288, *Maynaur Keyn* 1292, *Maynergeyn* 1395, *Maenor gain* c.1550
The n. may be elliptical for the lord's court at Llan-gain (q.v.) and they are apparently co-terminous.

Maenor Garllegan SN 2822
Meidrum
'administrative unit at Garllegan': *maenor*, ?rn. *Garllegan*
mainor garllegan 1584, *Maynor Garllegan, Maynor Garthllegan* 1592, *Maynor Garllegan* 1685
This is also likely to be *Kyllegan*' recorded in Meidrum in 1326 though it is a little difficult reconciling this form with Garllegan. The latter survives in two ho.ns. recorded as *Clift Side or Gerllegan* 1762, *Garllegan-fawr, ~ -fach* 1831, *Garllegan-fawr, ~ -fâch* 1889. These favour *garlleg¹* 'garlic' and suffix *an* possibly a lost n. for two small tributaries of Afon Dewi Fawr with the sense 'place *or* river where there is garlic'. Wild garlic typically grows on the shaded floor of old woodland. Both streams have small woods adjoining their upper courses. It is worth stressing, however, that historic forms are late and *garlleg¹* appears to be a very uncommon el. in pns. The more common term for wild garlic is *craf* as in Aber-craf, co. Brecon (DPNW xxxvii, 3, 134, 298) and it is possible that *Garllegan* is a reinterpretation of an unidentified el..

Maenor Glynystyn SN 5212
Llanarthne, Llan-non
'administrative unit at Glynystyn': *maenor*, Glynystyn
maynor Glynysty 1541, (manor) Mainor glin 1598, (manor) Glynne 1610, (manor) Glyn Ysti 1613
Glynystyn (SN 526125) is located in Llanarthne but the n. must have also applied to a place on the south side of Gwendraeth Fawr in Llan-non recorded as *Glan Asty* near the brook *Dynant* 1601 and *Glynystun alias Glynystyn* 1609. Dynant was one of several small streams that unite and join Gwendraeth Fawr (SN 517118) below Twmbl/Tumble leaving its n. in (hos.) *Dynant-fawr* (SN 225119), *Dynant-fâch* and *Dynant Colliery* 1880. The first el. is clearly **glyn** typically describing a narrow valley and the second el. may be ***estyn¹***, *ystyn¹* 'extension, a stretching, prolongation' though the precise sense is obscure. It could simply describe the area extending on both sides of the course of Gwendraeth Fawr. Melville Richards (WATU) misplaces Maenor Glynystyn in Llangyndeyrn, possibly because the former forest of Glynystyn extended from Mynydd Mawr (SN 541129) in Llan-non p. to the brook Carwe/Carway.

Maenor Grongar ?SN 5515
Llanddarog
'administrative unit at Grongar': *maenor*, ?pers.n. *Gryngar*
maynor gryngar c.1501, (manor) *Maynor Grongar, Maner Gryngar* 1609
Richards (WATU) gives *Gyngar* as an archaic form but this has not been traced. A pers.n. Cyngar is certainly recorded in medieval sources but this can hardly be linked to Maenor Grongar for which historic forms consistently have *Gryn-, Gron-*. Rees (SDL 174) maps it as an area in the south of Llanddarog extending towards Llanlluan. There is no connection with the hill and prehistoric site *Gron-gaer* (SN 574216) in Llangathen recorded as *the Grongaer* 1613, *Grongar Hill* 1811, *Grongaer, ~ Hill, ~ West, ~ East* 1831, probably 'round fort' (**crwn, cron, caer**).

Maenor Gwempa SN 4311
Llangyndeyrn
'administrative unit at Gwempa': *maenor*, Gwempa
(place called) *Maynor gwempa* 1597, *Maynor gwempy* 1611
Centred on Gwempa and Ffrwd (SN 455126).

Maenor Gwernolau SN 2123, 2124
Llanboidy
'administrative unit at Gwernolau': *maenor*, pn. Gwernolau
Maynore gwerne olei 1592, *Maynor Gwerne ole* 1592
Centred on hos. Wernoleu Fawr (SN 212241) and Wernoleu Fach (SN 215239) recorded as *Wernole fach, Wernoleu Fawr* 1692 (1758), *Wernolau* and *Wernolaufach* 1819, *Wern-oleu-fawr* and *Wern-oleu-fâch* 1889 and undistinguished as (mess.) *Gwernoe Oley* 1696, *Wernole* 1728. Gwernolau is 'bright alder-tree marsh' (**gwern, golau**) probably with the specific sense of 'sunlit alder-tree marsh'. Both hos. are on south-facing slopes.

Maenor Gwynfe SN 7221
Llangadog
'administrative unit at Gwynfe': *maenor*, Gwynfe
Maynor Wenuey 1317, *y vaenor ôinvei* 15th cent, (tmt. and lands in) *Manor Winvei* 1543
One of the four subdivisions of the cmt. of Perfedd, in Cantref Bychan (HCrm I, 235). Probably centred on Capel Gwynfe (q.v.) and identical in area to (ht.) *Gwinvey* 1705. Maenor Gwynfe is described in later sources as a single manor 'Gwynfe and Fabon' with Maenorfabon (SGStudies 24, 28-29).

Maenor Gynnwr ?SN 4219
Llangynnwr
'administrative unit in Llangynnwr': *maenor*, pers.n. **Cynnwr**
Maynorgonnor 1552, (manor) *Maynor Gonnor, Mannor Gonnor* 1609, *mannor Gonnor* 1634
Cf. Maenordeilo. The use of a saint's n. is also employed as a simple qualifier in Eisteddfa Gynnwr (on the site of a modern ho. Y Garth

SN 452184 in Nant-y-caws) recorded as *Eisteddfa* 1851, *Eisteddfa-Gwnwr* 1888-9, with *eisteddfa* 'seat, stool, chair'. The same el. occurs elsewhere notably in pns. in Eisteddfa Egwad (SN 535240, Llanegwad) and Eisteddfa Gurig (Llangurig, co. Montgomery), often possessing fanciful associations with the patron saint of the p. ch. (ELl 25). The antiquarian Richard Fenton states that Eisteddfa Egwad was 'an antient Mansion ... where the Saint of the Church ... might have lived' (Fenton 77).

Maenor Hengoed SN 5003, 5103
Llanelli
'administrative unit at Hengoed': *maenor*, **Hengoed**
(manor of) *Hengoed* 1542, 1571, (manor) *Maynor Hengoed, Mannor of Hengoed* 1609
Centred on Hengoed (q.v.).

Maenor Henllan Amgoed SN 1820
Llanfallteg
'administrative unit at Henllan Amgoed': *maenor*, **Henllan Amgoed**
Maynor Henllanamgoed 1592
Centred on the chp. or church. In Traean Morgan (q.v.).

Maenor Hernin see **Hernin**

Maenor Iddole see **Iddole**

Maenor Iscoed ?SN 3811
St Ishmael
'administrative unit at Iscoed': *maenor*, *Iscoed*
Maynor yscoed c.1501, (manor of) *Iscoid* 1504, (manor) *Maynor Iskoed* 1609, *Maynoyschoed* 1714 Probably centred on Iscoed otherwise known as Iscoed Morris.

Maenor Llanddeusant SN 7724
Llanddeusant
'administrative unit at Llanddeusant': *maenor*, **Llanddeusant**
Maynor Lanadeusant 1317, *y vaenor Lann y Deussant* 15th cent
One of the four subdivisions of the cmt. of Perfedd, in Cantref Bychan (HCrm I, 235)

Maenor Llanedi SN 5806
Llanedi
'administrative unit at Llanedi': *maenor*, **Llanedi**
(manor) *Llaneddy* 1529, (manor) *Maynor Llaneddy* 1609
Presumably centred on the p. ch. and village.

Maenor Llanfihangel SN 6331
Talyllychau/Talley
'administrative unit at Llanfihangel': *maenor*, **Llanfihangel** (Talyllychau/Talley)
(place) *Manor llanvihangell* 1584, *mainor llanvihangell* 1596
Later forms have not been identified.

Maenor Llanglydwen SN 1826
Cilymaenllwyd
'administrative unit in Llanglydwen': *maenor*, **Llanglydwen**
Maynor llanglodwen 1592
Located in Traean March.

Maenor Llangoedmor SN 5496
Pencarreg p.
'administrative unit at Coedmor': *maenor*, **Coedmor**
Maennaur Lancoedmaur 1303
The *maenor* does not seem to be recorded elsewhere but was probably centred on Coedmor.

Maenor Llangynin SN 2519
Llangynin
'administrative unit at Llangynin': *maenor*, **Llangynin**
(manor) *Llangynyn* 1570, *Maynor llangenyn* 1592
Centred on the p. ch. and village. Cf. Maenor Llanglydwen. In Traean Clinton.

Maenor Llan-non SN 5809
Llan-non
'administrative unit at Llan-non': *maenor*, **Llan-non**
(manor) *Llannon* 1529, *Llannon alias Maynor Llannon* 1562, (manor) *Maynor Llanon* 1609

Centred on the p. ch. and village. Cf. Maenor Llanglydwen.

Maenor Llansadwrn   SN 6931
Llansadwrn
'administrative unit at Llansadwrn': *maenor*, **Llansadwrn**
(land) *Maenaur Lansadurn* 1324, ('maynor' in) *Lansadour* 1335
Centred on the p. ch. and village.

Maenor Llansawel   SN 6236
Llansawel
'administrative unit at Llansawel': *maenor*, **Llansawel**
Maynahur Lansawil 1257, *Maynaur Lansawyl* 1265, *Maynaur Lan Sawyl* 1303
Centred on the p. ch. and village. in the southernmost part of the cmt. of Caeo,

Maenor Lleision   SN 3320
Meidrum
'administrative unit of (man called) ?Lleision': *maenor*, pers.n. ?*Lleision*
Maynor llaysshionen, Maynor llaysshyon 1592
The second el. is probably Lleision as Melville Richards supposes (WATU) but confirmatory evidence is lacking. The *maenor* was centred on Llysonnen (SN 332205) formerly Llysonnen-isaf adjoining Llysonnen-uchaf (*isaf, uchaf*). Historic refs. including *Llysonnenucha, Llysonnenissa* 1808 and *Llys-onen-uchaf, Llys-onen-isaf* 1831 are presumably reinterpretations as 'court at an ash-tree' (*llys, onnen*). Located in the lp. of Ysterlwyf.

Maenor Llys
Llandeilo Fawr
'administrative unit at the court': *maenor*, *llys*
Maynor llese c.1500, (manor) *Maynor Llys, Mannor of Lleys* 1609, *Mannor Lleyes* c.1619
Either an administrative unit under the direction of a seignorial court or more likely a specific location though this is unidentified.

Maenor Meddyfnych   SN 6213
Llandybïe
'administrative unit at Meddyfnych': *maenor*, *Meddyfnych*
mainaur med diminih before 830 x 850, *Maynor Veth Venigh, Manner of Methenvigh* 1609
The n. is taken from Meddyfnych shown as Myddyn-fych (SN 629134) on the OS Landranger as part of Rhydaman/Ammanford. This is (i) *Feddyfnych* c.1380 (c.1400) in an edited text, *tir Meddyfnych* 1651, *Mathinvach* 1740, *Muddunfych, Myddinfich* 1753, *Myddenfych* 1831, probably meaning 'place, land of (person called) Dyfnych': *ma-*, pers.n. **Dyfnych* (BBCS 7: 370). The pers.n. does not seem to be independently recorded.

Maenor Meibion Seisyll   ?SN 6440
Is Cennen ctf.; Caeo cmt.
'administrative unit of the sons of Seisyll': *maenor, mab* pl. *meibion*, pers.n. **Seisyll**
Maynaur filiorum Seysild 1257, *Mainor meyboncesyl, Mainor Meibon Scesyl* 1280, *Maennaur Meybon Seyssill* 1303
Seisyll was a fairly common pers.n. in medieval Wales, eg. in Seisyllwg (*-wg*). There seems to be no evidence to identify this person. The precise location is uncertain but it included Cwmblewog (see Cwmliog) and Penarth in Caeo p. in 1303.

Maenor Myddfai   SN 7730
'administrative unit at Myddfai': *maenor*, **Myddfai**
Maynor Methevey 1317, (rent) *Meynor Methevey* 1318, *Maenawr Vydduei* c.1400, *y vaenor Vyduei* 15th cent, *Maenawrvydhey* 1559
One of the four subdivisions of the cmt. of Perfedd, in Cantref Bychan (HCrm I, 235). See Myddfai.

Maenor Pencarreg   SN 5345
'administrative unit at Pencarreg': *maenor*, pn. **Pencarreg**
Maennaur Lancoedmaur et Penkarrec 1303
Centred on Pencarreg. See also Maenor Llangoedmor.

Maenor Penrhyn SN 4802
Pen-bre
'administrative unit at Penrhyn': *maenor*, pn.
Penrhyn
quarter of *Peynryn* 1361, (manor) *Penryn* 1529, (~) *Penryn* 1543, (~) *Penryne* 1555, (~) *Pennryn* 1571, (~) *Maynor Penryn, Mannor Penryn, Maenor Penrin* 1609, (ht.) *Penrhyn* c.1700, *Pendryn Hamlet* 1792
Centred on Penrhyn (SN 482021) referring to the projecting hill between Cwm Mawr and the coast at Pwll.

Maenor Pen-y-cnwc SN 2222
Llanboidy
'administrative unit at Pen-cnwc': *maenor*, pn.
Pen-y-cnwc
Maynor Pen y knowck 1581, *Maynor Penknock* 1592
Pen-cnwc (SN 222226) is (mess. etc) *Place Pen y Knwcke* (, 2 mess. etc *Troed yr Rhew* and *Drevach*) 1636, *Pen y Cnwch* 1738, *Pen-cnwc* 1819, *Penycnwc* 1739, (mess.) *Penyknwch* 1745 meaning 'top of the hillock' (*pen¹*, (*y*), *cnwc²*). Located in the division of Traean Morgan (q.v.) and later in Bach-y-ffrainc ht.

Maenor Rhiwtornor
Llanfihangel Abercywyn or Meidrum
'administrative unit at Rhiwtornor': *maenor*, ?pn. *Rhiwtornor*
Maynor Riwtornor 1592
Located in Ysterlwyf in 1592 but its precise position is undetermined. Further evidence is clearly needed but Rhiwtornor is likely to be a pn. or topographical n. containing *rhiw*; it is too hazardous to suggest what -tornor stands for.

Maenor Rhwng Twrch a Chothi SN 6541, 6641
Caeo
'administrative unit between (r.) Twrch and (r.) Cothi': *maenor*, *rhwng*, rn. **Twrch¹**, *a*, rn. **Cothi**
Mainiar between Turch and Kethy 1257, *Maynaur inter Turch et Kothy* 1265, *Maennaur Turch et Kothy* 1303

Later refs. have not been found and it is difficult to provide a precise location save that it applied to an area extending northwards between the two rs. which meet near Pumsaint and Dolaucothi. Glanville R. Jones (StC 28, 89-90) says that it was made up of four subdivisions or *gwestfâu* called Blaen-twrch, Maesllangelynnen, Cwm-twrch and Cwmcothi.

Maenor Tal-y-fan SN 1819
Llan-gan
'administrative unit at Tal-y-fan': *maenor*, pn. *Tal-y-fan*
Maynor tall y van 1581
Tal-y-fan is otherwise *Place tall y van* 1581 in Traean Morgan referring to a house or place (*plas*) in Tal-y-fan identifiable with the ho. Tal-y-fan (SN 18661925). The ho. is recorded as *Tall y van* 1701 and *Talvan* 1839. Located on the north-west slopes of a low hill between two stream probably meaning 'end of the peak' with *tâl²*, *y*, and *ban¹*.

Maenor Tegfynydd SN 1420
Llanfallteg
'administrative unit at Tegfynydd': *maenor*, pn. *Tegfynydd*
Maynor teg venydd 1592
Located in Traean Morgan in 1592. Tegfynydd is 'fair mountain' (*teg*, *mynydd* len. *fynydd*) recorded in (tmt.) *Tegvynydd* 1581, *Tegvynydd ycha* etc. *Tegvynydd Issa* 1614, (mess. etc) *Tegfynyth ycha* 1670, *Teggyvynydd, ~ Mill, Upper Tegyvynydd* 1774, *Tegfynydd* 1788 (*isaf*, *uchaf*, *mill*).

Maenor Tellwn
'administrative unit at ?': *maenor*, ?pn. ?
Maynor tallvom, Maynor Tellvon, Maynor Tellwn 1592
Apparently located in Traean Clinton which covered parts of Llangynin, Cilymaenllwyd and Llanfallteg in the western part of the barony of Sanclêr/St Clears but the apparent absence of later refs. prevents identification of a precise position.

Maenor Tre-dai SN 1519
Llanfallteg
'administrative unit at Tre-dai': ***maenor***, pn. *Tre-dai*
Maynor treyday, Maynor treydaye 1592, *Manor Treday* 1779
Centred on Tre-dai recorded as *Treday* 1819, *Tre-Dai* 1890, 1964 which appears on OS maps as *Brynglas* from c.1973. The first el. appears to be ***tref*** but *-dai* is uncertain. It seems unlikely to be the commonplace ***tai*** 'houses, buildings'. Located in Traean Clinton.

Maenor Tregelyn SN 2921
Meidrum
'administrative unit at Tregelyn': ***maenor***, pn. *Tregelyn*
Maynor Tregellin 1592, (manor) *Tre gelyn* 1602, (ht.) *Tregelin* 1697
Tregelyn is 'settlement near or among holly-trees' (***tref***, ***celynnen*** pl. ***celyn¹***) and was the area around Rhosyn-coch (SN 298212) near Drefach. Rhosyn-coch may be identified as the location of *Tithen capell kenock* 1602, *Tyr Cappell kynock* 1697, *Tyr cappel Kunnock* otherwise *Rhossin Coch* 1762 (***tyddyn***, ***tir***). Capel Cynog itself is *Cappell Kynock* 1661 (***capel***, ***Cynog***) and the scholar Edward Lhuyd records a well which reputedly cured 'the hag's fever' (*Kryd y wrach*) (***cryd***, ***y***, ***gwrach***). See also Merthyr Cynog.

Maenor Tre-goch SN 3018
Meidrum
'administrative unit at Tre-goch': ***maenor***, pn. *Tre-goch*
Maynor Trecoz 1592, ('maynor') *Trecoz* 1602, (mess. etc *Tyr Croes Eyno(n)* in) *the maynor of Trecoz* 1650
Centred on a fm. Croeseinon (SN 303184) recorded as *Crosseynon* 1841 (***croes***, pers.n. ***Einion***) otherwise known as *Lan-y-gors-uchaf* 1831-1953, *Lanygors Uchaf* 1971-2. Lan-y-gors Uchaf (*Lanygorse ucha*) is recorded on the tithe map of Meidrum in 1841 as adjoining *Lanygorse Issa* (SN 304182) (***glan***, ***y***, ***cors***, ***uchaf***, ***isaf***). Tre-goch is apparently 'red-brown settlement, settlement in an area of red or brown rocks or vegetation' (***tref***, ***coch***) but historic forms recorded 1592-1650 do not have len. (g for c). It is possible that Maenor Tre-goch 1592-1650 ch extended as far as Pentre-goch (SN 302218) which is recorded as *Pentre-goch* 1831 and *Pentre-côch* 1889 (***pentref***, ***coch***).

Maenor y Llan SN 4514
Llangyndeyrn
'administrative unit associated with the church': ***maenor***, ***y***, ***llan***
(place) *Maynor y llan* 1597
Probably (place) *the Llan* 1606 referring to the area around Llangyndeyrn p. ch. and village.

Maenor y Maes-gwyn SN 2023
Llanboidy
'administrative-unit at Y Maes-gwyn': ***maenor***, pn. **Maes-gwyn**
Maynor y maesgwyn 1592, 1597
Centred on Maes-gwyn (q.v.). The p. had a *Manor Pound* in 1844. Located in the division of Traean March.

Maenor y Merydd SN 1424
Cilymaenllwyd
'administrative unit at Blaenymerydd': ***maenor***, pn. *Blaenymerydd*
Maynor blaen y mehereith 1592, *Maynor blaen y merreith*, (ht.) *Maenor y Merydd* c.1700
Maenor y Merydd is evidently a reduced form of Maenor Blaenymerydd, 'headwaters of Merydd' (***blaen***, ***y***) with a rn. Merydd surviving in Rhydymerydd (SN 148249) recorded as (ho.) *Rhydymeridd* 1819 and *Rhyd-y-merydd* 1889-90 (***rhyd***). The scholar Edward Lhuyd mentions a brook here called *Nant Merydd* c.1700 (Paroch III, 65) (***nant***) and it is likely to be a former n. for the small stream east of Rhydymerydd now known Afon Wenallt. Later forms favour ***merydd¹*** adj. 'slow, sluggish' and 'stagnant'. In Traean March.

Maenor y Mynachdy SN 5022
Llanegwad
'administrative unit at Mynachdy': ***maenor***, pn. **Mynachdy**
maynor y manachty 1556, *y mynaghty* 1567

Probably corresponding with the ht. of Mynachdy (q.v.).

Maenor Ysgwyn　　　　　　　SN 5825
Llangathen
'administrative unit at Ysgwyn': *maenor*, pn. **Ysgwyn**
maynor yskwyn 1623
Probably corresponding with the ht. of Ysgwyn (q.v.) making up the northernmost part of Llangathen.

Maerdy　　　　　　　　　　SN 629204
Llandeilo
'demesne farm' or 'cattle farm, dairy farm': *maerdy*
Maerdy-bach 1831, *Merdy* 1852-3, *Maerdy, Maerdy-bâch* 1884-7
This may also be *Maherdyf* 1361 though the absence of intermediate forms prevents certain identification. *maerdy* originally meant 'steward's house' which developed secondary meanings including 'summer dwelling for the tending of cattle, dairy farm' and 'home farm' (DPNW 306). The n. is probably drawn directly from the fm. Maerdy-bach which is 'small, lesser' (*bach¹*) perhaps in contrast to Maerdy (SN 651279) north of Llandeilo.

Maes-gwyn　　　　　　　　　SN 2023
Llanboidy
'(the) white *or* fair field': (*y*), *maes*, *gwyn*
y Maesgwyn 1595, *Maesgwyn* 1692 (1758), 1736, *Mâsgwynn* 1774, (ht.) *Maesgwynne* 1844, 1851
Cited forms include refs. to the ho. now known as Maesgwynne (SN 204238) with unnecessary -ne, an affectation recorded from the 19th cent. Two hos. are recorded as *Maesgwyn issa* 1692 (1758) (*isaf*), *Maesgwyn vach* 1738 (1758) (*bach*). The ht. may correspond with Maenor y Maes-gwyn (q.v.).

Maesllanwrthwl　　　　　　　SN 654372
Caeo
'open land at (place called) ?Llanwrthwl': *maes*, ?pn. *Llanwrthwl*
(tmt.) *Tir Maes Llanwthwl* 1603, (mess. etc) *tyr maes llan wrthwll* 1613, (lands) *Maes Glan wrthwl ycha* and *Maes Glanurthwl yssa* 1663, (cap. mess. etc) *Maesllan wrthol* 1730, *Maesllanwrthwl* 1831
The ho. (see HCH 127) is in a small area of open land (*maes*) extending from Afon Cothi along Afon Annell. Llanwrthwl appears to be identical in meaning to Llanwrthwl, co. Brecon, 'church dedicated to (man called) Gwthwl' (DPNW 287), with **llan** and a conjectural pers.n. *Gwthwl*, later *Gwrthwl* with intrusive -r-. There seems to be just one ref. to a ch. or chp. at Maesllanwrthwl in 1836 but this is almost certainly a supposition (WSS 329) that -llan- stands for **llan**. It is worth noting that two forms appear to have **glan** 'bank (esp. of a river)' and although there is no r. Gwthwl or Wthwl on current maps they do raise further doubts with regard to **llan**. The final el. -*wthwl*, -*wrthwl* – whether it is a pers.n. or common noun – is otherwise found in a ho.n. recorded as *Maes Carn Wthiol* 1625 and *Tir Maes Coom wthwl* 1626. The second instance, with what appears to have **cwm** may be a scribal error for **carn** 'cairn'. A former prehistoric stone *Maen wrthwl* is recorded here c.1700 and is probably identical to Llech Wrthwl described in 1915 (**maen**, **llech**).

Maestreuddyn　　　Maestroyddin-fawr SN 631411
Caeo
'open land at Treuddyn': *maes*, pn. *Treuddyn*
Maystroethen c.1557, *Mays troythyn* c.1590, (mess.) *tyr Maes Troythin vawr* 1636, *Maestroyddyn* 1784, (ht.) *Maestroyddin* 1804, *Maestroyddyn-fawr, -fach* 1831, *Maes-troyddyn-fawr, ~ -fâch* 1887-8
Treuddyn (dialect Troyddyn) may describe the area around Maestroyddin-fawr and Maestroyddin-fach (SN 638419) extending to the stream Nant y Treuddyn and Glantreuddyn, possibly corresponding with Gwestfa Maestreuddyn (see also Faenor Isaf). It is evidently **trefddynn** probably with the particular sense 'homestead surrounded by a protective hedge' as suggested for Treuddyn, co. Flint (DPNW 475-6; PNF 190) or perhaps 'settlement in a raised area'. A small hill lies

between the two hos. marked by a wooded slope Allt Maes-troyddyn (*allt*).

Maes-y-bont SN 566166
Llanarthne
'open land at the bridge': *maes, y, pont*
Maes-y-bont 1844, 1883
Located where the B4297 crosses the r. Gwendraeth Fach. An adjoining ho. is *Hendrefaes* 1831 (*hendref*).

Maesycrugiau SN 473410
Llanllwni
'open land (or field) of the mounds': *maes, y, crug* pl. *crugiau*
(cap. mess. etc) *Maes y Cryggie* 1628, *maesycrigie* 1788, *Maesycrigie* 1795, *Maesycrugiau* 1831
The n. is taken from a ho. Maesycrugiau (SN 477405) (later a hotel) and is now applied to the area near St Llonio's ch. ('St Luke's' on OS maps, in Llanllwni). The shift of n. may have been prompted by the choice of Maesycrugiau for the railway station (SN 473407). The n. Llanllwni was transferred to the village (SN 487393) on the Caerfyrddin/Carmarthen-Llanbedr Pont Steffan/Lampeter road c.1970.

Maesyffynnon SN 752229
Llanddeusant
'open-field at the spring (or well)': *maes, y, ffynnon*
(mess.) *Maesyfynon* 1676, (ht.) *Maesyffynnon* 1804, (ht.) *Maes y Ffynnon* 1811
The n. of a ht. taken from a ho. on hill-slopes. A well is shown on OS maps near the ho. but any spring is likely to have related to two small streams a little to the west running down to Afon Sawdde. Surrounding fields have the irregular appearance of ancient enclosure.

Malláen
cmt. Cantref Mawr
'open land of (man called) Llaen': *ma-*, pers.n. *Llaen*
(land) *Mathlaen* 1210 (1323), *Machayn* 1257, (land) *Methlaen* 1282, (cmt.) *Mathlaen* c.1287, *Mallaen* c.1400, (cmt.) *Malleyn* 1439, Comm. *Mallaen* 1536-9, *Mallayne* 1559, 1585, *Mallân* 1811
Stressed on the second syllable. Historic forms with -*thl*- are AN attempts at representing -*ll*- [ɬ]. The pers.n. may be that found also in Porthdin-llaen, co. Caernarfon (DPNW 296; ETG 51). The n. of a former cmt. subdivided into Is Coed and Uwch Coed, 'above ~ , below a wood' (*is, uwch, coed*) recorded as *Mathlaen Iskoych* 1288, (cmt.) *Mathlaen Hughecoyt* 1290.

Mamog, Afon
rn. SN 320363 to Cuch SN 299356
?'mother (r.)': *mam, -og*
(r.) *Mamog* 1832
The rn. may otherwise have **mamog** 'dam, esp. in-lamb or breeding ewe, brood-mare'. In both cases, Mamog might be thought of as a larger r. in contrast to a smaller one and it is worth noting the ho. n. Babiog (SN 280367) (*babi* 'baby', *-og*) recorded as *Pabiog* 1851, on the north side of a small stream, a little over 1km to the north-west.

Marble Hall SN 515003
Llanelli
'building constructed of marble *or* likened to marble': E *marble, hall*
Marble Hall 1837, *Marble Hall*, (on) *Marblehall Road* 1880
Marble Hall is typically used to describe ornamented rooms and stylish buildings but it is also used facetiously for plain, unadorned buildings. The n. earlier applied to hos. on Penallt Road near its junction with Marblehall Road. An identical n. in Milford Haven, co. Pembroke, applied to two rows of hos. (SS 909064) constructed c.1900 on either side of a large building, later a community hall. Marble Hall (SN 117237), at Llandysilio, however, is a single ho. recorded as *Marblehall* and *Marble Hall* otherwise *Dyffrin* 1796, *Marble Hall* or *Ty Coch* 1853 and *Marblehall* 1889-90 located nearly opposite a ho. known as *Whitehall*. Charles also mentions Marble Hall (SN 022212), at Clarbeston, recorded in 1829 (PNPemb 407, 618).

Marlais[1] SN 4933-5330
rn. SN 478330 to Cothi SN 535300
'big, greater stream': **mawr, glais**
Marleys 1281, (*the fall of*) *Marlas* (*into Cothie*) 1584, *Marles* 1539, 1633, *Marles Brook* 1729, *River Marlais* 1887-9
See also Cwm-marles and cf. Abermarlais. Other instances of the rn. are found in the county, notably Marlais[2,-4]. See PNPemb 16.

Marlais[2], **Afon** SN 6117
rn. SN 571179 to Llwchwr SN 622132
'big, greater stream': **mawr, glais**
margles early 9th cent, *Marleys* 1556, *Marlaish River* c.1700, *River Marlais* 1878
The first ref. is recorded as one of the boundaries of Meddyfnych with Aberfferws (*aper ferrus*) (SN 621110) and *aper huerdic* (**aber**).

Marlais[3], **Afon** SN 6436, 5741
rn. SN 495334 to 570418 to Cothi SN 640368
'big, greater stream': **mawr, glais**
Marleys 13th cent (1332), *River Marlais* 1887
In Llansawel

Marlais[4], **Afon** SN 6831, 6929
rn. SN 675349 to Tywi SN 696289
'big, greater stream': **mawr, glais**
(r.) *Morlas* 1612, *Marlaish* c.1700, *River Marlais* 1887
In Llansadwrn. Rising above Cors Farlais (SN 675349) near Blaenmarlais (**cors, blaen**).

Marlais[5] see **Gronw**

Marros SN 207090
Eglwys Gymyn
'horse moor': **march, rhos**[2]
(mill at) *Marcros* 1307, *Marros, -e* 1516, *Marros* 1531-1819, *Marras* 1535-1760, *marchroes* c.1566, *Morisshe* 1630, *Marcroes* 1819
Loss of -ch- may be due to E influence, cf. Laugharne < Lacharn, Talacharn[2]. Marros is a small p. occupying a flattish hill above Bae Caerfyrddin/Carmarthen Bay, much of it unenclosed, on which horses probably ran free (DPNW 313).

Meidrum SN 289209
'middle ridge': **mei-, trum**
Meitrym 12th cent (13th cent), (ch. of) *Meydrym* c.1291-1831 (frequent), *Maidrym* c.1500, (manor) *Meidrym* 1535, *Meydrum, meidrim* c.1566, *Mydrym* 1602, *Midrym* 1612, *Mydrim in Elvett* 1663, *Midrim* 1685, *Mydrym* 1749, *Meidrym* 1831
The first el. is also found in Meifod, co. Montgomery (DPNW 315), and Meiros (q.v.). Melville Richards preferred 'ridge in the plain' (AMR) but it is located on the south-eastern slopes of a ridge and near the meeting place of this with three others. Meidrim (Llangeitho), co. Cardigan, is recorded as *Meydrim* (*and Sissilt and water-mill Rhyd y pandy mill*) 1804 (**rhyd, y, pandy**).

Meinciau SN461107
Llangyndeyrn
'banks, hillocks': **mainc** pl. **meinciau**
Mancha 1729, *Min Key* 1765, *Maingceu* 1811, *Maingieu, ~ -bach* 1831, *y Meincau ney y Fangalch* 1839, *Meincian* 1880-90, *Meinciau, Meinciau-mawr*, (ho.) *Meinciau-bâch* 1907
mainc generally means 'bench, long seat' and refers to its location on a low flat-topped hill among low hills and ridges (DPNW 315). Individual hos. are distinguished as 'small, smaller' (**bach**[1]) and 'large, larger' (**mawr**). Forms such as *Min Key* reflect dialect *Meinci*. Y Fangalch, 'the place of lime': **y, man**[1], **calch** nm. 'lime, chalk', properly refers to Van fm. (SN 453113) recorded as *Fan* 1880-90, 1953 on the west side of an area of limestone quarries and old limekilns.

Meiros SN 5026, 5126
Llanegwad
'middle moor': **mei-, rhos**[2]
(ht.) *Mairos* 1710, (mess.) *Pant-Gwyn otherwise Miros* 1737, (mess.) *Myros-vach, alias Pant-y-cwd* 1756, *Miros* 1811, *Hamlet of Meiros* 1844, (ht.) *Miros* 1851
Recorded by David Jones (Yr Haul 23, 259) as Maenor Miros. Centred on Cefnmeiros (SN 510261) located on a flattish hill-top recorded on the tithe map and in the apportionment

as *Cefen Miros* and *Cefenmiros* 1842 (no. 1658) (***cefn***). Cf. Meiros, in Llanharan, co. Glamorgan, which is *Myros* 1543, *Miros* 1621, 1726, *Meyros* 1651.

Melinddwr, Afon   SN 5937
rn. SN 562402SN 621361
'mill stream': ***afon, melin, dŵr***
Melyndwr 1509, *Velinddwr R.* 1729, *Afon Melinddwr* 1906
The persistence of lenited *-dd-* suggests stress on the first syllable rather than the penultimate which generally produces *Melindwr* with unlenited *-d-*. Melinddwr may have referred specifically to the stretch next to Llansawel village near the former water-mill. Recorded as *River Livy* 1887 but this seems to be unmatched and may be an error.

Merchon, Afon   SN 7341
rn. SN 712419 to Gwenlais SN 743414
?'little daughter river': ?***merchan***
River Merchon 1891, *Afon Merchon* 1907
It is difficult to be certain on the proposed meaning because recorded spellings are very late though Afon Merchon is a small tributary of Afon Gwenlais and might reasonably be thought of as a 'little daughter' stream. The r. rises on the south slopes of the hill Esgair Ferchon (q.v.).

Merthyr   SN 352208
'graveyard or shrine of a saint': ***merthyr***²
Mercher in Derclis 1313, *Merthier in Derthles* 1399, *Merther* 1417, *Merthir* 1535, *Merthyr* c.1566, *Merthir Elved* 1600
Occasionally distinguished as Merthyr Derllys and Merthyr Elfed from its location in the cmts. of Derllys (q.v.) and Elfed, both in the ctf. Gwarthaf. The dedication of the ch. to Martin/Marthin was probably inspired by its similarity to Merthyr and displaced an earlier dedication to an unknown saint. The ch. was otherwise known as Merthyr Mynach apparently after it was alienated to Caerfyrddin/Carmarthen priory in 1313 (CAS XXIV, 77) (***mynach***¹). A late 5th/early 6th cent inscribed stone found in the churchyard c.1875 indicates early medieval origins for the site (Corpus II, 267-9; HER 49310; EMESP).

Merthyr Cynog   SN 2409
Talacharn²/Laugharne
'graveyard or shrine of (St) Cynog': ***merthyr***², pers.n. ***Cynog***
Meirthir gvenauc 1222; *Mertherkenang* 1307, *Marconick* c.1592, (manor, towns, etc *Castle Lloyd* and) *Merckynock, Meterkynock* 1617, (lps. *Castle Lloyd* and) *Markenock* 1703 NLW Probate SD/1703/46
Lost but in the vicinity of Castell Llwyd (q.v.), Parc Cynog (SN 2500940 and Cynog's Well (SN 260099) recorded as *Park-cynog* (***parc***) and *Cynog's Well* 1831, *Park Kennock* and *Kennox Well* 1843 (***well***). Cf. Merthyr Cynog, co. Brecon (DPNW 318). Cynog may be one of the numerous progeny of Brychan Brycheiniog (EWGT 14, 15, 17-19).

Miawst   SN 549174
Llanarthne
'(land attended to in the) middle of August': ***mei-, Awst***
Miawst 1607-1851, *Myaust* 1650, *Myarost* otherwise *Nantyfran* 1797 (1833), (ht.) *Myawst* 1804
Miawst was a t. of Llanarthne p. and David Thorne has shown (CAS 53: 147-8) that Miawst also survives in fns. *Waun Miawst, Cae Miawst* and *Cae Miawst Uchaf* (***gwaun, cae, uchaf***) in the p. tithe apportionment and a br. Pont Miawst (***pont***) over a tributary of Afon Gwendraeth Fach. Miawst marks the end of the hay-harvest when animals were released on the land to graze on the aftermath. The precise sense is a little unclear but we know that the occasion was marked by festivities and it is possible that the n. refers to their location. This may explain why the ho.n. changes from Nant-y-frân (*Nantyfrân* 1813) 'stream of the crow' or 'stream called Crow' (***nant, y, brân***) to Miawst (1831-1974) and Derwen-deg 'fair oak-tree' (***derwen, teg***).

Middleton Hall see **Plas Middleton**

Mihartach, Afon SN 7822
rn. SN 785214 to Sawdde SN 778241
?'(r. rising near) a fold or yard': ?*buarth*, suffix *-ach*
Buarthach 1831, *River Mihertach* 1887, *Afon Mihartach* 1979
Historic forms are very late and further evidence is needed. The suggested meaning may be elliptical describing a r. polluted by dung in a farm-yard or one flowing from a farm-yard. Initial B- has a colloquial variant M- because they share the same nasal mutation and development of -art- in place of -arth- before -ch. Several pns.in Wales have colloquial -t in place of -th (EANC 4-5).

Milo SN 596178
Llanfihangel Aberbythych
'(place named from chapel called) Milo': chp.n. *Milo*
Milo 1870, *Milo Chapel* 1887
Named from an Independent chp. Milo was a part of biblical Jerusalem. Founded in 1828 (CYB 1933: 446; HEAC III, 515-6).

Moelfre[1] SN 3235, 3236
Cilrhedyn, Penboyr
'bare hill': *moel, bre*[1]
Moelfre 1887
A roundish isolated hill lacking woodland.

Moelfre[2] Moelfre Isaf SN 407
Llandyfaelog
'bare hill': *moel, bre*[1]
Moylvre 1590, *Moelvrey* 1591, *Molvre* 1609-1787, *Molvrey* 1628, 1648, *Molvere* 1700, *Moelfre* 1831, 1889, *Moelfre Issa* 1844
References above from 1889-1964 apply to a ho. now marked as *Moelfre Isaf* (*isaf*). Nearby Moelfre Uchaf (SN 403172) is *Llwyn Moelfre* 1831, plain *Molfre* 1844, *Moelfre-uchaf* 1889-1964 and *Moelfre Uchaf* from c.1970 (*llwyn, uchaf*). The hill extends from a point south of Moelfre Uchaf where it has no trace of woodland to Croesyceiliog and to a position south of Moelfre Isaf. Cf. Moelfre, Anglesey (DPNW 324).

Moelfre[3] SN 633341
Llansawel, Talyllychau/Talley
'bare hill': *moel, bre*[1]
Moylwre 13th cent (1332), (2 tmts.) *tithyn moylvre* 1535, (place) *moelvre* 1567, (hill) *Moelfre* 1831, (ho.) *Moelfryn* 1891
A distinctive, rough oval-shape hill (SN 629341) with woodland only on its lower slopes. Historic forms include *tyddyn*.

Morfa[1] SN 579127
Llandybïe
'moorland': *morfa*
(ho.) *Morfa* 1879-80
morfa typically applies to a sea-marsh (see Morfa[2]) but developed the broader sense of marsh or moorland and was applied to inland locations, particularly in cos. Carmarthen and Glamorgan. It is possible that this process was furthered by association with E 'moor' (*mōr*). The n. is taken from a ho. (SN 579126) and gradually passed to housing which developed along Black Lion Road leading to Gors-las and Gorsddu Terrace shortly before 1915. The 'moor' is also recorded in the ho.n. Pen-y-waun (SN 576129) recorded as *Pen-y-waun* 1831, 'head of the moor': *pen*[1], *y, gwaun*.

Morfa[2] SN 514988
Llanelli
'land by sea-shore, salt-marsh': *morfa*
(place) *Y Morva, Le Morva* 1556, *Morva* 1734
The n. applied to an area of flat marine marsh, mud banks and sands known as *Llanelly Marsh* 1813 extending from Llanelli into Llangennech p. including *Morfa Mawr* and *Morfa'r Ynys* 1831 (*mawr, ynys*). Industrial and housing development on marine flats from c.1870 centred on New Street north of Llanelly Works (tinplate) 1880 known in 1907 as Morfa Tin Plate Works and east of New Dock.

Morfa-bach SN 367136
Llansteffan
'little marsh by the sea-shore': ***morfa, bach***[1]
Morfa-bach 1831, 1888-9, 1964, *Morfa Bach* 1970
Located above the tidal estuary of r. Tywi.

Morlais SN 5308, 5405, 5502
rn. SN 511093 to Llwchwr SN 573018 Llannon, Llangennech
'big, greater stream': ***mawr, glais***
(r.) *Morles* 1520, 1550, (stream) *Morleis* 1564, the *Morlais* 1586, (brook) *Morleis* 1609, *Ryver Morleish* 1686, *Morlais River* 1891
Edward Lhuyd c.1700 describes *Morlais* as a 'little brook' ('cornant') and says that it rises in springs near *Byrgroes*.

Mountain SN 5825
Llangathen
'mountain, upland'
Mountain 1851, 1873
A former ht., not on OS maps, but probably centred on Penhill (SN 573251) and Lan (SN 581254) (***glan***), extending northwards towards Capel Isaac, and describing the upland part of Llangathen.

Mount Pleasant SN 509010
Llanelli
'(place or ho. on a) pleasant hill': ***mount, pleasant***
Mount Pleasant 1851, 1864, *Mount Pleasant Buildgs.* 1880
Mount Pleasant seems to first appear on OS maps in 1951 but was in local use in the mid 19th century. It is apparently taken from a ho. and the hos. Mount Pleasant Buildings. Llanelli also possessed *Mount Pleasant Inn* in 1856. Mount Pleasant is an exceptionally common ho.n. and chp.n. See PNGlamorgan 145 for examples in co. Glamorgan.

Myddfai[1] SN 772301
?'bowl-shaped land': ?***mydd***, ?***mai***
Meduey 1284, *Medvey* 1299, *Medevey* alias *Methevey* 1316, *Mothevey* 1492, *Methvey* 1535, *Llanyhangle muthuey* 1578, *Mothvey* 1600, *Mothvey* V. alias *Myddfai* 1763, *llanvihangell in mothvey* 1611, *Myddfai* 1831
Early forms with *Med-, Meth-* probably stand for ModW *Mydd-* because *-e-* was sometimes used to represent the middle vowel [ə]; cf. historic forms of Llandybïe and Llandyfaelog (q.v.). A pers.n. *Mai* has been proposed as the second el. but the suggested meaning fits the topography. Cf. Myddfai[2] and Myddfai (SN 6403, Llangyfelach), co. Glamorgan, recorded as (place) *Mithvay* 1585, (tmt. in) *Mothvay* 1650, *Mothvay* alias *Letty Thomas* 1764 (**llety**, pers.n. ***Thomas***)

Myddfai[2] SN 5119
Llanarthne
As Myddfai[1]
(village) *Myddfay* 1591, (place) *Tre Myddvey* 1639, *Myddfey* 1811, *Myddfai* 1811
The n. is not current but applied to the ht. west of r. Gwynnon and Middleton Hall towards the stream Ffinnant. The suggested meaning might apply to the short valley between Glasgoedfach (SN 521192) and Glasgoed-isaf (SN 517191). Sometimes recorded in association with Tre-clas (q.v.).

Myddyfi, Afon SN 6126
Llandeilo Fawr, Llangathen
rn. SN 630270 to Tywi SN 597220
Uncertain
(bet ar) ueithini c.1145, *The Modewie*, or some pronounce it *Motheuie* has 2 heeds joining above *Lanihangle* 1587, *River Myddyfi* 1887
The rn. survives also in a ho.n. *Glan-myddyfau* 1831, *Glanmythyfi* 1839, *Glan-Myddyfi* 1887 and br. *Pont Myddyfi* 1887 (***glan, pont***). R.J. Thomas (EANC 163 under Middifi) suggests that the first historic form is a miscopying (*n = u*) which would stand for *Meithifi* or *Meiddifi* in modern orthography with colloquial *i* for *ei*. Modern forms, however, favour Mydd- as on the OS.

Myhathan SN 4720, 4919
Llanarthne
'plain, open-land of (man called) Cathan': ***ma-***, pers.n. ***Cathan***

Maghhatau 1291, *Maghhatan* 1339, *Meyhaythen'* 1576, (t.) *Myhathen* 1620, (place) *Tre myhathan*, *Tremihathan* 1633, (mess. etc *hendre wenyn* at place) *Tre Myhathan* 1639, *Gweyn Myhathan*, (t.) *Myhathan* 1699, (ht.) *Myhatham* (and *Trefreyn*) 1804, *Meathan* 1851

With aspirate *-chathan* after **ma-**. Melville Richards suggests (BBCS 23: 324-5) that if this contains a pers.n. **Cathan** then there may be a connection with Llangathen and Catheiniog (q.v.). Cited historic forms include **tref** and **gwaun**.

Mynachdy[1] SN 506229
Llanegwad
'monk's house, monastic grange': **mynachdy**
y mynaghty 1567, *manachdye* 1613, (ht.) *Mynachty* 1710, *Manachdy* 1811, *Monachty Hamlet* 1851
See also Maenor y Mynachdy. There is no evidence of any monastery as supposed by David Jones (Yr Haul 23: 99).

Mynachdy[2] SN 6246, 6345, 6347
Llan-y-crwys
'monk's house, monastic grange': **mynachdy**
Mynachty Hamlet 1851
There is no specific place bearing this n. but it is possible that the monastic gr. (belonging to Talyllychau/Talley abbey) is recalled in the ho.n. Llwyn-cwrt (SN 644454) recorded as *Llwyncwrt* 1841, *Llwynycwrt* 1851 and *Llwyn-cwrt* 1888 (**llwyn**). The gr. lands were concentrated between the rs. Camddwr and Twrch (CStudies 116-7).

Mynydd Du, **Black Mountain** SN 6917, 7517, 7719, 8222
'dark mountain': **mynydd**, **du**, E **black**, **mountain**
minÿd du 1129 (c.1170), *ymynÿd du* c.1170, 15th cent, (hill) *Mynyth Duy* 1541, *Mynydh duy* 1584, *Y Mynydd Dy* 1596, (mtn.) *Mynydd Dy* 1609, *the Blak Montayne, ~ Mountayne* c.1538, *the Blacke Mountaines* 1559
'Dark' perhaps from the shadow which it casts or from its wildness. A large area of open mountain pasture covering parts of Llangadog and Llanddeusant ps. extending into Breconshire, reaching its highest points in Bannau Sir Gaer (SN 8121) and Fan Brycheiniog (SN 8221) (**ban**[1] pl. *bannau*).

Mynydd Figyn SN 5830-6030
hill Llanfynydd, Talyllychau/Talley
'mountain at the marsh': **mynydd**, (**y**), **mign**
Mynydd y figin 1668, *Mynydd Figin* 1831, *Mynydd Figyn* 1888
Centred on crossroads (SN 594306) with extensive wet, rough pasture extending north and eastwards.

Mynydd Llanllwni SN 5037, 5138
Llanllwni
'mountain in (p. of) Llanllwni': **mynydd**, **Llanllwni**
Mynydd Llanllwni 1888
Cf. Mynydd Llanybydder.

Mynydd Llanybydder SN 5339, 5439
Llanybydder
'mountain in (p. of) Llanybydder': **mynydd**, **Llanybydder**
Llanybudder Mountain 1729, *Mynydd Llanybyther* 1888-9
Simply a mountain in Llanybydder as opposed to Mynydd Llanllwni (q.v.)

Mynydd Malláen SN 7344, 7444
Caeo
'mountain in (commote of) Mallaen': **mynydd**, **Mallaen**
Mynidd Mallaen 1554, *Malhan Mountain* c.1700, (hill or bank on) *Mallan Mountain* 1777, *Mallane Mountain* 1818, *Mynydd Mallaen* 1831
A broad area of hills and moorland between rs. Cothi and Tywi defined on the north by Gwenffrwd.

Mynydd Myddfai SN 8029
Myddfai
'mountain in (p. of) Myddfai': **mynydd**, **Myddfai**
Mynydd Myddfai 1831
An area of mountain land extending to the border with co. Brecon.

Mynydd Pen-bre SN 4503
'mountain in (p. of) Pen-bre': ***mynydd***, **Pen-bre** (Pembrey)
Penbray Mountain 1729, *Mynydd Pembrey* 1813, *Mynydd Penbre* 1830, *Mynydd Pen-bre* 1880, *the Pembrey Mountain* 1851
A ridge extending from the village to Penymynydd (q.v.).

Mynydd Pencarreg SN 5742, 5843
Pencarreg
'mountain in (p. of) Pencarreg': ***mynydd***, **Pencarreg**
Pencarreg Mountain 1729, *Mynydd Pencarreg* 1831
The mountainous part of Pencarreg.

Mynydd Sylen SN 514080
Llanelli
?'mountain named from (man called) Sylen': ***mynydd***, ?pers.n. ***Sylen***
Mynidd Sylen 1609, *Mynyth sellen* 1618, *Mynydh Coese Cawr a Common als Mynydh Syle* c.1700, *Mynydd Sylan* 1811
The most prominent hill in the area north of Llanelli. Sylen may perhaps be a variant of the pers.n. Sulien.

Mynyddygarreg SN 427081
Cydweli
'mountain at the stone': ***mynydd, y, carreg***
(commons) *Mynith y Garrege* 1609, (pasture) *Mynith y garreg* 1620, *Mynidd y Garreg* 1753, *Mynydd Garreg* 1831
The pn. is also found in (land) *Parke Mynith y Kerregge* 1564 (***parc***). Named from the hill, heavily quarried for limestone extending north-eastwards towards Meinciau. The n. need not refer to a specific stone but be elliptical for a hill characterised by very large stones and rocks.

Mynys, Afon SN 7235
rn. SN 711389 to Tywi SN 724318
Uncertain
river Mynys c.1700, (r.) *mynys* 1785, *Mynys River* 1887, *Afon Mynys* 1907
Probably derived from a pers.n. ***Mynys***. The rn. is also recorded in Gwestfa Ystradfynys (q.v.) and ho.ns. *Cwmynys-fach* and *Cwmynys-fawr* 1831 (***cwm***). The pers.n. is best known in Maesmynys (SO 017476), co. Brecon (PNBrec 116).

Myrtle Hill[1] SN 505062
Llanelli
'(place or ho. on) a hill where myrtle grows': E ***myrtle, hill***
Myrtle Hill 1851, 1857, (ho.) *Myrtlehill* 1880, *Myrtle Hill* 1906
Myrtle is an evergreen plant widely used as an ornamental garden plant.

Myrtle Hill[2] SN 766308
Myddfai
(ho.) *Myrtle Hill* 1854, 1886
On the north side of a small road running northwest from Myddfai opposite Berllandywyll. Simply a ho.n. and it is unclear why the n. has been given prominence on recent OS maps. The ho. is shown but not named on the OS map 1831. Myrtle Hill (SN 409190), Caerfyrddin/Carmarthen, is also a ho.n., recorded in 1840 (HCH 136). The ho.n. also occurs at Llanrheithan, in co. Pembroke (PNPemb 234).

Nant Aeron to Newton

Nantgaredig. Looking north from Heol yr Orsaf/Station Road c.1920.

Nant Aerau SN 4530
?'stream of battles': *nant*, ?*aerau* pl. of *aer¹*
Nant Aerau 1889
With a ho. Cwmaerau (SN 460298) recorded as *Cwm-aerau* 1889 (*cwm*). Historic forms are too late for certainty but they might describe a r. which fights its way over rocks and rapids; cf. Cadnant, cos. Montgomery and Caernarfon (*cad¹* nf. 'battle, conflict', *nant*). Nant Aerau is only 4km from Nant Aeron (q.v.). Note also Brynaere (SN 639436), Brynaire (SH 897038), in Llanbryn-mair, co. Montgomery, recorded as *Brinaure* 1676, *Brynaire* 1783 and possibly *Brynarey ycha* 1615, and Brynaerau-uchaf (SH 436520) and Brynaerau-isaf (now Plas Brynaerau), at Clynnog, co. Caernarfon, recorded as *bryn yr Aure* 1639, *Bryn Aera* 1770, *Brynaerau* 1834 and *Bryn-eira-isaf, ~ Uchaf* 1841. Both might describe a hill (*bryn*) associated with battles or disputes.

Nant Aeron SN 4130
rn. SN 417336 to Gwili SN 418292 Llanpumpsaint
?'stream (in area) notable for berries': *nant*, ?*aeron*
Nant Aaron 1889, *Nant Aeron* 1906
The n. need not be taken literally; it might simply describe an area notable for its fertility. Historical forms are so late that it is impossible to be certain that *aeron* is the correct second el., however. A ho. Brynaeron (SN 127212), in Llandysilio West, is recorded as *Brynayron* 1793, *Brynaeron* 1819 and *Bryn-aeron* 1890 (*bryn*) possibly a former n. of Afon Rhydybennau as a nearby ho. is *Llan-aeron* 1890. It is 1890

(?*glan*). It is tempting to associate these with the r. Aeron, Ceredigion, generally thought to be linked to the Celtic goddess of battles **Aeron** (DPNW 1) but evidence is lacking.

Nant-bai SN 775446
Llanfair-ar-y-bryn
'destructive stream': *nant*, ?*bai¹*
(gr.) *Nantbey* 1291, (~) *Nantbay* 1535, *Nauntvey* 1550, *Nant y Bai* 1809, (chp.) *Nant y Bai* (in ht. Rhandir Abbot) 1811, *Nant-y-Bai* 1888-1970
Named from a stream (SN 782454 running to Tywi SN 772442). The first el. is *nant* but -bai is uncertain though plausibly *bai¹* nm. 'fault, failing; transgression' with the extended sense of a r. which is destructive. It is difficult to find comparable ns., however. The def.art. *y* is intrusive (after the plosive b) and does not appear in early forms. It was dropped by the OS after 1970. Ystrad Fflur/Strata Florida abbey possessed a gr. (*cwrt¹*) possibly at or very near Bryn-y-cwrt (SN 774448) recorded as *Bron-cwrt* 1834 with (*bryn*) which may be 'the miserable cot' where services were held in 1809. The gr. may have had its origins in a grant before 1202 with an additional chp. at Capel Peulin (q.v.), possibly the original chp. The area of monastic property is said to have been co-terminous with Rhandir Abad (WCist 202, 204, 309).

Nant Bargod SN 3636
rn. SN 371340 to Teifi SN 348405
'stream forming a boundary or edge': *nant*, *bargod¹*
(r.) *bargod* 1553, (stream) *Bargod* 1580, (brook) *bargod* 1688, (r.) *Bargoed* c.1700, *Nant Bargod* 1889
Rising near Blaenbargod (SN 376341) recorded as *Blaen Bargod* 1831, *Blaenbargoed* 1832 (*blaen*), and marking the boundary between Penboyr and Llangeler (EANC 49). The n. may be compared with Bargoed (PNGlamorgan 10-11). See also Maenor Aberbargoed (above).

Nant Brechfa SN 4225
rn. SN 439262 to Gwili SN 420248 Llanpumsaint
'stream at (area called) Brechfa': *nant*, *brech*, *-fa*
Nant Brechfa 1889
Cf. Brechfa. Brechfa here has not been identified but may have referred to an area with variegated soil or vegetation or more likely speckled by stones. A small hill here is recorded as *Cerig Fawr* 1889 (SN 425254) and nearby ho.ns. include Danycerig (SN 428254) and Cwm-dan-cerig (SN 425260) (*cerrig* 'stones')

Nant Corrwg SN 4527
rn. SN 466281 to Gwili SN 437288
'stream which is small (like a dwarf)': *nant*, *cor*, suffix *wg*
(r.) *Corrug* 1670, *Corrwg R.* 1729, *Nant Corrwg* 1889 OS 1:2,500
The n. is also found in Ystradcorrwg (SN 454281) recorded as *Ystrad-corwg* 1831. An unlocated *Glyncorrwg* is recorded here in 1743, possibly a ref. to the fm. Glancorrwg (SN 434284). *cor* is also found in *corrach* 'dwarf, pygmy', *cornant* 'little brook'. Corrwg may have been regarded as small in comparison with Afon Gwili.

Nant Crychiau SN 4424
rn. SN 470261 to Gwili SN 432214 Abergwili
'stream which ripples *or* bubbles': *nant*, *crych* pl. *crychiau*
Nant Crychiau 1889
The ho.n. Cwmrheiddol (SN 464253) presumably draws its n. from the uppermost part of the valley.

Nant Cwm-merydd SN 5031
rn. SN 495324 to SN 503305
Uncertain
Nant Cwm-marydd 1889, 1978-81
The first two els. are clearly *nant* and *cwm* but historic forms are too late for certainty with regard to merydd. It is possible that the rn. is wrongly divided and that *Cwm-marydd* properly stands for *Cwmarydd* with *arydd* nm. 'ploughman' with the sense of a r. which

cuts through the landscape. The current form seems to have been associated with **merydd**[1] adj. 'slow, sluggish' (see Maenor y Merydd) or **merydd**[2] nm. 'fat beast, animals'. Neither seem appropriate for a valley or a small torrent.

Nant Cynnen SN 3622
rn. SN 381240 to Afon Cywyn SN 334210
?'battling (r.)': ?*cynnen*
Nant Cynnen 1889
Perhaps describing a r. forcing its way through its landscape and comparable with other rns. such as Cedi (*cad*[1], *-i*); see R.J. Thomas in EANC 110-1. He adds that it may otherwise be a pers.n. *Cynein* (or *Cynen* in local dialect) and notes that at Y Traws-mawr, near a tributary of the r., there is a stone bearing the inscription *Cunegni*, possibly a variation of *Cunagni* (gen.) which gives ModW *Cynan*. Cynnen might also derive from *Cynien*, OW *Cyngen*, *Congen* < Brit *Cunogenos*. The rn. occurs in Cwmcynnen recorded as *Cwmcunnen* 1764, 1771, *Cwm-cynnen* 1831 (*cwm*) and rises near Blaencynnen (SN 378227) recorded as *Blane Diffrin cunnen* 1700 and *Blaen-cynnen* 1831 (*blaen, dyffryn*).

Nant Dâr SN 7144
rn. SN 720450 to Cothi SN 697442 Caeo
'stream where there are oak-trees': *nant, dâr*
Nant Dâr 1887
Identical in meaning to Dâr (Dare), in co, Glamorgan (PNGlamorgan 3), which has the same el. as that found in **derwen** and **deri**. A ho. Esgair-dda (SN 702442) on the hill (*esgair*) north of the stream is recorded as *Tir Eskeir Ddaer* 1603, *Eskeir daer* 1678 and *Eskerddar* 1724 (*tir*). Blaen-dâr is *tir blaen dar* 1584, *Tir blaen daer* 1652 (*blaen*).

Nant Eiddig SN 5845
rn. SN 589439 to Teifi SN 570471
'voracious stream': *eiddig*
Nant Eiddig 1888-9
The n. might describe a stream which was notably powerful and destructive. R.J. Thomas (EANC 186) argues that the meaning may be closer to *aidd* nmf. 'ardour, zeal' but this is a back-formation from *eiddig*. The n. also survives in Coedeiddig (SN 585459) recorded as *Coed-eiddig* 1831 (*coed*). Identical to Eiddig, in Crucadarn, co. Brecon.

Nant Felys SN 4224
rn. SN 437254 to Gwili 419236
'sweet *or* pleasant-tasting stream': *nant, melys*
Nant Melys 1889-90, *Nant Felus* 1906
Possessing water suitable for human and animal consumption.

Nantgaredig SN 494217
Abergwili and Llanegwad
'kindly brook' or 'kindly valley': *nant, caredig*
(meetinghouse called) *Nantcaredig* 1792, *Nantgaredig* 1808-9, *Nant-garedig* 1831
The first recorded form indicates that **nant** is a nm. in the sense 'valley' but all of the evidence is late. A very small stream passes through the ht. near the crossroads on the A40 becoming little more than a ditch marked by field-boundaries passing near Rwyth farm (SN 490211) on the Tywi flood-plain.

Nant Garenig SN 6712
SN 663105 to Aman SN 667135 Betws
'(r. associated with the) crane': *garan*[1], *-ig*
Nant Garenig 1878
The development *-an-* > *-en-* is through affection of the following vowel *-i-*. The rn. is also recorded in *Abergarenig* 1831 and *Cum ŷ garrenigg* 1725 and *Cwm-garenig* 1831 (*aber, cwm*).

Nant Gochen SN 3529
rn. SN 343329 to Afon Duad SN 373274
?'red-coloured stream': *nant,* ?*cochyn* fem. *cochen*
Nant Gochen 1889
See also Afon Fawr. Red or red-brown perhaps because of iron pyrites leaking into the water or from local vegetation. *cochyn* usually describes a red-haired person.

Nant Gwennol SN 8436
rn. SN 868387 to Gwydderig SN 822335
'stream swift like a swallow': *nant, gwennol*
Nant Gwennol 1887-8

A stream fancied as flowing quickly, dipping and weaving like a swallow.

Nant Gwythwch　　　　　　　　SN 6717
rn. SN 688178 to Llwchwr/Loughor SN 664174
'stream which behaves like a wild pig': *gwythwch*
Gwythwch Brook 1890-1, *Nant Gwythwch* 1907
A stream which rushes and roots like a pig; cf. Twrch. The rn. is recorded in a lost ho.n. (place) *Blaen Gwythwych* 1609 (***blaen***).

Nant Hust　　　　　　　　　　SN 5241
rn. SN 544398 to Teifi SN 505427
?'stream which hushes *or* sounds like hushing': ***nant***, ?***ust***, *hist*
Nant Hust 1877-89
ust in the sense 'a hush' or 'silence' is inappropriate to this hill-stream and the name is better understood as a stream that is imagined as 'hushing, urging silence' or making a sound like the word 'hushing'. Historic forms are so late, however, that these suggestions must be regarded as tentative.

Nant Melyn　　　　　　　　　SN 7346
rn. SN 727493 to Gwenffrwd SN 746465
?'mill stream': ***nant***, ?***melin*** displaced by ***melyn***
Nant Melyn 1888
Most earlier forms of a ho.n. Blaennantmelyn (SN 732475) suggest that ***melyn*** 'yellow, brown' has replaced ***melin*** 'a mill'. Historic forms include *tir blaen nant melyn* 1693, *Blaennant Melin* 1782, *Blaennantmelin* 1812, *Blaen Nant Melin* otherwise *Blaen Nant y Velin* (... watergrist mill *Velin Troed Rhiw yr hwch* and mess. *Glan Croythir*) 1758, *Nant Melin* 1784, *Blaen-nant-melyn* 1888 (***tir***, ***nant***, ***blaen***), and is about 1km above a former woollen mill Melin-y-coed (SN 734466) (***y***, ***coed***).

Nant Pedol　　　　　　　　　SN 6915
rn. SN 717181 to Aman SN 688134
'horse-shoe shaped valley': ***nant***, ?***pedol***
Speddul 1609, *Pedwl* 1734, *Pedol* 1831
There are doubts with this explanation because local pron. is 'pedwl' (CAS VI, 11), borne out by the 1754 form. No better explanation can be offered, however, and 'horse-shoe' could describe the middle reaches of the valley above Pantyffynnon (SN 698153) or perhaps the shape of an adj. hill recorded as *Y Foeldegarbedol* 1831, shown as Foel Deg-arbedd (SN 705159) on current maps, probably 'the fair bare-hill over against Pedol' (***y***, ***moel***, ***teg***, ***ar***, rn.).

Nant Pen-y-cnwc　　　　　　　SN 4623
rn. SN 476250 to Crychiau SN 445227 Abergwili
'stream at Pen-y-cnwc': ***nant***, ho.n. *Pen-y-cnwc*
Nant Pen-y-cnwc 1889
Named from Pen-cnwc (SN 463231) recorded as *Pen-y-cnwc* 1889, 'top of the hillock': ***pen***[1], (***y***), ***cnwc***[2]. Cf. Maenor Pen-y-cnwc.

Nant Pibwr　　　　　　　　　SN 4418
rn. SN 498188 to Tywi SN 404175
'stream which squirts': ***nant***, ***pibwr***
Peebwr c.1693 (late 18th cent), *Pibwr R.* 1729, (r.) *the Peebwr* 1804, *Nant Pibwr* 1891
pibwr might describe a mountain stream which makes a piping sound or more likely one which squirts up from springs sometimes with unpleasant connotations such as the colour of its waters or unpleasant odours; cf. Cachan, co. Monmouth (EANC 44). The rn. is also recorded in that of the lp. of Pibwr (*Pybour* 1467), a ho. (*Pyboure* 1572), a ho. Aberpibwr recorded as *Aberpibor* 1618 (***aber***), the br. Pont Pibwr (*Pont Pybor* 1798, *Pybor Bridge* 1792) (***pont***), a mill (*Pybors Mylle* 1609, *Felinpibwr* 1831) (***mill***, ***melin***) and compounded in the ns. of several hos. including Pibwr-wen (*Pibwr Wen* c.1735) (SN 404175), 'white Pibwr' (***gwen***) and Pibwrlwyd (q.v.).

Nant Rhydw　　　　　　　　　SN 4314
rn. SN 458164 to Gwendraeth Fach SN 435127 Llandyfaelog, Llangyndeyrn
?'filthy stream': ***nant***, ?***rhyd-***, ***-wy***[1]
Nant Rhydw 1888
This could describe a r. which was exceptionally muddy or perhaps one which was foul; cf. Bawddwr. ***rhyd-*** is the same el. found in ***rhwd***[1] 'rust, filth, mud' (R.J. Thomas in BBCS 8: 39). The r. rises more than 2km above the ho.n.

Blaen-rhydw (SN 440153) recorded as *Blane Rhwddw* 1569, *blane Rhyddwy* and meadow ground *gwaine Rydig* 1668, *Blaenrhydw* 1788, *Blane yr hydw* 1804 and *Blaenrhydwfach* 1811 (*blaen*, *bach¹*, *gwaun*).

Nant Tawe SN 6041
rn. SN 594433 to Marlais 608385
?'flowing one, stream': ***nant***, ?Br ****tam-***, ***-wy¹***
Nant Tawe 1887
The obvious comparison is with Tawe, co. Glamorgan (PNGlamorgan 213), earlier Tawy, but historic forms are very late. It is better recorded in the n. of an Independent meeting-ho. as *Escair Dowie Meeting House* 1756, *Escerdowe* 1765, *Esgairdawe* 1768-9, 1826, *Eskir Dave* 1779 and *Esgair Ddewi* 1831 (***esgair***) and a ho.n *Cwm Ddewi* 1831 (***cwm***). All historic forms display great uncertainty with regard to sp. which raise suspicions that Tawe/Tawy has influenced an earlier and unidentified rn. Some very late forms have been influenced by the pers.n. Dewi.

Nant Thames SN 6140
rn. SN 615423 to Tawe SN 608401
Nant Thames 1887
The n. may be mocking for a small insignificant stream and possibly inspired in part by the nearby fm.n. Llundain-fechan (SN 614406) meaning 'little London' and comparable with ho.ns. Llundain-fach in Llanelli and Llansawel, Llundain-fach (Nantcwnlle), co. Cardigan, and Little London otherwise Llundain-fach (in Llandinam), co. Montgomery (pn. *Llundain*, *bechan*, *bach¹*). Note also Nant Thames recorded in 1887 applying to a small stream which rises in or near Llyn Taliaris (SN 634283) entering Afon Dulais near *Glanthames* 1841 (***glan***) (SN 642268).

Nant Treuddyn (Nant Troyddyn) SN 6342
rn. SN 616429 to Cothi SN 648413
'stream called Treuddyn' or 'stream in area called Treuddyn': ***nant***, *Treuddyn*
Nant Troyddyn 1887-8
The ho.n. *Glan-troyddyn* 1887-8 (***glan***) tends to favour the first explanation but it does not make good sense since Treuddyn, 'protected homestead' (***trefddyn***) would apply to a place not a r. Cf. Maestreuddyn (q.v.). *Troy-* in OS sps. represents local dialect.

Nant Tridwr SN 8438
rn. SN 855389 to Gwennol SN 850369
'stream composed of three waters *or* streams': ***nant***, rn. *Tridwr*
Nant Tridwr 1888 OS 1:2,500
The three waters (***tri***, ***dŵr***, *dwfr*) may refer to the area where Tridwr is joined by two unnamed tributaries (approx. SN 845379). Cf. Abertridwr, co. Glamorgan (PNGlamorgan 6).

Nant y Bai see **Nant-bai**

Nant-y-caws SN 458182
Llangynnwr p.
?'stream in fertile land', ***nant***, ***y***, ***caws***
(ho.) *Nant y Caws* 1729, (mess. etc) *Nant y Caws* 1771, (hos.) *Nant-y-caws*, ~ *-isaf* 1831; (stream) *Nant y Caws*, (ht.) *Nant-y-caws* 1888-9
Literally 'the cheese stream' perhaps a stream with clouded waters or one with an unpleasant smell but comparable ***llefrith*** 'milk' was also used to describe land which was fertile and it is possible that ***caws*** is used in a similar sense here. The first explanation is more likely with the former fm. Nant-y-caws (SN 783463), in Llanfair-ar-y-bryn, which adjoined a small stream in rocky terrain. This is recorded as *Nant y Caws* 1629-30, *Nant y Caws* alias *Cawse* 1692, *Nant-y-caws* 1834. A stream in Cydweli, near Pont Sbwdwr (SN 434059), is recorded as *Nant y Caws* 1609 and cf. Nant-y-caws (in Oswestry), Shropshire, which is *Nantcause* 1755, *Nant-y-caws* 1837.

Nant y Dresglen SN 8434, 8534
rn. SN 843364 and SN 865375 to SN 834328
'stream of the thrush (or mistle thrush)': ***nant***, ***y***, ***tresglen***
Nant y Tresglen 1887-8, 1964, *Nant y Dresglen* 1978-81
From the abundance of this bird or an allusion to the short repeated rattling call of the mistle thrush.

Nant-y-ffin SN 554321
Llanfihangel Rhos-y-corn, Llanybydder
'the boundary stream': *nant, y, ffin¹*
Nantfyn, (tmt.) *Nant y Ffyne* 1584, *Nant-y-ffin Woollen Factory*, (br.) *Pont Nant-y-ffin* 1888-9, 1906, *Nant-y-ffin* 1952-3
Located on the B4310 which passes over a small stream Nant y Ffin here recorded as *Cledagh Wen* c.1762 (see Clydach¹). The stream (SN 556366 to Cothi SN 556320) defines part of the p. boundary of Llanfihangel Rhos-y-corn and Llanybydder.

Nant y Rhaeadr SN 7543
rn. SN 740 to Tywi SN 766415
'stream with a waterfall': *nant, y, rhaeadr*
Nant Rhaeadr 1888, *Nant y Rhaiadr* 1905-6, *Nant y Rhaeadr* 1979-80
Notable for a waterfall at Craig y Rhaeadr (SN 755436) (*craig*).

National Botanic Garden of Wales see **Neuadd Middleton**

Nenog, Afon SN 5336
rn. SN 541386 to Clydach SN 529345
'(r. in) elevated area': ?*nen, -og*
Nenog 1584, *Nant Nenog* 1887-9, *Afon Nenog* 1906
Without more evidence it is difficult to be certain but the rn. may perhaps describe a r. in an elevated area. The alternative is that it is derived from an unrecorded adj. **haenennog* 'stratified, layered' of *haenen* a dim. form of *haen* describing perhaps small slabs in the r. bed, ie. *Haenennog, Henennog* > *Nennog*. See Llidiardnenog.

Neuadd Middleton, Middleton Hall SN 521182
Llanarthne
'hall of Middleton family': *neuadd*, family n., E *hall*
Middleton hall 1676, *Middleton Hall* 1699-1809
The original ho. was built by Henry Middleton (d. 1644). Sold to William Paxton in 1789 who commissioned the architect Samuel Pepys Cockerell to build a new ho. 1793-5 (HCH 132-134). This was largely gutted by fire in 1931 and the great glasshouse was built on its site. The museum opened in 2000.

Newcastle Emlyn see **Castellnewydd Emlyn**

Newchurch see **Llannewydd**

New Inn¹ SN 635248
Llandeilo Fawr
'new inn': E *new, inn*
New Inn 1678, 1783, *New Inne* 1712, *New Inn, ~ Shop* 1886
The n. applies in 1831 and 1887 to a ho. at SN 635246 but has shifted in recent years northwards to what is *Efail-newydd (Smithy)* 1887, 'new smithy' (*gefail¹, newydd*).

New Inn² SN 472367
Llanfihangel-ar-arth, Pencader
'new inn': E *new, inn*
New Inn 1765, *the New Inn; New Inn. A new house lately opened* 1804, (fair at) *New Inn* 1818, (four fairs at) *New Inn* 1831
The inn otherwise known as the 'Traveller's Rest' was reputedly built in 1711. Richard Fenton (Fenton 5) noted 'three or four houses' and a country shop 'on a very large scale' at *the New Inn* 1804. The ht. expanded later around a Calvinistic Methodist chp.

New Mill SN 261132
Talacharn²/Laugharne, Llanddowror
'mill near Newton' or 'new mill': E pn. *Newton, new, mill*
the New Mill c.1765, *Newton Mill* 1831, *New Mill* 1888-9
Newly-built or new from its proximity to Great Newton (SN 261129) recorded as (old mansion) *Great* or *Lower Newton* 1820, *Lower Newton* 1831, *Great Newton* 1843, 1888-9 and see Newton².

Newton¹ see **Drenewydd**

Newton² SN 2612
Talacharn²/Laugharne, Llanddowror
'new farm or settlement': E *new*, ME *-ton*
Neuton 1307, *Newton* 1532-1745, *Upper Newton* [= Newton], *Lower Newton* [= Great Newton] 1831

Probably in the sense 'newly-established' rather than new in contrast to some specific older settlement.

Pantarfon to Pysgotwr

Pen-dein/Pendine. Viewed westwards along seashore c.1910.

Pantarfon SN 568254
t. Llanfynydd
?'hollow over against the bottom (of hill-slopes)': ***pant***, ?***ar***, ***bôn***[1]
Pantarvon 1577, 1778, *Pantarfon* 1799-1871, (ho.) *Pant-yr-afon* 1831, *Pantyrafon* 1841, *Pant-Arfon* 1887-8
The ho.n. probably gave rise to that of the t. and is located near the bottom of hill-slopes. ***bôn***[1] also has the meaning 'tree-trunk, stump' in ref. to a particular fallen tree. Earlier evidence, however, is needed.

Pant-gwyn SN 591253
Llangathen
'fair hollow': ***pant***, ***gwyn***
(mess. etc) *Tir y Pant gwyn* 1637, *Pant Gwynn* 1682, 1788, *Pant-gwyn* 1831-1964
The hollow may be the wooded dingle to the east.

Pant-y-caws SN 150262
Cilymaenllwyd, Llanglydwen
Pantycaws 1794, *Pant-y-caws* 1887, 1964
?'hollow in fertile land': ***pant***, ***y***, ***caws***
For the use of ***caws***, cf. Nant-y-caws. The small settlement also has two hos. known as *Pant-y-menyn* (***ymenyn***, ***menyn***) and *Pant-y-ffynnon*

141

(*ffynnon*) (known as Kensington Villa from 1907).

Pantyffynnon SN 624114
Llandybïe
'hollow of the well', *pant, y, ffynnon*
Pant-y-ffynnon 1906, *Pantyffynnon* 1953
The n. now applies to a part of Rhydaman/ Ammanford which developed between the Llanelly & Llandilo and Garnant & Brynamman branch railways near the former Dynevor tinplate works and Pantyffynnon Colliery. The n. is ultimately drawn from a mill and Pantyffynnon farm (SN 619106) in Llanedi p. recorded as *Syddin Tir pant y ffynnon* 1668, *Pant y ffonnon* 1699, *Pant y ffynnon* 1751, *Pantyffynnon Farm* 1809 (*syddyn, tir*). The n. was first transferred to the railway junction before 1860 and then to the colliery.

Pant-y-llyn SN 606169
Llanfihangel Aberbythych
?'hollow at the lake': *pant, y, ?llyn*
Pant-y-llyn 1831, 1906, *Pantyllyn* 1835, *Panty llyn* 1841, *Pant-y-llwyn* 1878
No lake is shown here on OS maps. The 1878 form suggests association with *llwyn* 'grove. copse'.

Parc Howard SN 508013
Llanelli
'park named from Howard (family)': *parc*, family n. *Howard*
Parc Howard 1913
The n. appears in 1913 as a description for the municipal park, and later the museum and art galley, established at the ho. Brynycaerau (recorded 1836) and was gradually extended to include housing constructed during the 1930s and immediately after World War II. The ho. and park were gifted to Llanelli by Sir Stafford and Lady Howard in 1912.

Parc-y-rhos SN 5874560
Pencarreg
'enclosure at moorland': *parc, y, rhos²*
(mess.) *tir park y rhose* 1649, (close of lands) *parck y Rhose* 1670, *Park y Rose* 1714, (tmt. etc)

Llaingoy or *Parkyrhose* 1784, *Parc-y-rhos* 1831, *Parc-y-rhôs* 1888
Former moorland lies south-east of the settlement. Earlier refs. appear to apply to Parc-y-rhos House. The alternative n. in 1784 stands for Llain-gou, 'enclosed strip (of land)': *llain, cau¹* dialect *cou, coy*.

Pedair-heol, Four Roads SN 448094
Llangyndeyrn
'(meeting of) four roads': *pedwar, pedair, heol*; E *four, road*
Pedairheol 1811, *Pedair-heol* 1831, (inn) *Four Roads* 1875, *Pedairheol* 1879
At the crossing of the Meinciau-Cydweli road and a minor road.

Pemberton SN 530005
Llanelli
Pemberton 1916
Pemberton recalls a family of coal pioneers and the n. is also commemorated in the former *Pemberton Arms* (P.H.) 1880 in Station Road. Pemberton was an area of housing which developed from the late 19th cent in the fork between what is now Gelli Road and Pemberton Road (SN 52860043). The n. is drawn from a second Pemberton Arms (described in 1870 as at Llandafen) at the junction but does not become fixed until shortly before 1916. Pemberton has partly displaced both Cwmcarnhywel (q.v.) on the 1880 OS map and hos. on Pemberton Road near the White Lion PH (formerly at SN 53110033). This n. was briefly extended to include the hos. on the eastern side of Pemberton Road on the 1907 OS map but had shifted back to its earlier location by 1916. By 1959 the n. applied to hos. constructed in the 1950s around Heol Gwili and Heol Elfed.

Pembrey see **Pen-bre**

Penallt SN 518007
Llanelli
'top of wooded slope': *pen¹, allt*
Penallt Terrace, ~ Road 1916, *Penallt* 1950-1
Apparently a new n. first applied to a terrace of housing built near Box Colliery shortly before

1907 recorded as Penallt Terrace in 1916. Penallt Road replaced the n. Marblehall Road about the same time. The n. was extended before 1950 to housing on both sides of Penallt Road.

Penarth see **Gwestfa Penarth** (Caeo)

Pen-boyr SN 360363
Llangeler
Probably 'hill of (r.) Beyr': *pen¹* and ?rn. *Beyr Penbeyr* 1222 (1239), *Penbeyr* 1337, 1349, *Pen beyr* c.1566, *Penbeher* c.1291, 1296, *Penbeir* 1487-1563, *Penbyere* 1532, *Pen boyr* 1590-1, 1752, *Penboer* 1627, *Penboyr* 1624, 1831, *Pemboyr* 1750
The rn. is conjectural but is supported by a ref. to *Rhydvoyr in Pemboyr* 1754 and a ho. recorded as *Rhyd-y-foyr* 1831, *Rhyd-foyr-uchaf, ~ -isaf* 1889 (*rhyd*). The conjectured rn. could apply to a stream marking part of the western side of the terrace on which the village is located. The meaning of *Beyr* is uncertain but appears to have been disyllabic, possibly a pers.n. or perhaps a compound of *pau* 'district' and *hir* (DPNW 362). Later forms with *-boyr* have dialect -oy- for -ey-; cf. Coynant ('kɔinant) for Ceunant.

Pen-bre (Pembrey) SN 428012
'end of hill': *pen¹*, *bre*
Penbray 1141, c.1160, *Penbrei* c.1145, *pennbre* c.1170, *Penbrey* c.1291-1602, *Pembray* 1378, *Penbre* 1535, *Pembrye* 1618, *Pembre* 1747, *Pembrey* 1833
Local pron. 'Pem-bre' explains the non-standard *Pembrey* (ETG 52, 249; DPNW 360-1). Cf. colloquial 'Demby' for Denbigh and 'Llambed' for Llanbedr Pont Steffan, co. Cardigan. The village is located at the western end of a hill Mynydd Pen-bre extending eastwards to Penymynydd (q.v.).

Pencader SN 446360
Llanfihangel-ar-arth
'hill shaped like a chair': *pen¹*, *cadair* colloquial *cader*
Pencadeir c.1191, c.1400, (chp. of) *Penkadeir* 1595, (chp.) *Penkadeyre* 1637, *Pencader-Capel* 1680, *Pencader* 1725-1831

cadair is used in a topographical sense for a chair-shaped hill or simply an elevated place (DPNW 362-3). The first suggestion seems likelier here and it is worth noting that the ref. to *Pencadeir* c.1191 is glossed *cathedrae caput* 'head of a chair' (Latin *cathedra, caput*). The n. probably refers to the site of the castle Dinweilir (HER 1785) described by the antiquarian Fenton in 1804. The chp. recorded in 1595 and 1680 was ruinous c.1700 and in 1763. Stones had apparently been taken away to use in building a meeting ho. and school by the churchyard on which there was a datestone 'July 1705'.

Pencarreg SN 535451
'stone hill or headland': *pen¹*, *carreg*
Penkarreck c.1291, *Pencarrik* 1331, *Pencarrek* 1377, 1535, *Penkarok'* 1544, *p.kareg* c.1566, *Pencarreg* 1565, 1763
Located on slopes of a low hill. A ho. Blaencarreg (SN 545450) recorded as *Blaencarreg* 1782 may be employed elliptically to describe the upland area of Pencarreg.

Penceiliogi SN 531006
Llanelli
?'hill where there are male (wild-) birds': *pen¹*, ?*ceiliog* pl. *ceiliogau*
Penceiliogy 1813, *Pen-ceiliogau* 1830, *Pencilogi* 1851, *Pen-ceiliogi* 1880-1953, *Penceiliogi* 1953
There is a small hill a little to the east centred on the ho. Llys-y-bryn, formerly known as Uplands (SN 539008). The historic form for 1830 supports the preferred interpretation though one might expect *ceiloge, ceilioge* in local dialect unless there was an unrecorded variant pl. *ceiliogi*. Instances of *ceiliog* in placenames are numerous (see Glossary for examples). Llanelli also has a ho. Penceiliogwydd (SS 539986) a little over a mile to the south recorded as *Penclackwydd* 1841 and *Pen-ceiliogwydd* 1879, 1964, meaning 'hill notable for ganders' (*pen*, *ceiliogwydd* variant *clacwydd*). The ho. is next to former salt-marshes. This may also be *Pengwythfawr* 1785 and *Pengwyddmawr* alias *Penychlackwydd* 1810, with *mawr* len. *fawr* qualifying 'hill notable for geese' (*gŵydd*),

though it is a little difficult separating these references from those of a former ho. Pen-gŵydd (approx. SS 540984) recorded as *Pengwydd* 1745 and *Pen-y-gwydd* 1830.

Pen-cwm SN 4419
Llangynnwr
'top of the valley': ***pen¹***, (***y***), ***cwm***
Penycwm 1843, *Penycwm, Pencwm* 1845
Recorded on the tithe map 1842 and apportionment 1843 as part of Hendy (SN 446195). The n. may refer to a small valley north of the site of *Penycwm* or to the valley of Nant Pibwr to the south.

Penddeulwyn SN 4519, 4619
Llangynnwr
'head (or hill) of the two groves': ***pen¹***, (***y***), ***dau, deu, llwyn***
(mess., etc) *Tythin att Pen y ddoylwyn* 1633, (two messuages, etc) *Sythyn pen y ddoylwyn* and *Tythin att Penyddoylwyn* 1668, (ht.) *Pennyddoylyne* 1724, *Pen y ddaulŵyn* 1728, *Penyddoyllwyn* 1738, *Penddoilwn* 1788, *Penddoylwn, Penddailwyn* 1845
The n. survives in that of several hos. recorded as *Pen-ddaulwyn-fâch*, *Pen-y-ddau-lwyn-ganol* and *Pen-y-ddau-lwyn-uchaf* 1887-8 (***bach¹, canol, uchaf***). Cf. Pendeulwyn (Pendoylan), co. Glamorgan (PNGlamorgan 163-4).

Pen-dein, Pendine SN 229088
'headland by a ?fort': ***pen¹***, ?***din***
Pendyn 1307, *pendyn* c.1566, *Pendyne* 1594-1633, *Pendeyn* 1664, *Pendine* 1596-1831, *Pendein* 1710, 1815, *Pendŷn* 1811, *Pen-dain* 1814
The current pron. [pen'dain] is longstanding as the W forms *Pen-dein* and *Pen-dain* indicate. The suggested meaning cannot be regarded as conclusive but there is a promontory fort (SN 2283076) above Gilman Point (HER 3287). If the origin lies with ***din*** then the development *din* [dɪn, di:n] > *dain* [dain] *dein* is probably through E influence; cf. Penrice < Pen-rhys, co. Glamorgan (PNGlamorgan 169). Pentywyn (cf. Pentywyn²) has been in use as the W form of Pendine since the publication of a gazetteer in 1957 (GazWPN) and is based on a misidentification dating at least as far back as 1893 (LL 415). Heather James has shown (CBeyond 158) that several early refs. to Pentowin (q.v.) have been mistakenly identified with Pen-dein/Pendine. Melville Richards evidently had doubts in this respect because he collects evidence under 'Pendyn' (AMR).

Pengwern SN 4307
Cydweli/Kidwelly
'top of alder-trees': ***pen¹***, ***gwern***
Pengwerne 1408, *Pengwern* 1607, (ht.) *Pengwern* 1692, *Pengwerne* 1611, *Pen-y-wern isaf* 1830, ~ *uchaf* 1831, *Penywern Issa*, *Pengwern*, *Penywern ucha* 1840
This is also likely to be *Pengwern* 1319. *Penywern Issa* and the adjoining *Pengwern* (SN 436071) on the tithe plan 1840 are omitted on the OS map 1880. *Penywern Ucha* changed its n. to *Cae-gwyllt* (SN 438075) before that date (***isaf, uchaf, cae, gwyllt***). All three hos. appear on 19th cent maps next to a narrow wooded valley Cwm Teilo followed by Capel Teilo Road. There is no obvious marshy ground to justify the alternative sense of 'alder-marsh' for ***gwern***; the def. art. ***y*** is a later intrusion.

Peniel SN 435241
Abergwili
'(place named from chapel called) Peniel': ***Peniel***
The Independent chp. was built in 1809 (HEAC III, 466-7) and is recorded as *Capel penuel* 1831 and *Capel Peniel* 1889 (***capel***). Peniel or Penuel was the biblical city on the east side of the r. Jordan. The village expanded from the 1960s and is recorded as plain Peniel from 1971.

Pennant (= Fforest) SN 6445
Llan-y-crwys
'head of the streams': ***pen¹***, ***nant*** pl. *naint*, variant *neint*
(forest) *Pen Neint* 1303, 1584, *Forest of Pennent* 1577, (forest) *Penneynt* 1353-1580, (forest *Glyncothy* and) *Peynient* 1432, (*fforestes of Glincothie* and) *Penneynt* 1584, *Peneint* c.1600

The forest in 1584 covered most of Llan-y-crwys centred on Fforest (q.v.) ht. and lay around the headwaters of Afon Twrch and Camnant.

Penrherber SN 290390
Cenarth
'above the lodging *or* shelter' or 'above the arbour': **pen**¹, def.art. **yr**, **herber**² or **herber**¹
Pen'r-herber, *Dan-yr-herber* 1831, *Penrherber* 1843, 1906, (inn at) *Penreherber* 1875, *Pen-r-herber* 1891
An inn is shown at the crossroads 1891 which tends to favour the first interpretation. Daniel E. Jones (HPLlangeler 152-3) cites Rhŷs in Cymru on 'Pen-yr-herber'. **herber**² is derived from ME *herberge* (Fr *auberge*) but is sometimes confused with **herber**¹ (< ME *herber(e)*, *-our*) 'herber, arbour' (EEW 117). Cf. Penyrherber (in Castell Caereinion), Montgomeryshire.

Penrhiw-goch SN 557180
Llanarthne
'top of red-brown slope': **pen**¹, **rhiw**, **coch** len *goch*
Penrhingoch otherwise *Gwaingoch* 1797 (1833), *Pen-rhiw-goch* 1831, *Penrhiwgoch* 1839
A ho. on the north side of the ht. is *Garn-gôch* 1887 apparently taken from a 'red-brown cairn' (**carn**), the redness describing either the colour of the vegetation or geology. The first historic form appears to contain a scribal error *-rhin-* for *-rhiu-* or *-rhiw-*; the alternative n. is 'red moor' (**gwaun**) applied on OS maps to a ho. (SN 557178) as *Waen-gôch* 1887, *Waun-gôch* 1906.

Pen-rhos Pen-rhos Uchaf SN 562270
Llanfynydd
'head of the moor', **pen**¹, **rhos**²
(mess. etc in) *Penrhose* 1585, (place) *Penrhose* 1604, (mess. etc) *Tythin pen y rose* 1669, *Pen-y-rhos-uchaf* 1831
The n. of two fms. and a t. centred on a small hill east of the village of Llanfynydd.

Penrhyn¹ SN 482021
Pen-bre
'promontory', **penrhyn**
(manor of) *Penryn* c.1500, 1545, *Penrhyn* 1605, (ht.) *Penrhyn*, (ruined chp.) *K. Kynnor* c.1700, (ht.) *Pendryn*, ~ *Hamlet* 1792
The former chp. is Capel Cynnor otherwise Capel Cynnwr (q.v.).

Penrhyn² (Penrhyn Deuddwr)
Llangynog, Llansteffan
'promontory' **penrhyn**
(cmt.) *Penryn* 1292, *Penryn* 1332, 1586, (cmt.) *Penryndudour* 1391, *Comm. Deilis à Penryn* c.1537
Penrhyn Deuddwr in full, 'promontory between two waters': **dau**, **deu**, **dŵr**, *dwfr*. The n. of a former cmt. in Cantref Gwarthaf between the rs. Taf and Tywi corresponding with the lp. of Llansteffan. Penrhyn is described as a manor associated with Llansteffan ferry (*la Verie*) in 1462 and in 1622 (**ferry**) where it is said to lie in the parish of Llangynog (SN 339163) but the lp. actually applied to the entire promontory including Llandeilo Abercywyn, Llansteffan and Llan-gain, as well as Llangynog. *Deilis* refers to Derllys (q.v.).

Pen-sarn SN 419189
Llangynnwr
'head of the causeway': **pen**¹, **sarn**
Sarnes end 1609, *Pensarne* 1623, *Pensarn* 1729, *Pen-sarn* 1831, 1887-90
The first cited form has E *end* as a purely literary translation of **pen**¹. The causeway runs southwards from the br. of the t. Tywi at Caerfyrddin/Carmarthen across the floodplain known as *Morva Pen y Sarne* 1609 (**morfa**) to Pen-sarn where the road reached drier land and forked, one road running southwards to Pibwrlwyd and the other towards Llanelli and Pontarddulais. Antiquarians such as Richard Fenton (Fenton 1) identified this road as the Roman road running between Caerfyrddin/Carmarthen and the fort at Casllwchwr/Loughor, co. Glamorgan. The first stretch of this road, now known as Roman Road, was replaced by a turnpike taking a graded route around Mounthill. Proof of a Roman origin is lacking.

Pentowyn see **Pentywyn**

Pentowin SN 293190
Meidrum
'hollow of ?sand-banks: *pant*, ?*tywyn*² variant *towyn*
Pantowin 1716, 1737, *Pantowyn* 1755, *Pentwyn* 1811, *Pentywyn* 1831, *Pen-tywyn* 1888-9, *Pentowin* 1906
The first el. has been misassociated in late sources with *pen*¹ 'head, top of' – a very common el. in local ho.ns. The second el. is apparently *tywyn*² but Pentowin is 6km distant from the salt-marshes, mud banks and sand-dunes of the estuaries of Afon Taf and Afon Cywyn and is located among low hills adjoining Afon Dewi Fawr. It is conceivable, however, that *tywyn*² was adopted locally to describe sand-banks adjoining a r. in inland locations; cf. *morfa*. It is unlikely that *tywyn*² has displaced some other el. such as *tŵyn* in view of the historical evidence.

Pentre SN 282164
Llanfihangel Abercywyn, Sanclêr/St Clears
'village': *pentref*
Pentre 1680, *Pentre* 1727-1868, *Pentrey* 1748
The n. survives in Pentre Road which was once that part of the main Caerfyrddin/Carmarthen to Haverfordwest road (A40) between the brs. over Afon Cynin and Dewi Fawr.

Pentre-bach SN 812334
Llanfair-ar-y-bryn, Llywel, Myddfai
'small village': *pentref, bach*¹
(ho.) *ty Pentre bach* 1746, (mess.) *pentre Bach* 1760, *Pentre-bach* 1832, *Pentre-bâch* 1887
A scatter of hos. along Brecon-Llandovery road (A40) on both sides of Afon Gwydderig, partly in co. Brecon, and 'small' perhaps in contrast to Pentre-tŷ-gwyn (q.v.).

Pentrecagal SN 339403
Penboyr
'sheep-dung village': *pentref, cagl* variant *cagal*
Pentrecagl 1760-1, *Pen-tre-cagyl* 1831, *Pentrecagal* 1862, *Pentre-cagl* 1888-9
A small settlement on the A484 describing a road along which animals were once driven.

The n. might also be disparaging for a place which was thought to be of little consequence. It occurs elsewhere, notably Pentre-cagl (SN 287307), in Tre-lech, which is *Pentrecagal* 1861 and *Pentre-cagl* 1891. Other instances are found in Coychurch (*Pentre Cagal* 1720) and Ystradyfodwg (*Pentre Caegil* farm 1781), co. Glamorgan. There is a Tregagle (SO 526079) (*Tregagallt* 1812; 1831; *Tre-gagle* 1890) (*tref*) in Pen-allt and Caggle Street (SO 3617) (*Caggle street* 1831) (*street*) at Llanwytherin/Llanvetherine, both in co. Monmouth.

Pentrecŵn SN 604214
Llandeilo Fawr
'village of dogs': *pentref, ci* pl. *cŵn*
Pentre coon 1729, 1769, *Pentrecwn* 1791, 1972-6, (place) *Pentre Cwn* 1809, *Pentref Cŵn* 1811, (ht.) *Pentrecwm* 1804, *Pentre-cwm* 1884-6
The westernmost ht. of Llandeilo Fawr. There are identical ns. elsewhere, notably Pentrecŵn (SN 604214) at Llandyfaelog, recorded as *Pentrecwn* 1788 and *Pentre Cwn* 1795, and Pentrecŵn (SN 718382) at Cil-y-cwm, which suggests that the n. may have applied to a fm. or ht. with the responsibility of raising dogs for the hunt, etc. Cf. Tre-cwn (*tref*), co. Pembroke (PNPemb 83, 184, 226).

Pentre-cwrt SN 389388
Llangeler
'village at the grange', *pentref, cwrt*¹
Pentre Court 1841, *Pentrecwrt* 1852, 1875, *Pentre-cwrt* 1891
Located on the site of the gr. Maenor Forion (q.v.).

Pentrefelin SN 597236
Llangathen
'mill village': *pentref, melin*
(hos.) *Pentre-felin* 1887
The ht. is shown as a small group of buildings near Pont Myddyfi over Afon Myddyfi including *Pant-y-bas* and a smithy on the OS map 1887. Pentrefelin properly refers to the hos. (SN 597238) near *Pont Pentre-felin* 1887 (*pont* 'bridge'). The mill is *Melinbrynhafod* 1813, *Melin-bryn-hafod* 1831 and *Brynhavod Mill* 1835,

'mill near (or belonging to Brynhafod' (*melin*). Brynhafod (SN 592239) is 'hill at the summer-dwelling' (*bryn, hafod*) recorded as *Brynhavod* 1626, *Brin-Havod* 1678, *Bryn-hafod* 1831.

Pentregwenlais   SN 608163
Llandybïe
'village at (r.) Gwenlais': ***pentref***, rn. *Gwenlais Pentre Gwenllaise* 1725, *Pentregwenlais* 1760, *Pentre Gwenlas* 1778, *Pentre-gwenlas* 1831, *Pentregwellas, Pentregweles* 1851
Pronounced 'Pentregwilesh'. For the rn., cf. Gwenlais, 'fair stream' (***gwyn, glais***). The p. has a well Ffynnon Gwenlais (SN 600161) (***ffynnon***). Nant Gwenlais rises a little to the west of Glangwenlais (SN 601161) (***nant, glan***) and meets Llwchwr at SN 619160. The village developed near limestone quarries.

Pentre Morgan   SN 410255
Llannewydd/Newchurch
'village of (man called) Morgan': ***pentref***, pers.n. **Morgan**
(building) *Forge Mill* 1889, *Forge Mill (Corn)* 1906
A new name, apparently adopted after 2016, probably recalling Robert Morgan, iron entrepreneur, who resided at nearby Cwmdwyfran farm and at Furnace House (now Carmarthen Library). He acquired the iron forge in 1786 but it was evidently re-built c.1840. The n. applies to hos. extending north along the A484 road from Forge Mill to Forge Quarry.

Pentre-poeth   SN 507016
Llanelli
'burnt village *or* burnt farm': ***pentref, poeth***
Pentre-poeth 1830-1992
poeth generally means 'hot' and might describe a settlement in a warm, sheltered location but it also seems to have developed the sense 'burnt' and it may be significant that it is located only a short distance from Ffwrnes/Furnace[1] (q.v.). The development began in the early 19th cent near small coal-works. Orchard Street, in Caerfyrddin/Carmarthen, is recorded as *Pentrepoth* 1871 and *Pentrepoeth* 1931. The pn. is common with other examples in Llangyndeyrn, in Gelligaer, Llangyfelach (PNGlamorgan 171), co. Glamorgan, Machen, co. Monmouth, Dolbenmaen, co. Caernarfon, and Llansilin, co. Denbigh

Pentre-tŷ-gwyn   SN 816355
Llanfair-ar-y-bryn
'village at Tŷ-gwyn': ***pentref***, *Tŷ-gwyn*
Pentretygwyn 1740-1867, *Pentredywyn* 1781, (meeting ho. at) *Pentretygwyn* 1799, *Pentre-ty-gwyn* 1832, 1891
The location of an Independent chp. (HEAC III, 396). Tŷ-gwyn is 'white house' (***tŷ, gwyn***) probably for a white-washed ho.

Pentre-wyn   SN 324130
Llansteffan
?'white or fair settlement':? ***pentref***, ?***gwyn***
(lp., manor or fm. of) *Kentregwyne* and *Hendre* 1601, *Centrewyne* 1633, (churchway from) *Pentrewin* (to *Llanstephan*) 1718 and 1739, *Pentre wyn* 1767, *Pentre-gwyn* 1889, *Pentre-wyn* 1906
There are some doubts with this interpretation because the two earliest forms suggest that *Pentre-* may have displaced *Cefntre-, Centre-* (***cefn, tref***) under the influence of nearby Pentrenewydd (SN 322136). Later forms see-saw between *-gwyn* and *-wyn* perhaps through association with **ŵyn** pl. of *oen* 'lamb'. Hendre (SN 327137) is a nearby fm. otherwise recorded as (place) *yr Hendreye* 1610, *Hendre* 1700, *Hendre* 1799 and *Hêndref* 1831 meaning 'old farm or settlement' (***hen, tref***). Pentre-wyn may be the location of *Egluys Treffwenyn* 1308 though this has also been identified with Pentwyn (CBeyond 158). *Treffwenyn* appears to stand for Trefwenyn perhaps meaning 'township associated with bees': ***tref*** and ***gwenyn***, one where honey was gathered. This may be compared with Tir Cwmgwenyndy (Llanfynydd) (***tir, cwm, tŷ***) and Brynygwenyn (Llan-non), in co. Carmarthen, Brynygwenyn, in co. Monmouth, (***bryn***), Esgairygwenyn (in Llanwrin), (***esgair***) and Gwenynog Isaf and Gwenynog Uchaf (in Llanfair Caereinion), in co. Montgomery.

Pen-twyn SN 561115
Llan-non
'hillock top': *pen¹*, **twyn**
Pentwyn 1790, 1831, (hos.) *Pen-twŷn* 1879-80
Located near the top of a lane climbing up steep slopes from Cross Hands. The Calvinistic Methodist chp. adopted the n. Capel Pen-twŷn. A very common ho.n. and pn. in south Wales with other examples in Betws, Llanfair-ar-y-bryn and Llangadog.

Pentywyn¹ (Pentowyn) SN 320106
Llansteffan
'head of sand-dunes': *pen¹*, **tywyn²**
Lannteliau penn tiuinn c.1145, *lanteliau penn tiwin* c.1250, *Pentewy* c.1135, *Pentewi* c.1170, (gr.) *Pontowyn* 1535, *Pentowin* 1676, *Pant tewin* alias *Pant Owen* 1700, (mess.) *Pantowen* 1704, *Pentowen* 1729, *Pentywyn* 1831
Close to the estuary of Afon Taf. Early refs., formerly identified with Pen-dein/Pendine (q.v.), stand for Llandeilo Pentywyn, 'church dedicated to Teilo at Pentywyn (**llan**, **Teilo**). No physical traces of a ch. or chp. have so far been found here but Heather James has identified three fields on the 1841 tithe map east of the ho. recorded as *Parc Sinshill fach*, *Parc Sinshill issa* and *Parc Sinshill ucha* (SN 324107), apparently standing for 'saint's hill' (**sant** pl. **saint**, **hill**, **bach¹**, **isaf**, **uchaf**). Pentywyn was the location of a gr. belonging to Caerfyrddin/Carmarthen priory (CBeyond 158-163).

Pentywyn² see **Pen-dein**

Pen-y-banc¹ SN 616240
Llandeilo Fawr
'top of the bank': *pen¹*, *y*, *banc¹*
Penybank 1724, *Penbank* 1753, *Pen-y-banc* 1831, 1886
Located at crossroads and on hill-slopes above an unnamed stream and Nant Gurrey-fach. The name is drawn from that of a ho. The village developed around the Greyhound PH, Gate Inn, and Capel Seilo Independent chp. built c.1830 (HEAC III, 528) or 1833 (CYB 1933: 446). A very common ho.n. and pn. with other examples in Abergwili, Caeo, Cil-y-cwm, Llanarthne, Llanboidy, Llanegwad, Llangadog and elsewhere.

Pen-y-banc² SN 614116
Llandybïe
'top of the bank': *pen¹*, *y*, *banc¹*
Penybank 1852, *Pen-y-banc* 1878-80, 1965, *Penybanc* 1977
The n. referred to the area (SN 617119) around the Golden Lion PH at the junction of A483 and a lane running northwards uphill. By 1906 the n. was extended westwards to housing along the A483.

Pen-y-fan SS 514998
Llanelli
'top of the hill', *pen¹*, *y*, *ban¹*
(place) *Penvan* 1519, *Penyvan* 1597, (ho.) *Pen vann* 1613, *Penyvann* 1619, *Penyevan* 1729, *Pen-y-fan* 1931
An area of housing centred on Penyfan Road developed from the late 1930s and especially after 1945. *ban¹* in its common sense of 'summit, peak' is inappropriate since the ho. lay near the foot of a low hill centred on Bigyn (q.v.). It is possible that the n. has been transferred from some other ho. in Llanelli.

Pen-y-garn SN 574318
Llanfynydd
'(place at) top of (stream called) Y Garn': *pen¹*, *y* and rn. *Y Garn*
Pen-y-garn 1831, 1887-8
The rn. is also recalled in (ho.) *Blaen-nant-y-garn* 1887-8 (SN 570319) (**blaen**, **nant**) evidently referring to an unnamed tributary of Afon Cothi which it meets at SN 564322. Presumably the rn. recalls an unidentified cairn (**carn**).

Pen-y-groes SN 587133
Llandybïe
'top of the crossroads', *pen¹*, *y* and **croes**
Penagroes 1680, *Pen-y-groes* 1831
The n. initially applied to hos. at the junction of the B4556 and B4297. The village developed from the late 19th cent near California Colliery (SN 579136) abandoned c.1900 and the n.

was extended southwards to include the area around the B4297 towards Capel Hendre after the opening of Emlyn Colliery 1892-3 (SN 583134).

Penymynydd SN 464038
Pen-bre
'top of the mountain': *pen¹, y, mynydd*
Pen-y-Mynydd 2012
Located where the B4308 crosses Mynydd Pen-bre. The name appears to be modern and applies to housing built in the 1960s incorporating existing hos. Brynmelyn, Brynhyfryd and Ffald.

Perfedd
Cantref Bychan
'middle (commote)': *perfedd*
le Commod P(er)ued 1203, *Comoth Perueth* 1265, *Commot Pervet* 1287, *Comot' Perueth, Comot de Perueyth* 1317, (cmt.) *Perveyth, Perveth* 1317, *y Kymwt Perued* c.1400, *y K6mw6t perved* c.1500
'Middle' from its location between the cmts. of Hirfryn and Is Cennen in the ctf. of Cantref Bychan. The cmt. (*cwmwd*) covered the parishes of Llanddeusant, Llangadog and Myddfai. Perfedd, co. Cardigan, has similar forms. Cf. also Perfeddwlad, 'the middle country', between the ancient kingdoms of Gwynedd and Powys (HW I, 239; PNF 184).

Peuliniog
Cantref Gwarthaf
'land of (man called) Peulin': pers.n. *Peulin, -iog*
Peunlyok c.1285, *Pelbiniock* 1282, *Pulinog* 1308, *Pellunyauc, Pellinnyauc* 1332, *Pellonyok* 1382, (*a*) *Phelunyawc* c.1400
A cmt. in Cantref Gwarthaf covering Llangynin, Llanwinio and Sanclêr/St Clears. The territorial suffix is fairly common in the names of cmts. and ctfs. coupled with a pers.n., eg. Anhuniog, co. Cardigan, Brycheiniog, co. Brecon, Cyfeiliog, co. Montgomery, Rhufoniog, co. Denbigh. Peulin may be an unknown secular lord. The ch. at Llan-gors, co. Brecon, is dedicated to an ecclesiastical Paulinus.

Pib, Afon SN 4831
rn. SN 479326 to Marlais SN 526302
'river with a narrow channel like a pipe': *pib*
river Peeb 1668, *Nant Pib* 1889
A narrow, fairly straight valley with only short tributaries, now almost entirely in forestry plantation; cf. Nant Pib. The area at the head of the r. was evidently Blaen-pib recorded as *Blayn Pyb* 1331, *Blaen Peeb* 1668 (*blaen*).

Pibwr see **Nant Pibwr**

Pibwrlwyd SN 413183
Llangynnwr
'grey (house near) Pibwr': rn. **Pibwr**, *llwyd*
Court Pibour 1609, *Pibwr Llwyd* 1753, (ho.) *Pibwr Court* 1831, *Pibwrlwyd* 1833, 1872, 1990-2, *Pibwr-lŵyd* 1887-90, 1969
The n. in historical sources referred to a large ho. (HCH 160-1) a little to the north of Nant Pibwr (q.v.). Pibwr-wen (SN 405174), recorded as *Pibwr Wen* 1754, *Aberpibwr* alias *Pibwr Wen* 1766 is by contrast 'white, fair' (*gwyn* fem. *gwen*) – probably a whitewashed building. The variant refers to its location near the confluence (*aber*) of Pibwr and Afon Tywi.

Pigwn, Y SN 828312
Llywel, Myddfai
'the cone, peak': *y, pigwn²*
Pigwn 1832, *Y Pigwn* 1854
Referring to the hill described in 1854 as *Mynydd y Begwns* 'the Beacons mountain' on which are located several Roman military camps.

Pinged SN 423037
Pen-bre
?'gates to a pound': ?ME *pund*, ?ME ?*gate(s)*
Penggetys 1408, *the Pyngetts* 1492, *the Pynget* 1554, 1591-1907, *Pinged Marsh* alias *Morva Pingett* 1711, *Pinget Marsh* 1830
Possibly Pingate as a variation of E *pingot* 'small enclosure' under the influence of *gate, gates*. Recorded instances of *pingot* are, however, mainly in Cheshire (NDEFN 331, 489) and north Shropshire (SFN 11; PNShr 5: 271, 279). In neither case are there strictly comparable parallel examples in the E dialect

of neighbouring Pembrokeshire or Gower where there has been an E population since the 12th cent. An area of small irregular enclosures occupying a low promontory above former marshland.

Piodau SN 612153
Llandybïe
?'(places associated with) magpies': **piodau* ?pl. of *pi²*
Piode 1609, 1704, *Peode, Pyode* 1666, *Piodey* 1675, *Piodeu* c.1700, *Piode Hamlet* 1734, *Piodau, Piodau-fach* 1831
*piodau appears to be an unrecorded double pl. but it is more likely that it was coined to describe two hos. Piodau and Piodau-fach (SN 619146) (***bach¹***). **Pioden** and *pi²* pl. *piod* are common pn. el.s often associated with **llwyn** with examples in Llandybïe (SN 632166), Llanarthne (*Llwyn y Piod* 1713), Llandeilo Fawr (*Llwynypiod* 1755) and Sanclêr/St Clears (*Llwyn y Piod* 1783) where it may be taken as literally 'grove inhabited by magpies' or sometimes disparaging as in the case of Tafarnypiod (SS 580986), co. Glamorgan, 'the magpies' tavern' (**tafarn**), for a run-down inn.

Pistyll Pistyll-ganol SN 629172
Llandybïe
'spout, well': ***pistyll***
(ht.) *Pistyll* c.1700, 1811, *Pistill Issa* 1774, *Pistill-uchaf, -ganol, -isaf* 1831
The n. of a ht. and three hos. distinguished by ***canol, isaf, uchaf***. Springs are marked on OS maps dating from 1879 at SN 629172. Gomer Roberts (HPLland 280) prefixes the def.art. ***y*** but it seems to be otherwise uncorroborated.

Pont Abram (Pont Abraham) SN 576072
Llanedi, Llan-non
'bridge of (man called) Abram': ***pont***, pers.n. ***Abram***, *Abraham*
(br.) *Pont Abraham* 1964
Recorded as *Gwyly Bridge* 1851, *Gwili Bridge* 1879-1921 carrying the Fforest-Cross Hands road over Afon Gwili constructed c.1780-c.1800 but the n. is subsequently dropped on OS maps. Local pron. is 'Pont Abram'. The ht. is reputedly named from a local but unidentified person (HPLlan-non 22). The old br. was dismantled when the M4 motorway and services were constructed during the 1970s. Often pronounced as if it were the E form Abraham.

Pontaman SN 640126
Betws
'bridge over (r.) Aman': ***pont***, rn. **Aman**
Amman Bridge 1821, *Pontamman* 1831, 1891, *Pont-aman* 1879
The village developed in the mid 19th cent near *Pontamman Chemical Works* 1861.

Pontantwn SN 441130
Llangyndeyrn
'Anton's bridge': ***pont***, pers.n. **Antwn**, *Anton*
(ho., mill, etc) *Pont Anton* or *Gwendrath vach* 1664, (mansion ho., etc) *Pont Antwn* 1673, *Pont Anton* 1760, *Pontantwn* 1765, 1875, *Pont-anton* 1831, *Pont-Antwn* 1891
The pers.n. Antwn – probably derived from Anthony – is fairly common in co. Carmarthen and is otherwise found in lands known as *tir Res Anton'* 1561 (***tir***, pers.n. **Rhys**) in p. Llan-non and the ho.n. *Plas Antwn* 1880 (***plas***) (SN 430089) near Mynyddygarreg. It is possible that there is some family connection with Anthony Jones to whom the ho. was mortgaged in 1673. The br. crosses the r. Gwendraeth Fach and is recorded as a common and ancient br. in c.1791. Pontantwn has been confused with *Ponthanan'* (SDEA 66) but this was near Crymlyn Bog, co. Glamorgan (PNGlamorgan 185).

Pontarddulais SN 592036
Llandeilo Tal-y-bont/Llanedi
'bridge on (r.) Dulais': ***pont, ar*** displacing *aber*, rn. **Dulais³**
(*Lanedi nigh*) *Brige end, Pen y bont aber Duleis* 1550, *Ponte artheleys* 1557, (place) *Bridge end* 1589, *Penybont ar ddylays* c.1670, *Pontardilash* 1674, *Pont ar Ddylais* 1765, *Pontardulais* 1740, *Pontardulais* 1832
The old village (SN 587038) is located in Llanedi, next to a br. called Y Bont Fawr, 'the large bridge' (***y, pont, mawr*** len. *fawr*) over Afon

Llwchwr. The n. also extends to the large urban development on the eastern side of Llwchwr in Glamorgan extending along the Dulais valley and the road leading towards Swansea. Pontarddulais is now the favoured official form based on the supposition that the pn. contains the prep. *ar* 'on' which causes len. of Dulais to *ddulais*. This is an historical contraction, as Deric John has shown (PNBont 18-23), of an earlier Penybont Aberdulais, '(place at the) end of the bridge in Aberdulais'; *aber* refers to the area around the confluence of Dulais and Llwchwr just below the br. Historic forms see-saw between minor variations of Pontarddulais and Pontardulais.

Pontargothi SN 506217
Llanegwad
'bridge on (Afon) Cothi': *pont, ar*, rn. **Cothi**
Pont-ar-Cothi 1309, (gr.) *Pontcothy* 1535, *pont cothe* 1540-1, *Cothy Bridge* 1824, *Pont-ar-Cothi* 1831, (2 fairs at) *Pont-ar-Gothy* 1831
The monastic gr. is Maenor Frwnws (q.v.); see Llandeilo'r-ynys and Mynachdy[1]. The ht. developed at the meeting point of several roads and the br. carrying the main Caerfyrddin/Carmarthen-Llandeilo road over the r.

Pontarllechau (Pont-ar-llechau) SN 728244
Llanddeusant, Llangadog
'bridge built on slabs of rock': *pont, ar, llech* pl. *llechau*
Ponte lleche 1610, *pont ar lleche* 1728, *Pontarlleche* 1824, *Pont-ar-llechau* 1887
The br. (recorded in 1610) crosses Afon Sawdde. Llechau occurs elsewhere as a rn., eg. Nant Llechau (Llanwynno, co. Glamorgan), a tributary of Rhondda Fach, recorded as *nant llecha* 1633 and *Lleche* 1852 (*nant*).

Pont-ar-sais (Pontarsais) SN 442284
Llanllawddog
'the Englishman's bridge': *pont, y, Sais*
pont ar Sais or *Tyr pen pont ar sais* 1731, *Pontarsais* 1744, 1953, *Pontysais* 1783, *Pont-yr-Sais* 1831, *Pont-ar-Sais* 1889
There are some doubts with this explanation because of the early and persistent forms with *ar* 'on' but it may have replaced *y* through association with other pns. such as Pontarddulais and Pontardawe (DPNW 387). The OS 1831 and 1889 show that the br. (SN 442287) crosses Afon Gwili and there is no rn. here to justify a translation 'bridge over r. Sais' as Melville Richards suggests (AMR). The stream on the south side of the village is Nant Corrwg.

Pontbrenaraeth SN 662237
Llandyfân
'foot-bridge on (r.) Araeth', *pontbren*, rn. *Araeth*
Pontpren Areth 1791, *Pontprenaraith* 1810, *Pontbrenaraeth* 1831, *Pontbren Araeth* 1887, 1980
The r. Araeth rises on the western slopes of the hill Trichrug (SN 700230). Pontbren Araeth properly refers to the br. not the small ht. which developed immediately around it extending up the valley to Bethel chp. The rn. is poorly recorded and the only comparable rn. appears to be *Aryth* recorded in *Blaen Aryth* 1668 (**blaen**) in Llanfihangel-ar-arth. It could be a variant of *araith* 'speech or address, speech', etc perhaps a r. which sounds like talking; cf. Ieithon, co. Radnor (PNRad 63).

Pont-henri SN 476094
Llanelli
'Henri's bridge': *pont*, pers.n. **Henri**, *Henry* (mess.) *tyr pont henry*, (br.) *Pont Henry* c.1627, (~) *Pont Henrie* 1642, *Pont Hendry* 1729, *Pont Henry* 1765, 1891
Henri/Henry has not been identified (DPNW 388). The br. crosses Gwendraeth Fawr but the focus of the village has shifted to the crossing (SN 480091) of the Pontyates-Pontyberem road and the Pont Henri-Cynheidre road.

Pont-iets, Pontyates SN 462111
Llanelli
'bridge of (man called) Yates': *pont*, surname **Yates**
Pont Yates 1729, 1831, *Pontyeats* 1765, 1875, *Pont-Yates* 1891

A br. over Gwendraeth Fawr. The surname Yeates is recorded in the neighbouring p. of Llangyndeyrn in 1646 and a Walter Yates of Llangyndeyrn 1666. Later forms suggest association with *iet¹* 'gate' (with E pl. -s). This el. was certainly in use in this area and further afield. Reputedly named from the n. of the first builder of the br. (CAS I, 51; DPNW 393).

Pont Sbwdwr   SN 434059
Pen-bre
'bridge of (man called) Sbydwr' or 'bridge of the hospitaller': ***pont**, (**y**), **ysbydwr***
Ponte Rees Powdwr 1499, *Spoders Brydge* 1592, (br.) *Pont y Spowder* 1609, *Pont Yspydor a Stone Bridge over Gwendraeth Vawr* c.1700, (*i*) *ben pont Rees bodor* 1741, *Spudders Bridge* 1792, (br.) *Pont y Spwdwr* 1796
The epithet or descriptor is first recorded in the n. of Thomas *Spodur*, a juror of Cydweli, in 1292 (BBCS 13: 218) and according to Melville Richards (BBCS 25: 423-4) may derive from *ysbydwr* '(Knight) Hospitaller, hospitaller'. He also suggests that the 1499 sp. with the pers.n. *Rhys* may be by popular etymology although it is curious to see that it recurs in 1741. The el. also seems to occur in a fn. *Cae Sputtre* in Llangyndeyrn p. in 1771 (***cae***) and is found in the ns. of David *le Spodur* c.1281 at Bonvilston, Thomas *le Spodur* 1291 at Tythegston and Adam *Spudur* 1492 and 1542 at Roath, Glamorgan (AC 1920: 85).

Pont-tyweli   SN 412402
Llanfihangel-ar-arth, Llangeler
'bridge over (r.) Tyweli': ***pont**, **Tyweli***
Pont-taf-wili 1831, (place) *Pontyweli* 1841, *Pont-Tyweli* 1889, 1974, *Ponttweli* 1899
Named from a br. over r. Tyweli (SN 412401). The n. applied to the ht. which developed either side of it along the road (now Station Road and Lewis Street) to Llandysul Bridge.

Pontyates see **Pont-iets**

Pontyberem   SN 501111
Llanelli, Llangyndeyrn
'bridge at ?confluence of (Nant) Berem (and Gwendraeth Fawr)': ***pont**, ?**aber**,* r. *Berem*
(br. called) *Pont y berran* 1606, (~) *Pont y Beran* 1609, *Pont y Berem* c.1700, *Pont Beram* 1729, *Pontyberem* 1783, *Pont-y-berem* 1831, 1889
Pontyberem is probably a contraction of an earlier *Pontaberberan in which the first of the repeated syllables -ber- has been dropped in common speech; cf. Abérthin (< Aberberthin), Glamorgan, and unstressed a- has been associated with the def.art. ***y***. The br. is immediately below the meeting point of the stream Nant Berem (so spelt 1880) with Gwendraeth Fawr, also recorded in the ho.n. Cwm-berem recorded as *Cwm Beren* 1847 and *Cwm-Berem* 1880 (***cwm***). Beran, Berem may contain ***ber*** 'short' and suffix ***-an*** with the common variation of -n/-m and perhaps influenced by association with ***berem***, *berm* 'yeast' (DPNW 393).

Pontyfenni   SN 2317, 2417
Llanboidy, Sanclêr/St Clears
'bridge of (Afon) Fenni': ***pont**, **y*** and rn. ***Fenni***
Pont y Venny Bridge c.1700, *Pont Fenny* 1729, *Pontfenni* 1739, (br.) *Pont y Venny* 1792, *Pontifenny* 1819, *Pont-y-fenni* 1888-9
The rn. is recalled in Plas y Fenni (unlocated, in Llangynin) recorded as *Plas y veny otherwise Tythyn y werne Lloyd otherwise Park y Gwenyth* 1621 which may be interpreted as 'mansion near Y Fenni' (***plas***), 'small farm at the grey alder-marsh' (***tyddyn, gwern, llwyd***) and 'the wheat field' (***parc, gwenith***).

Porth Tywyn, Burry Port   SN 446010
Pen-bre
'harbour at the sand-dune': ***porth³, tywyn²***; 'port on Burry estuary': rn. *Burry/Byrri*, E ***port***
Berry Port 1833, *Burry Port* 1841, 1867, *Porthladd Porth Tywyn* 1870-9
The n. derives from the northern part of Burry inlet recorded as *Burrey a Creeke* (for Llanelli) 1566 and *the creek of North Burry* 1636. The modern town developed on sand burrows described as *Twynbach* in 1911 (HMSG 73)

(*tywyn, bach¹*). The harbour (*porthladd*) was initially known as the New Pembrey Harbour and re-named in 1835 as Burry Port Harbour 1835 (SGStudies 126, 130-1). Pembrey Harbour is a distinct construction made in 1819. Burry/Byrri is ultimately drawn from a rn. in Gower, co. Glamorgan, recorded as (r.) *Borry* 1318, *Byrri Auon* early 14th cent, *Burrey water* c.1670 (see PNGlamorgan 25-26). The n. was extended in error to the whole estuary of the r. Llwchwr/Loughor as *Burry Inlet*, now dropped on recent OS maps. The n. may be drawn from OE *byrig* 'fort', possibly referring to a small fort at North Hill Tor (SS 45309381) in Gower (DPNW 57). It has been suggested that the rn. contains *bwr* adj. 'fat, strong, big', with a suffix *-i* but Burry/Byrri is a small stream passing through marshes and mudbanks. There seems to be little support for the suggestion that the rn. is related to E *burrow* (OE *beorg*) in the sense of 'sand-dune' since there are no sand-dunes immediately around the stream in Gower.

Porth-y-rhyd¹ SN 711378
Cil-y-cwm, Llanwrda
'gateway at the ford': ***porth²**, **y**, **rhyd***
Tyr porthe yr hyd 1612, *Tyr Porth y Rhyd* 1677, *Tyr Porth y Rhyd, ~ issa* 1716, *Porth y Rhyd ycha and Porth y Rhyd issa otherwise Rhyd yr Evel* 1735, *Porthyrhyd* 1831, *Porth-y-rhŷd* 1891
Located at and near several crossroads adjoining Afon Mynys (DPNW 398). The ford may have preceded the br. taking the road running east-west over Mynys through the village.

Porth-y-rhyd² SN 519160
Llanddarog
'gateway at the ford': ***porth²**, **y**, **rhyd***
(two closes) *tir porth y rhyd* 1612, *Porthyrhud* 1740, *Porth-y-Rhyd Gate* 1813, *Porth-y-rhyd* 1831
Located at a road junction; no ford is shown on the OS 1:10,560 in 1891 but it may have been replaced by Pont Lan Lucis (SN 521161) north of the village on the lane (to Llandeilo'r-ynys) crossing Gwendraeth Fach. Other instances of Porth-y-rhyd are recorded in Llansawel and Llan-y-crwys.

Pump-heol, Five Roads SN 490054
Llanelli
'(meeting of) five roads': ***pump**, **heol***; E ***five**, **road**, -s*
Pump-heol 1833, *Five roads* 1875, 1880, *Pumpheol* 1884, *Pumheol* 1886
The W form is variously given as *Pum Heol* 1960 and *Pump-hewl* by Elwyn Davies (GazWPN). The n. also applies to *Five Roads Inn* 1880, 1960 which may be the *Farmers Arms* 1875.

Pumsaint SN 656406
Caeo
'five saints': ***pump**, **pum**, **sant** pl. **saint***
(chp.) *Lanympymseint* 1331, *Capel llanpymsent* 1578, *Lana Pinsent chappell* 1587, (village) *Pymsaint* 1704, *Pumsaint* 1831, *Pumsant* 1836, *Pumpsaint* 1888
The n. indicates either a well, ch. or chp. dedicated to five saints as in the case of Llanpumsaint (q.v.) The first ref. in 1331 records it as one of the chps. of Cynwyl Gaeo (see Caeo). Pumsaint has also been identified with *lann teliau pimseint caircaiau* mentioned in a papal bull c.1170 (LL 62). If the identification is correct then there must have been a church here dedicated to both St Teilo and the five saints. One of them Gwynno is recalled in Ffynnon Gwenno (***ffynnon***), a cave in the Roman goldworkings at nearby Dolaucothi. No ref. to a chp. has been found after 1587. A standing stone known as Carreg Pumsaint (***carreg***), at the workings, has been identified as a mortar stone and circular marks on the stone made by industrial processing were explained in legend as the marks of the five saints stuck to the stone by the devil until released by Merlin.

Pwll SN 480010
Pen-bre
'pool' or 'pit': E ***pool**, **pwll***
Pool 1726, 1729, *the Pool* 1792, *Pwll* 1813, 1830
Early forms favour E ***pool*** perhaps a coastal pool in view of the early refs. Traces of early coalworking suggest that it may have been associated with a pit (***pwll***) used for the extraction of minerals. The n. is applied in 1880 to the area near Pwll Colliery (SN 474009) but

extended eastwards before 1907 to include Pen-llech and the area around Holy Trinity ch. (SN 480010). Coal was exported from Pwll Colliery by way of its own quay.

Pwll-trap SN 2616
Sanclêr/St Clears
'pit used as a trap' or 'pit near (a place called) Trap': *pwll*, *trap¹*
Pwlltrapp 1682, (water corn grist mill) *Pwll Trap Mill* 1688, *Pultrap* 1729, *Pwll trap* 1786, *Pwll-trap* 1839, *Pwlltrap* 1868
The significance of the n. here is uncertain. There is now no mill pool or suitable place for a fish-trap in the close vicinity. It is possible that *trap¹* is employed in a like manner to Trap (q.v.).

Pysgotwr, Afon SN 724537, 766482
rn. to Afon Doethïe
'fish water, water notable for its fish': *pysgod*, *dŵr*
Pescotter flu: 1578, *Pescover* 1612, (rs.) *Pyscottwr vawr*, ~ *vach* 1690, *Pyscottwr* 1815
Rising as Nant y Bryn and known as Pysgotwr Fawr down to a tributary Afon Pysgotwr Fach (SN 717508, 747498) (*mawr*, *bach¹*).

Ram to Roche Castle

Rhydaman/Ammanford. Looking southwestwards down Wind Street c.1910.

Ram　　　　　　　　　　SN 580468
Pencarreg
(area at) Ram (inn)
Ram 1851, (inn) *Ram* 1875, (area) *Ram, Ram Inn* 1888
A PH once having the sign of a ram (**ram**) shown at the junction (SN 590468) of the A482 and an older road running south-eastwards to meet the A482 at the former ho. at Crosshands (SN 613437) (DPNW 406).

Red Roses see **Rhos-goch**

Regwm (Y)　　　　　East Regwm SN 226175
Llanboidy
Uncertain
y regwm 1609, (lands) *tyr yr eywm* 1613, *Regum* 1620, *Regwm* 1670, *Rhegum* 1751, *Regum* 1811
A lost rn. proven by refs. to *Aber yreggom* 1561 and *Aberregum* 1680 (**aber**). The rn. seems to be *Yregwm* or *Eregwm* later misdivided as *Yr Egwm, Regwm* through association with the def.art. *y, yr, 'r*. The unfamiliarity of the n. explains the misguided suggestion that it is a 'corruption' of Latin *terra regwm* 'kings' land' (**terra, rex** pl.gen. **regum**) because Regwm contained a small forest common to Crown tenants (AC 1975 108-9). East and West Regwm (SN 217177) lie either

side of a short tributary flowing southwards to Afon Taf. In the 1670 ref. it is described as in the gr. of Iscoed (*Yscoed*) which David H. Williams identifies as the final site of Hendy-gwyn/Whitland abbey (SN 208182) (WCist 182-3, 313).

Rhandir Abad SN 775446
Llanfair-ar-y-bryn
'abbot's shareland': *rhandir*, (*yr*), *abad*
Randir Rabad 1671, *Rhandir Abad or Abbotts hamlet* 1710, *Rhandir-Abbot* 1804, *Rhandir yr Abad* 1809
The area was held by the abbot of Ystrad Fflur/Strata Florida and *abad* distinguishes it from nearby Rhandir-mwyn (q.v.). Ystrad Fflur/Strata Florida possessed the chp. at Capel Peulin and nearby Nant-bai (q.v.).

Rhandir Cil-y-gell see **Cil-y-gell**

Rhandir Ganol SN 8040
Llanfair-ar-y-bryn
'middle shareland': *rhandir*, *canol*, cenol
Randir genoll 1574-1609, *Rhandir ganol, or middle hamlet* 1710, *Rhandir-ganol* 1804
The central part of Llanfair-ar-y-bryn around Cynghordy. Cf. Rhandir-isaf and Rhandir-uchaf and Rhandir Ganol (Myddfai) recorded as *y Rhandir Genol* 1608.

Rhandir Isaf SN 8135
Llanfair-ar-y-bryn
'lower shareland': *rhandir*, *isaf*
y Rhandir issa 1602, (place) *y Randir Yssa* 1637, *Rhandir Isha Hamlett* 1671, (ht.) *Rhandir isa, or lowest hamlet* 1710, *Rhandir issa* 1739, (ht.) *Rhandir-Isaf* 1851
One of the four hts. of Llanfair-ar-y-bryn (CarlisleTD) and 'lower' in contrast to Rhandircanol and Rhandiruchaf (q.v.).

Rhandir-mwyn SN 784437
Llanfair-ar-y-bryn
'shareland at the mine': *rhandir*, lost def.art. *y*, *mwyn*²
Rhandir y Mwyn 1814, *Rhandirmwyn* 1822, *Rhandir-mwyn* 1834

The n. need not refer to a specific mine and probably describes an area associated with lead-mining since it occurs in company with the ho.n. *Nantymwyn* 1831 and the hill Pen Cerrig-mwyn (SN 793442). The OS map 1891 shows *Nant y Mwyn Lead Mine (Deep Boat Level)* at SN 781434. Rhandir-mwyn applied in 1834 to buildings at SN 785446 but the n. has shifted south. It seems to be a relatively late pn. and it is significant that it is not named as one of the four administrative ts. (Rhandir Abad, Rhandir Ganol, Rhandir Isaf and Rhandir Uchaf) of Llanfair-ar-y-bryn.

Rhandir Rhydodyn SN 6334
Llansawel
'shareland at Rhydodyn': *rhandir*, pn. **Rhydodyn**
(place) *Randir Rid odin* 1558, (ht.) *Rhandir Rhydodyn* 1603
Centred on Rhydodyn (q.v.).

Rhandir Uchaf ?SN 8135
Llanfair-ar-y-bryn
'upper shareland': *rhandir*, *uchaf*
Rhandir Ycha 1671, (ht.) *Rhandir ucha, or highest hamlet* 1710, *Rhandir-ucha* 1804
'Upper' in contrast to Rhandir Isaf and Rhandir Ganol (q.v.). The precise area of the ht. is not established but it probably included the area north and east of Rhandir-mwyn.

Rhiw-las Rhiw SN 610259
Llandeilo Fawr
'blue-green slope': *rhiw*, *glas*
(place) *y Riwlas* 1611, (3 messuages) *Rhywlas* 1680, mess. etc) *Rhulas* 1683, (ht.) *Rhiwlas* 1724, (ho.) *Rhiw-lâs* 1831
On the western slopes of a small hill Cefn Rhiwlas overlooking Afon Myddyfi. A common pn. and ho.n. with further examples in Llanfynydd and Pen-bre.

Rhiw'radar SN 596231
Llangathen
?'the birds' slope': *rhiw*, *yr*, *adar*
Arauder 1390, *Riwoireder* 1523, (tmt.) *rhyw r addar* 1665, *Rue Raddor* 1675, *Rhiw'radar* 1743, *Rhiw'r Adar* 1807, *Rhiw-yr-adar* 1831

There are some doubts with this interpretation in view of the first historic form. Located on the eastern side of a low hill over which a small lane leads down to Afon Myddyfi. Rhiw'radar was one of the meeting places of courts for Catheiniog (q.v.) (HCrm I, 226) and Richards (WATU) identifies it as a manor in Llangathen coupled with Cilsân (q.v.).

Rhos SN 385355
Llangeler
'moor (in Llangeler)': ***rhos**²*
Rhôs 1844, (ho.) *Rhos-Geler* 1891, (fm.) *Rhosgeler* 1906, *Rhos*, (fm.) *Rhos Geler* 1947
The ht. is shown on OS maps as *Sarn-fach* 1831 and *Sarn-fâch* 1891 (**sarn**, **bach**¹) but this gave way to *Rhôs* before 1906, a n. taken from the fm. (enclosed from moorland) and applied to the area extending along the Llangeler-Cynwyl Elfed road running south from *Tŷ-isaf* (**tŷ**, **isaf**) and *Glasbant* (SN 379355) (**glas**, **pant**) to St James's ch., Bryn Saron chp. (opened 1877), and *Pen-ffordd-newydd* 1906 (**pen**¹, **ffordd**, **newydd**). This may be the location of a fair at *Castell newydd yn Ros* (*a Chynwilgaio*) 1612 and (~) *Castell Newydd yn Rhos* 1831 (**castell**, **newydd**). The boundary with the parish of Caeo or Cynwyl Gaeo lies a little over a mile south of Rhos. The longer form *Rhosgeler* simply records its location in Llangeler p.

Rhosaman SN 7314
Llangadog p.; Cwarter-bach
'moor near (r.) Aman': ***rhos**²*, **Aman**
(area) *Rhos-amman* 1831, *Rhôs-aman Colliery* 1877, *Rhôs-amman Colliery* 1906
The small ht. developed near a ford taking what is now the A4088 over the r. Aman and is described as *Rhyd-gwyn* and *Rhŷd-wen* 1906 and 1917. See Rhyd-wen.

Rhos-goch, Red Roses SN 203118
Eglwys Gymyn, Talacharn²/Laugharne
'red moor': ***rhos**²*, **coch** re-interpreted as E **red**, **roses**
Rosgoch 1307, great *Rosegough* 1653, *Rose Gough* 1655, *Great Rhos-Goch Fawr* 1780, *Rhosgochfach* and *Rhosgochfawr* 1819, *Red Roses* 1843, (fms.) *Red Roses, Rhosgochfawr* 1875, *Red Roses (Rhos-goch)* 1970-1
The ht. developed after 1819 near the Llwyn-gwair Arms and a Calvinistic Methodist chp. Red Roses is probably a re-interpretation of the nearby ho.ns. Rhos-goch-fach (formerly SN 205124) and Rhos-goch-fawr (SN 208127) (**bach**¹, **mawr**). The latter is now plain Rhos-goch after the abandonment of the former ho. shortly before 1907. Association with roses may have been encouraged by confusion with **rhos**¹ 'roses, rose bushes' especially as an heraldic device (DPNW 407).

Rhos-maen SN 640239
Llandeilo Fawr
?'moor (or promontory) of the stone': ***rhos**²*, **maen**
Rosmayne 1532, (vill) *Rosemaen* 1542, *Rosemayne* c.1575, *Rosmaen, Rosemane* 1640, *Rhos y Maen*, *Bryn y Maen* 1760, *Pentre Rhosmaen* 1820
The n. must be taken from the hill above the ht. but this is not typical moorland and there is little or no regularity of field boundaries to indicate late enclosure. It is possible that **rhos**² here has the alternative sense of 'promontory'.

Rhwng Twrch a Chothi see **Maenor Rhwng Twrch a Chothi**

Rhydaman, Ammanford SN 630124
Llandybïe
'ford on (r.) Aman', E **ford**, W **rhyd**, and rn. **Aman**
Ammanford, Rhydaman 1882, *Rhydaman (Amanford)* 1891
The ford over Aman was at Betws br. Recorded as *Cross Inn* 1807 and on OS maps down to 1891 and as *y Gwesty-croes (Cross Inn)* 1835 in ref. to an inn (**gwesty**, **croes**). Ammanford and Rhydaman were adopted as new ns. by the Post Office in 1882 because post was straying to other places called Cross Inn. The railway station changed its n. about the same time. Some W-language sources continued to use Cross Inn, eg. James Morris 1905 (in ESeion) in preference to Rhydaman. The area around the inn developed from the late 18th cent and was

the location of a fair in 1831 (Coflyfr 74). The town developed rapidly with the expansion of coal mining in the late 19th and 20th cents.

Rhydargaeau SN 438263
Llanllawddog
'ford at the weirs': *rhyd*, *argae* pl. *argaeau*
rydargaye otherwise *pensyr* 1601, (mess.) *Tyr Rhyd ar Gaye* 1747, *Rhydargaue* 1784, (ht.) *Rhŷd yr Gaue* 1811, *Rhydargaie Chapel* 1812, *Rhyd-y-caeau* 1831, *Rhydargaeau* 1839
Located on Nant Brechfa but there are now no weirs in the close vicinity and this may have contributed to the supposition that the pn. contains *y* and *caeau* pl. of *cae*. The form *pensyr* refers to a ho. *Pantyseiry* 1811, *Pant-y-seiri* 1889 (SN 440271), *Pantseiri* 1979-80 a little to the north, 'hollow of the ?paved path or causeway' with *pant*, *y*, and ?*seri¹* associated in later forms with *seiri* pl. of *saer* 'carpenter' etc. A former t. in Llanllawddog but part of the modern village – along the road between Horeb Baptist chp. and Bethel Calvinistic Methodist chp. – is in the historic p. of Llanpumpsaint.

Rhydcymerau SN 578389
Llansawel, Llanybydder
'ford of the confluences': *rhyd*, lost def.art. *y*, *cymer¹* pl. *cymerau*
Rhydkymere, Rhydkimere 1584, *Rheed y Cum(m)ere* 1729, *Rhyd-cymmerau* 1891
The ford (now bridged) is located where the Llanybydder-Llansawel road crosses over Afon Melinddwr. There is a second ford over a small tributary Nant Moelen immediately before the first. A ridgeway – now largely a bridleway – runs north-west from Llansawel to Rhydcymerau where it meets the B4337. Shown as *Capel* (*capel*) on the OS map in 1831 referring to the Calvinistic Methodist chp.

Rhydodyn, Edwinsford SN 632346
Llansawel
'ford near a lime-kiln': *rhyd*, *odyn*, E pers.n. *Edwin, ford*
Ridodyn 1394, (*o*) *ryd odyn* late 15th cent, (land) *tythin rydodyn* 1526, *Rhyd Odyn* 1542, *Rhydodyn* 1591, *Rhyd Odwyn* 1655, *Edwins foord* 1711, *Edwinsford* c.1735, 1809
A ford over r. Cothi replaced by a br. Rhydodyn was reinterpreted from c.1710 as the possessive form of an E pers.n. *Edwin* and *ford* partly prompted by occasional W forms containing inorganic -*w*- which seems to first occur mid 17th cent. Use of the reinterpreted form Edwinsford was reinforced by its use for the large mansion (SN 63124577).

Rhydowen SN 194285
Eglwys Fair a Churig
'ford of (man called) Owen': *rhyd*, pers.n. *Owain* variant *Owen*
Rhydowen 1873, *Pont Rhyd-Owen, Rhyd-owen Station* 1889-90, 1907
The n. is taken from a ford over Afon Taf which marks the boundary between Llanfyrnach, co. Pembs., and Eglwys Fair a Churig, replaced by a br. (*pont*) before 1826 (when it was re-built). The n. took root owing to the construction of the station on the Whitland & Cardigan Branch railway opened in 1873. The closure of the railway station in the 1960s was followed by addition of new hos.

Rhydsarnau SN 571100
Llan-non
'ford at the causeways': *rhyd*, *sarn* pl. *sarnau*
Rhyd-sarnau 1831, 1879, (mess. etc called) *Rhydsarne* 1837
The ford gave passage for the road – now the A48 – running between Cross Hands and Pontarddulais over an unnamed tributary of Afon Gwili. The br. here was constructed before 1857.

Rhyd-wen SN 731140
Llangadog (Cwarter Bach)
'white ford': *rhyd*, *gwen*
(place) *y Rhyd wen ar Amman* 1610, (place) *y Rhyd Wen ar Ammon* 1764, *Rhyd-gwyn* 1877, 1898, *Rhŷd-wen* 1906, *Rhyd-wen* 1953
A ford, white perhaps from foam, is shown here on the OS maps 1877 and 1898. Now part of Rhosaman. Identical ns. are found in Cynwyl Elfed recorded as *Rhyd Wen* 1787

and in Llandysilio recorded as *Rhŷd-wen* 1729, *Rhydwenn* 1774 and *Rhydwen* 1819 (SN 143217).

Rhydwilym SN 113249
Llandysilio
'Gwilym's ford': **rhyd**, pers. **Gwilym**
Rhyd Willym 1694, *Rhyd Wylym* 1704, *Rhid willim* 1767, *Rhydwilim* 1819, *Rhydwilym* 1981-2
Located on Cleddau Wen/Eastern Cleddau. Best known for Capel Rhydwilym Baptist chp. (erected 1701 RC I, 369.

Rhydybennau, Afon SN 1320
rn. SN 128227 to SN151195
'r. near Rhydybennau': **afon**, ho.n. *Rhydybennau*
Afon Rhyd-y-bednau 1890
The ho. is recorded as *Redbenny* 1631, *Rhyd y benno* c.1700, *Rhydybennau* 1819, *Rhydybenne* 1851 and *Rhyd-y-benau* 1890 meaning 'ford of the carts, a ford wide enough for carts': **rhyd, y, ben**[1] pl. *benau* located near a ford over the r. in p. Llandysilio. The rn. may properly be Aeron with hos. *Llan-aeron* (SN 129214) (**glan**) and *Bryn-aeron* (SN 126211) (**bryn**). Cf. Nant Aeron.

Rhydyceisiaid SN 240211
Llanboidy, Llangynin
'ford of the tax-collectors *or* catchpoles': **rhyd, y, ceisiad**[2] pl. *ceisiaid*
Read y kyssed 1609, (Protestant ch. at) *Ryd y Ceisied* 1723, *Rhydyceisiad* 1742-3, *Rhydycaished* 1819, *Rydyceisiaid* 1845, (br.) *Pont Rhyd-y-ceisiaid* 1907

The location of an Independent chp. (SN 243210) established c.1723 and re-built in 1777 and 1888 (HEAC III, 366-8; CYB 1933: 448, says 1707) but the n. no doubt originated with a ford preceding the br. which carries the Llanboidy-Sanclêr/St Clears road over Afon Gronw. **ceisiad**[2] has a number of meanings including 'sergeant of the peace, treasurer, seeker' and its precise significance here is uncertain.

Roche Castle SN 295103
Talacharn[2]/Laugharne
'castle of the de la Roche family': Fr family n., E **castle**
Roch late 13th cent, *Roze* 1327, *Rupa* 1545, *Roche alias Machrellswalles* 1573, *the Roches* 1584, *Roch Castle* 1798, *Roach Castle* 1831
The family n. appears in the ns. of Gilbert de Rupe late 13th cent, Thomas de la Roche otherwise Thomas de Rupe 1307, and his son William de Rupe, lord of Macrels (q.v.) with Latin **rupa** 'a rock' corresponding with Fr **roche**. Thomas de Rupe may be identified with Thomas de Roche who witnesses a charter of Guy de Brian granted to Talacharn[2]/Laugharne. A David Roche also occurs in connection with property in Talacharn[2]/Laugharne in 1565. The castle – or fortified ho. – was in ruins in the early 17th cent (HCH 174). Many historic sps. bear close comparison with those for Roch (Y Garn), co. Pembroke (PNPemb 622) and Roche, in Cornwall (CDEPN 503).

Salem to Sylgen

Sanclêr/St Clears. Looking northwards up High Street from the Butcher's Arms towards Capel Mair c.1900.

St Clears see **Sanclêr**

St Ishmael see **Llanismel**

St Peter's see **Carmarthen**

Salem SN 623266
Llandeilo Fawr
Salem 1973-4
Named from the Independent chp. recorded in 1887. The area is *Heol-galed* 1831, 1887, *Heolgaled* 1854 meaning 'hard road': *heol*, *caled* describing the road running north-south through the village. The chp was established 1817 on land sold by Lord Robert Seymour, Taliaris (HEAC III, 573-4). Salem is an Old Testament n. for Jerusalem.

Sanclêr, St Clears SN 2715, 2716
Llanfihangel Abercywyn
'(church dedicated to) St Clair *or* Cleer': *sant*, variant *san*, and *Clêr*, E and Fr *saint*, pers.n. *Clear*,
(*de*) *Sancti Clari castello* c.1191, (*in*) *Sancto Claro* 1230, (*de*) *Sancto Claro* c.1285, *St. Clare*, *Seint Cler* 1325, (lp.) *Seintclere* 1382, *castell Seint Cler*

c.1400, (cmt.) *Seynclere*, (manor) *Senclere* 1435, (vill) *St. Clair* 1515, *Clare Castel hard by Saint Cleres Chirch* c.1537, *St. Clere* 1584, *Sainct Cleeres* 1610, *sain kler* c.1566

Sanclêr/St Clears as a n. seems to have applied initially to the Cluniac priory established between 1147 and 1184 as a daughter ho. of St Martin-des-Champs, Paris. Clair or Cleer has been associated with a 9th cent E saint also dedicated at St Cleer, Cornwall, whose main cult was at St-Clair, Normandy (CornPN 69). Cleer was also dedicated at a chapel near Somerton, co. Somerset, and at two chapels in Hartland, Devon (LBS II, 15). At Sanclêr/St Clears it is likely that the dedication was an AN import applied to the priory and later to the castle and town. At some point the priory acquired an additional dedication to Mary Magdalen (SGStudies 242) since it is described in 1279 as *prioratus beate Marie Magdalene, alias Sancti Clari in Vallis*. The description of the priory as 'in the valley' (*in Vallis*) may allude to its location in the valley of Afon Taf. The town church is described as dedicated to St Mary in 1417. Dual dedications are known elsewhere in Wales often through the addition of a continental or biblical saint to a local one but there seems to be no good evidence of any dedication to Cleer or a local saint here. Local pron. is *Singclêr* or *Sangclêr* (DPNW 431). The W form is clearly derived from E forms such as *Sencler* c.1386-7 and *Senclere* 1435. The priory was not dedicated to St Clare of Assisi (1194-1253) as has been supposed.

Sandy SN 498006
Llanelli
'sandy place': E **sandy**
Sandy 1803, 1879, *y Sandy* 1931
From its location near the Loughor estuary.

Sannan, Afon SN 5526
rn. SN 558297 to Dulas SN 560232
?pers.n. *Sannan*
River Sannan 1887-9
The pers.n. is thought to be that of a saint in Llansannan, co. Denbigh, but R.J. Thomas also cites secular instances (EANC 85-6). The base is said to be Latin *sanctus* 'saint' but it is more likely to be **sant** 'saint, holy' with suffix **-an** and refers directly to the r. This rises just above Blaensannan (**blaen**) and reaches Dulas below Abersannan (SN 561240) recorded as *Aber Sanan* c.1555, *Abersannan* 1605-1809 and *Aber-Sannan* 1891 (**aber**). Thomas also locates *Abersynnann* 1331 here but this appears to be a slip for *Aber y Ffynnaun* (**ffynnon**).

Sardis[1] SN 584080
Llanedi
Named from a Baptist chp. **Sardis**
Sardis 1873
The chp. was built in 1849 and is located near a ho. *Heol-dwr* 1830, *Heol-y-dwr* 1879, 'the water road (**heol**, **y**, **dŵr**) leading southwards from Llanedi to Fforest and the r. Loughor/Llwchwr. Sardis was a notable city in western Turkey mentioned in Classical writings and the Bible.

Sardis[2] SN 745289
Myddfai
Named from an Independent chp. **Sardis**
Sardis 1799, 1836
An Independent chp. 1792, rebuilt 1827 (HEAC III, 579-81).

Sarnau SN 339189
Meidrum
'causeways': *sarnau* pl. of **sarn**
Sarne 1766, 1817, *Sarnau* 1831, 1875, (ho.) *Sarnau, Sarnau Farm* 1889
Located on the old Caerfyrddin/Carmarthen-Haverfordwest road which crosses the floodplain of Afon Taf. Meidrum has another Sarnau (SN 313209) recorded as *Sarnau* 1757, 1831 on the Meidrum-Caerfyrddin/Carmarthen road (B4298) with a nearby Sarnbwla (SN 317205, 318202) recorded as *Sarn-y-bwla* 1831 and 1889 (**bwla**, *bwly* 'bull, castrated bull').

Saron[1] SN 601125
Llandybïe
Named from a Baptist chp.: **Saron**
Eglwys Saron 1839, *Saron* 1856, (village) *Saron* 1878-9
The Baptist chp. was built in 1824.

Saron[2] SN 373372
Llangeler
Named from an Independent chp.: **Saron**
Saron 1839, 1852, Capel Saron (Ind.) 1889
The Independent chp. was built in 1792 to replace an older, smaller chp, and rebuilt again in 1897-8 (HPLlangeler 205, 210; HEAS 16). The earlier chp is also said to date to 1790 (HEAC III, 415-9). Biblical Saron was an area of beauty and fertility.

Sawdde, Afon SN 8021, 6928
rn. SN 802218 to Tywi SN 694281
'sinking river': **sawdd**[1], **-ai**
Sawdai (= *Sawddai*) late 15th cent, *Sawthey flu: 1578,* (r.) *Sawtheie 1586, Sawthy 1612, Sawthey 1623, Sawdde 1831*
The r. flows over sandstone, as R.J. Thomas notes (EANC 32-33), and the n. probably alludes to pot-holes formed in deep crevices in the rocks in the last three miles of its course. Sawdde has two branches, the main one rises in Llyn y Fan Fach, formerly called Llwch Sawddai (***llwch***), running by Blaenau and Twynllanan to the r. Tywi. The other, Sawdde Fechan, rises on Mynydd Du/Black Mountain (q.v.) close to Blaenllynfell (SN 766184), the source of Llynfell, meeting Sawdde at Pont Newydd (SN 737235). See also Blaensawdde.

Seaside see **Glanmôr**

Sïen, Afon SN 2526
rn. SN 239288 to Cynin SN 274238
?pers.n. *Sïen*
Afon Sien 1889
Sïen may be a variant of the recorded pers.n. Sïan (not Siân). R.J. Thomas (EANC 123) suggests that it describes the natural sound of its waters, i.e. 'r. which babbles *or* murmurs' with ***si***[1], ***-en*** but recorded forms are very late. Rises near Blaenffynnon (**blaen, ffynnon**).

Siloh SN 741371
Cil-y-cwm
'(place at chp. called) Siloh: **Siloh**
Pentre-Siloh, Capel Siloh (Calv. Meth.), Shop-Siloh 1889, Siloh 1979-80
Named from a Calvinistic Methodist chp. opened 1860 (HMSG 170) taking its n. from Biblical Siloh or Shiloh.

Soar SN 620285
Llandyfeisant
Named from a Baptist chp. called Soar: **Soar**
Pantsoar 1845, Pant-Soar 1888, 1947, Soar 1977
Named from a Baptist chp. near Ffos-ddu. Capel Soar (***capel***) is now Soar Cottage. The n. is consistently Pant-Soar, 'Soar hollow' (***pant***) down to recent OS editions.

Strade (Stradey) SN 491015
Llanelli
'vales': **ystrad** pl. *ystradau*
Straddye 1574, Parke Estrade 1577, Istradey 1611, Strade 1632, Stradey 1677-1891, Strada 1792
The n. probably refers to the flattish land where the valley of Afon Dulais (*Dulais* 1880) merges with Afon Cwm-mawr which flows down from Penymynydd (q.v.) to meet the coastal land. The anglicised form Stradey is found in Stradey Park rugby ground, the home of Llanelli RFC. Stradey was also the n. of a former mansion demolished when its replacement Stradey Castle was built (HCH 176-7).

Swan Pool Drain SN 4002, 4101
SN 426008 to Gwendraeth Fawr SN 412056
'drain at Swan Pool': E ***swan, pool, drain***
Swan Pool Drain 1879-80
The OS applies the n. to the stretch from SN 410017 to SN 412014, the northern part shown on a plan c.1681. It is evidently named from *Swan pool* recorded c.1700 by Edward Lhuyd near Pen-y-bedd (SN 414028) and Twyn-mawr (approximately SN 397037). The pool was notable for eels, ducks and 'sometimes wild Swans or Elkes and Wild Geese'. The drain was part of a system constructed to drain the marshes between Pen-bre and Cydweli (SGStudies 152-4).

Swiss Valley see **Glyn y Swisdir**

Sylen SN 511070
Llanelli
Sylen 1880
The n. is taken directly from the former Baptist chp. Capel Sylen recorded as *Capel-sulen* 1833 and *Capel-sylen* 1880 (SN 519067) otherwise known as Capel y Mynydd (***capel, y, mynydd***). The chp. should not be confused with a much older chp.-of-ease which may be identified as Capel Dyddgu (q.v.). The n. is drawn ultimately from the hill Mynydd Sylen (q.v.). Sylen also occurs in a fm.n. (SN 503071) recorded as *Glyn Sylen* 1868, *Glynsylen* 1880 (***glyn***).

Sylgen, Afon SN 3033
rn. SN 325336 to Cuch SN 293337
Uncertain
Afon Sylgen 1889-90
The apparent absence of early evidence prevents confident explanation. The main stream rises near Blaen-trench (SN 325336) (?***transh***, *trensh*) but the ho. Blaensylgen (SN 317329), in Cilrhedyn, recorded as *Blaen-sylgan* 1831 (***blaen***), is near a tributary Nant Fach and it is possible that Sylgen was a two-headed r.

Tachlouan to Tywi

Trimsaran. Looking northwestwards from the bridge over Afon Marlais towards Sardis Independent chapel and Bryncaerau c.1930.

Tachlouan SN 5927, 6027
Llandeilo Fawr
?'hill-slope of Cloufan': ?*telych* and pers.n. *Cloufan*
Telichclouman, Telich clouuan c.1145, (place) *Tal ach Loyan* 1605, *Talychlayan* 1611, (ht.) *Tachloyan* 1741, *Tachloian* 1746, *Tach Lleuan* 1811
Telych Cloufan is the form preferred by Richards (WATU) based on the 12th cent forms in the Book of Llandaf. The first el. *telych*, qualified by a pers.n. Cloufan, seems to survive in other parts of south Wales in Bryntelych (Caeo), Bryntelych (Llandeilo Tal-y-bont), Cefntelych (Llanddingad/Myddfai) recorded as *Keventellich* 1682, and Cefntelych-hen (Glynrhondda, co. Glamorgan). Its coupling with the pn. els. *bryn* and *cefn* may be behind the suggestion that it is Ir *tulach*[1] 'low hill, hillock' or a loan from that language. It is worth noting that Tachlouan was centred on a ho. Ffynnondeilo (SN 594279) recorded as *Ffynnon Dilo alias Tochloyan* 1873 located on the south slopes of a hill. See further under Telych.

Taf, Afon SN 1622, 2515
rn. SN 184340 to SN 3308
'flowing (r.)': Brit *****tam-**, **-i**
Tam, taf c.1150; *Taf* c.1170, *Taph* c.1191, *Taua, Taue* c.1538, *the Tave* 1586, *Afon Tâf* 1890

Identical in meaning to Taf (PNGlamorgan 211) and belonging to a group of rns. including Thames, Tawe, etc which were once thought to contain IE *tam-* in the supposed sense 'dark'. The gen. form *Tam-i* developed into *Tyf* by *-i* affection and probably survives in Fforest Gaerdydd/Cardiff Forest (q.v.) and Cardiff (PNGlamorgan 33, 125-7), co. Glamorgan.

Talacharn[1]
Cantref Gwarthaf
Uncertain
?*tâl*[2], ?pn. ?*Acharn*
Talacharn c.1191, *Talechar* c.1236, (barony) *Thalagharn* 1276-7, (barony, castle, *villa*) *Tallauhern* 1307, (cmt.) *Lagharn* 1439, *Comm. Talegarne* 1536-9, *swyδ lacharn* c.1566

The cmt. and lp. of Talacharn (covering Eglwys Gymyn, Talacharn[2]/Laugharne, Llan-dawg, Llanddowror, Llansadyrnin and Pen-dein/Pendine) occupies a wedge-shaped promontory. The n. was attached to the castle and small medieval borough of Talacharn[2]. Historic forms of Talacharn and Laugharne are interchangeable from c.1191 down to the early 17th cent. From that point on Lacharn and Laugharne (reduced forms of Talacharn) are increasingly confined to the town. *tâl*[2] is employed in the sense of 'end of' but *-acharn* or *-lacharn* are unexplained. It is possible that the cmt. and lp. were called Acharn since there is mention of *gorthir Acharn* in poetry of Iolo Goch late 15th cent (GIG 8.64n (pp. 225-6). This would appear to mean 'upland of Acharn' with **gorthir** nm. 'uplands, highland' but it is uncorroborated. There are similar problems with *-lacharn* which makes little convincing sense unless it represents a compressed form of **llachar** adj. 'bright, shining' coupled with **carn** 'cairn' (DPNW 211).

Talacharn[2], **Laugharne**   SN 301106
Talacharn c.1191, (ch. of) *Sancti Michaelis de Talachar* 1223, *Lauchern* 1288, *Tallach(a)rn* c.1291, *Talagharn* 1292, *Locharn'* 1323, *Thalazarn*, *Thallacharn* 1331, *Talacharnn* c.1400, *Lagherne* 1403, *Lagharn* 1439, *Lacharn* 1535, *Tallaugharne, otherwise Laugharne* 1603, *Laugharne* 1606

Talacharn is taken from that of the medieval cmt. (above) occupying the promontory between the r. Taf and Bae Caerfyrddin/Carmarthen Bay. The town is located at the point where Afon Corran (which rises at SN 271125) reaches Taf, hence its earlier n. Abercorran recorded as *Abercorran* c.1194, *Aber Corran* c.1400 and *Aberkorran* 1559 (**aber**). This has prompted the suggestion that Talacharn contains the rn. Corran but historic evidence makes this highly unlikely. Corran is recorded as *Corran* 1623 and *Coron* 1680 probably meaning 'dwarf (r.)': **cor** nm. 'dwarf, pigmy' and suffix **an**.[2] Glancorran (SN 283129) and Abercorran (SN 276120) are modern hos. The latter is a misnomer since it is two miles from the outfall of Corran. The official W n. of the town is Talacharn but historic evidence in W sources favours *Lacharn* and *Llacharn* from c.1535 down to the 19th cent. The anglicised form Laugharne developed from Lacharn > *Lagharn, Laghern(e)* > Laugharne with loss of fricative *-gh-* probably in the 16th or early 17th cent producing '*Llaugharne* (pronounced Larne)' 1860. Loss of unstressed *Ta-* may be a result of confusion with Latin and Fr *de* 'of, from' and finds parallels in some of the historic forms for Tafolwern, co. Montgomery (PNMont 161).

Talhardd   SN 621201
Llandeilo Fawr
'end of a hill': *tâl*[2], **ardd**[1]
Talharth 1331, *Talardd* 1776, 1831
The n. survives in two hos. Talhardd and Talhardd-bach (**bach**[1]). The first developed as a fortified mansion ho., apparently on the site of a castle, near the foot of a small hill and looks over the valley of Afon Cennen towards the second, occupying a hill-top near the A483. Some forms may have been influenced by **hardd**.

Taliaris   SN 640280
Llandeilo Fawr
'top of a place notable for hens': *tâl*[2], **iares**
(land) *Taleyares* 1324, *Talliares* 1356, 1634, *Taliaris* 1693, *Taliarus* 1710, (hill and park) *Taliaris, Taliaris Chapel* 1831

The likeliest explanation is that it describes a raised area or a hill where hens, particularly wild birds, gathered. There are comparable ns. elsewhere such as Branas, earlier Branes, co. Merioneth (***brân*** 'a crow', suffix *-es²*) (HESF 33-34) and cf. ***buches²*** 'herd of cattle' (see Tomos Roberts in ADG², 92). The chp. is apparently a rebuilding c.1600 (HCH 177-8) of an earlier chp., recorded later as *Trinity Chappell* 1719 and *Holy Trinity, or Talliaris* 1763, and located (SN 654283) on the B4302 about 2 km east of the ho. Maerdy (SN 653279) is *Merdytalleris* 1859 and plain *Maerdy* 1831, 1891 (***maerdy***).

Talley see **Talyllychau**

Talog¹ SN 3225
Aber-nant
'dirty ford': ***rhyd, halog***
Rhyd Talog 1754, *Rhydtalog Mill* 1763, *Rhyd talog Mill now called Havod Halog* 1816, *Talog* 1831, 1875
Talog probably derives from Rhytalog, a form produced through colloquial pron. of the cluster -dh- as -t-; cf. Rhytalog, co. Flint (PNF 172-3). The same development is seen in many historic forms of Coedtalog (SJ0511), co. Montgomery, such as *Coytalauk* 1278, *Coythalauc* 1291, *Coedtallog* 1774. Loss of Rhy- may have been partly prompted by association with the def. art. ***y, yr*** and partly through association with ***talog*** 'jaunty, cheerful'. Hafodhalog does not appear on OS maps and refs. to it have not been found before 1816 but it appears to have ***hafod*** 'summer dwelling' and ***halog*** 'polluted'. The mill may be confidently identified with *Talog Mill (Corn)* 1889. The ford is probably that at the small br. over Afon Cywyn in the middle (SN 332255) of the ht.

Talog², Afon SN 4637
rn. SN 506390 to Tyweli SN 444363
'jaunty river': ***afon, talog***
River Talog 1888, *Afon Talog* 1905
Pont-ar-talog (SN 462376) (***pont, ar***) is a former woollen mill located where the New Inn-Llanfihangel-ar-arth road crosses the r. and is recorded as *Pont ar Talog* 1730, *Pont-ar-Talog Woollen Factory* 1888 (***pont***), *Talog Fulling Mill* 1905.

Talsarn SN 779260
Llanddeusant
'end of the causeway': ***tâl², (y), sarn***
Talysarn 1737, *Talsarn* 1824-2016
Located on a ridgeway and once busy road running westwards from Brecon through Trecastell, co. Brecon, to a point beyond Twynllanan where it split into several roads linking to Llangadog, the Aman valley, Trap and Llandeilo. The location of a Calvinistic Methodist chp. from 1800 (HMSG 351-3).

Tal-y-bryn see **Dol-y-bryn**

Tal-y-foel (Tan-y-foel) SN 5846
Pencarreg
?'end of the bare hill': ***tal, ?(y), moel***
Talevoel 1599, *Tal y voel* 1722, (ho.) *Dalfoel* 1842, *Tanyfoel* 1881, *Tan-y-foel* 1888
The ho. is at SN 587462) and the bare hill must be the unwooded promontory between the A482 and the older road leading southeast from Ram to open land above Mountain Gate (SN 599456). Later refs. have been misassociated with ***tan¹***.

Talyllychau, Talley SN 632327
'end of the stone slabs': ***tâl², y, llech¹*** pl. *llechau*
Talelech' 1222, *Taletheu* 1223, *Talelek'* c.1222, *Tallechev* 1281, *Talelezeu* c.1291, *Talley* 1382-1765, *Tal y Llecheu* c.1400, *Talallecheu* 1495, *tal y lleche* c.1566, *Taley or Tâl y Llycheu* c.1700
The n. may refer to a causeway, ford or road. Talley ['tălī] is an anglicised form of Talau, a contracted form of Talyllychau, represented as *Talley* 1382, *Tallay* 1395, *Tallagh* 1383 and similar forms, and in specifically W sources as *Talau* 1740 and 1859. Talley/Talau is neither 'an abbreviation or corruption' (OPemb IV, 381) nor 'an abomination' (CStudies 112). The supposition that Talyllychau contains *llychau* sg. ***llwch²*** seems to begin with the scholar Edward Lhuyd ('from the two Lakes .. or Fishponds') and was repeated by others such as Richard

Fenton in 1809. It is true that *e* was sometimes used to represent the middle vowel [ə] as in the def.art. *y* but this will not account for forms such as *Talilazhau* c.1348, *Tal y Llecheu* c.1400 and *tal y lleche* c.1566 taken from W sources. The monastery was established c.1184-9 by Rhys ap Gruffudd as a ho. of canons of the order of Prémontré (Laon) following the rule of St Augustine (MOSW 35).

Tanerdy   SN 422209
Caerfyrddin/Carmarthen
'tannery': ***tanerdy***
Tanerdy 1811-1890, *Taner-dy* 1831
This is also likely to be *Tandery* 1812 recorded under Abergwili. ***tanerdy*** is found elsewhere in co. Carmarthen, eg. (piece of ground called) *yr ardd* alias *yr ardd wrth y Tannerdy* 1767 (***gardd***) in Myddfai and (mess.) *Tannerdy* 1783 in Llangadog.

Tanglwst   SN 310340
Cilrhedyn
pers.n. **Tanglwyst**
Tanglwys 1882, *Tan-glŵys* 1890, *Tanglwst* 1904, *Capel Tan-glŵys (Independent)* 1906
The n. is drawn from the ho. recorded as *Gelli Danglws* 1842, now plain Gelli (SN 306348) (***celli***), and may be compared with Tanglwyst (SN 351480), in Troedyraur, co. Cardigan, formerly Esgair Tanglwst (and similar forms) which has lost prefixed ***esgair*** 'ridge'; see PNCrd 154. In the latter Wmffre shows that Tanglwst derives from Tanglw(y)st, a metathesised form of the pers.n. Tangwystl found in medieval pedigrees (EWGT 36, 98, 106). Both pns. are comparable with Tŷ Tanglwyst (SS 822809), in Pyle and Kenfig, co. Glamorgan, recorded as (gr.) *Tangluscland* 1547, *Tanglusland* 1623, *Tanglust-land* 1685, *Ty Tanglust Farm* 1731 *Tyr Tanglwst* 1738, *Ty tan-glwys* 1833, and Llety Tanglwst, in Defynnog, co. Brecon, recorded as (place) *llettuy Tangloust* 1589 (***tŷ***, ***llety***).

Teifi, Afon
rn.
?'flowing one': Br **tam-*, rn. suffix *-i*
Tuerobios fluvius c.150, *Teibi* c.830 (11cent), *Teuwi* anno 1136 (12cent), *amnem Theibi* 11cent, *(fluvio) Teyvi* c.1191 (c.1214), (water) *Tywy* 1241, *aqua de Teivi* 1302, *Teywy, Teiwy, Deiwy* 1184 (1285), *Teivy, Teivye* 1585, *the Teifie* 1587, *river Tivie* 1606
The first recorded form (latinised from original Greek) taken from Ptolemy, mathematician, 1st cent is almost certainly geographically accurate but is thought to be corrupt. It is impossible to trace any development directly to Teifi without amending it to a dubious **Tuegobis* (PNRB 480). The rn. is better compared with Tawe and Taf (DPNW 450, 458). A fuller discussion and additional historic forms are given by Iwan Wmffre (PNCrd 1294-5). The r. rises in Llyn Teifi (*Llin Tyue* c.1537, *Llyn Teyvy* 1603) (***llyn***) and flows through the valley Dyffryn Teifi (*difrin teiui* c.1145) (***dyffryn***) forming the greater part of the border between cos. Carmarthen and Cardigan. The rn. is frequently spelt *Teivi* and *Tivy* down to the 19th cent reflecting E orthography.

Telych   SN 7833, 7932
Llandingad
?'hill-slope': ***telych***
(lands) *Teleigh* 1609, *Telith Hamlett* 1671, *Telich* 1710, *the Hamlet of Keventelich* 1798
The n. is not current but applied to the easterly part of Llandingad centred on a ho. Cefntelych (SN 793327) recorded as *Keven-tellich* 1682, *Cefntelyrch, Keventelyrch,* (ho.) *Cefn-telych* 1886 (***cefn***). Penlantelych (SN 786334) is recorded as *Penylan* 1831, *Pen-lan-telych* 1886 (***pen¹***, ***glan***). The el. ***telych*** is clearly very old since it is found in *Tir Telih* c.840 identified with Bryntelych (approx. SN 648389) (***bryn***) and it has been suggested that it is Ir ***tulach¹*** 'low hill, hillock' or a loan from that language though this is unproven and cf. ***tyle*** and Tachlouan. Rhŷs (AC 1895: 28) suggested that *Telih* is a pers.n. but there seems to be nothing to substantiate this.

Temple Bar   SN 589175
Llanfihangel Aberbythych
'(place named after) Temple Bar'

Temple-bar 1763-4, (ho,) *Temple bar* 1778, *Temple Bar* 1831, *Temple Bar (P.H.)* 1887
A n. belonging to a group of pns. transferred from London streets and landmarks and typically applied to public hos. and turnpike gates. Sometimes they appear to have been used in an ironic sense for small insignificant places (DPNW 458-9). There is also a ho. (SN 660831) in Caeo recorded as *Temple Barr* 1728, *Temple Bar* 1851 and *Templebar Cottage* 1888 but no evidence of a nearby turnpike gate. The n. may have been influenced by that of nearby Crug-y-bar (q.v.).

Tigen, Afon SN 1926
rn. SN 205253 to SN180268
Obscure
Tigen c.1700, *Afon Tigan* 1889
R.J. Thomas (EANC 124-5) notes a possible origin in *Ticīnā* related to *Ticīnos*, *Ticīnus* in rns. in Gaul. The r. rises at a place called *Blaen Tigen* c.1700 (AC 1975: 109) though many OS maps misapply Tigen to the tributary Dyflyn which rises near a ho. Blaendyflin (SN 215270) and joins Tigen at SN 205262. The rn. is also recorded in *Abertigen* 1670-1, 1757 and *Abertegen* 1828 (**aber**).

Tiresgob SN 6226
Llandeilo Fawr
'bishop's land': **tir, esgob**
(ho., tmt. of lands) *Tire Iscobe* 1664, *Tier Escob Hamlet* 1671, *Tir Esgob* 1811, 1851
Possibly *Patria de Lanteyllowe* 1326. The n. preserves the memory of the ecclesiastical manor of Llandeilo (HW I, 268). An area between Afon Dulais and Afon Myddyfi. Tir Esgob was also an alternate n. for the bishop's manor of Llangadog recorded as *Patria de Langadok* 1326.

Tir Rhoser SN 606160
Llandybïe
'land of (man called) Rhoser': **tir**, pers.n. **Rhoser**
(mess. etc) *tyr Rosser* 1633, *Tir Rosser* c.1700, *Tir Rosser Hamlet* 1734, *Tir Rhôs Hir* 1811, (ho.) *Ty-Rosser* 1878, 1964-5

Re-interpreted in 1811 as if it were **rhos²** and **hir**, 'long moor'. Tŷ-Rosser 1878 and 1964-5 refers to a ho. substituting **tŷ** 'house, dwelling' for **tir** in the area now generally called Pentregwenlais (q.v.).

Tir-y-dail SN 625128
Llandybïe
'land in the leaves': **tir**, (**yn**), **y**, **dail**
(tmt.) *Tyr yny Dayell* 1597, (mess.) *Y ty yn y dail* 1693, *Ty'n'-y-dail* c.1700, 1831, *Tirydail* 1859, (ho.) *Tir-y-dail* 1878, 1906
The n. might describe a place in an area which is densely wooded or one associated with a particular species of plant. Note the use of **dail** in ns. such as *dail cawl* 'colewort', *dail y cwrw* 'laurel-leaves' and *dail y fendigaid* 'St John's-wort'. Three sps. have a variant form *Ty'n-y-dail* (**tŷ, yn**). The n. was transferred from a ho. (SN 625124) before 1906 to housing constructed from c.1900 on its northern and eastern sides.

Tor-y-coed Torcoed-fawr SN 480141
Llangyndeyrn
'slope of the wood, slope with a wood': **tor²**, **y, coed**
Skebor Toragoed 1561, (tmt.) *Tarrakoed* 1609, *Toricoed* 1620, *Tor y Coid* c.1735, *Tor-y-coid-fawr*, ~ - *fach*, ~ -*ganol*, ~ -*llipry* 1831
tor² is often difficult to distinguish in pns. from **tor¹** nmf. 'a breaking; gap, breach' but the suggested meaning makes better sense here. The hos. Torcoed-fawr (HCH 181-2) and Torcoed-fach (loss of unstressed **y, mawr, bach¹**) and a small intervening wood occupy the northern slopes of Mynydd Llangyndeyrn. This is presumably *Tor-y-coid-llipry* 1831 with the particular sense of 'drooping wood' (**llipryn**). The first ref. has **ysgubor** 'barn'. An identical n. Tor-y-coed (ST 066827), Llantrisant, co. Glamorgan, recorded as *Torre Ycoide* 1541, *Tor y Coad* 1768 and *Tor-y-coed* 1885, occupies a similar position on hill-slopes above Afon Clun.

Towy see **Tywi, Afon**

Traean Clinton
barony of Sanclêr/St Clears
'third belonging to Clinton': *traean*, pers.n. ***Clinton***
Trahan 1390, *Treyne Clinton* 1451, *Traneclyton* 1546, *Dominium de Clinton, Lordshippe of Clinton* 1592, *Trayne Clynton* 1608
The n. applied to one-third of the lp. of Sanclêr/St Clears which was an AN amalgamation of older W administrative units including Amgoed and Peuliniog (q.v.). After the death of William Braose in 1230 without male heirs the lp. was divided between three of his daughters Maud (d.1301), Eleanor (d. 1251) and Eva (d.1255) (HCrm I, 11). The *traean* known as Traean Clinton descended from Eva through the Cantilupe and Hastings families down to 1389. Clinton probably recalls William de Clinton (d. 1354), first earl of Huntingdon, who married Juliana Hastings after 1325. Its history after 1390 is unclear since the three thirds were often undifferentiated in historical sources and can sometimes only be identified by context. Clinton first appears as a qualifier in 1450. The following year Traean Clinton was granted to Margaret, wife of Henry VI, drawing rents from Sanclêr/St Clears by way of its own portreeve mentioned in 1592 and c.1700) with Llangynin (SN 1419), Castelldwyran, Bachsylw and Cilymaenllwyd (SN 1523). Traean Clinton is known to have contained Maenor Bachsylw (Cilymaenllwyd), Maenor Capel Martin (probably in Sanclêr/St Clears), Maenor Llangynin and Maenor Tredai (Llanfallteg).

Traean March
barony of Sanclêr/St Clears
'third belonging to (the earl of) March': *traean*, title (earls of) ***March***
Traynemarch 1425, (lp.) *Trayne March* 1515, (lp.) *Tranemerch* 1584, (manor) *Tranemershe* 1590, *Manerium de Traynemarche et St. Clere* 1597, (lp.) *Trayne Marche* 1608, (portreeve of) *Train March* c.1700
Traean March was that third of the lp. of Sanclêr/St Clears acquired by Roger Mortimer (d. 1282) on his marriage to Maud Breos (d.1301), one of the heirs of William Breos (see Traean Clinton and Traean Morgan) (HCrm I, 11). This descended to their son Edmund Mortimer (d. 1304) and grandson Roger Mortimer II (d.1330) (LSMW 53-56). During the turbulent reign of Edward II (1307-27), Traean Mortimer was taken from him and granted to Sir Rhys ap Gruffudd in 1322. Roger recovered the third in 1328 when he returned to royal favour and became first earl of March. March almost certainly refers to his shortlived possession of the *traean* as earl though the qualifier is first recorded in 1425. Roger's title lapsed until 1355 when Edward III conferred it on his grandson Roger Mortimer. Traean March stayed more or less continuously in the custody of the Mortimer family down to Edmund, fifth earl of March, who died in 1425. The *traean* then passed, along with Narberth, into the custody of his widow Anne and her new husband John, earl of Huntingdon. The historic form for 1590 (above) is probably an attempting at representing March [martʃ] in W form as Mersh [merʃ] though this has not been generally adopted in modern sources. The suggestion (OPemb IV, 365) that Traean March was centred on a fomer ho. Cefn-march recorded as *Cefn-march* 1831 (approx. SN 247169) can be dismissed. This is an unrelated n. meaning 'ridge where horses are seen' (*cefn, march*) in ref. to the hill crossed by the Sanclêr/St Clears to Pontyfenni road. Traean March covered most of Llanboidy (Maenor Bach-y-ffrainc, Maenor Cilgynfyn, Maenor Cilhengroes, Maenor Cilhernin, Maenor Cilnawen, Maenor Gwernolau and Maenor Maes-gwyn) and parts of Cilymaenllwyd (Maenor Cilellyn, Maenor y Merydd), Llandysilio (Maenor Castell Dwyran) and Llanglydwen (Maenor Llanglydwen) and possessed rents in Sanclêr/St Clears gathered by its own portreeve recorded in 1592 and c.1700.

Traean Morgan
barony of Sanclêr/St Clears
'third belonging to (man called) Morgan': *traean*, pers.n. ***Morgan***

Trayne 1404, (lp.) *Trayne Morgan* 1592, 1608, *Traney Morgan* 1613, (lp.) *Trayn Morgan* and *St. Cleers* 1692

Traean Morgan was inherited by Eleanor (d. 1251) who married Humphrey Bohun (1204-75), second earl of Hereford. It subsequently passed to Eleanor (1241-90), wife of Edward I, by exchange. After her death Edward granted it with Boduan, co. Caernarfon, in 1299 to Morgan ap Maredudd (d. 1329) from whom it took its n. Morgan's father Maredudd (d. 1270) was ruler of Machen, Edlogan and Llefnydd, in Gwent. We know that Morgan left a daughter and heir Angharad and that she married Llywelyn ab Ifor. It has been said that Angharad brought with her estates at Tredegar and Cyfoeth Maredudd though there is no mention of these at her father's death. Traean Morgan probably descended as an unspecified third of Sanclêr/St Clears to a grandson Llywelyn ap Morgan who is thought to have joined the revolt of Owain of Glyndyfrdwy against Henry IV before 1404. Lands in the town and lp. of Sanclêr/St Clears were granted with lands in Newport and Cardiff to his widow Margery in 1413 who had already re-married. We do not know when *Morgan* was added as a qualifier but the fuller form was probably in use from the mid 14th cent in order to differentiate the pn. from Traean Clinton and Traean March in the same barony. Traean Morgan covered part of Egrmwnt (Maenor Egrmwnt), Llanboidy (Maenor Pen-y-cnwc), Llandysilio (Maenor Castell Madog, Maenor Clun-tŷ), Llanfallteg (Maenor Henllanamgoed, Maenor Tegfynydd), Llan-gan, (Maenor Castell Draenog, Maenor Tal-y-fan) and Meidrum (Maenor Cilau) in 1592 with rents collected by its own portreeve in 1592 and c.1700, and Clog-y-frân (q.v.) in 1313 in Sanclêr/St Clears.

Traethnelgan   SN 6436
Talyllychau
'marshy place associated with (man called) Elgan': **trallwng**, pers.n. **Elgan**
Trailneygan, (mill of) *Trathleneygan* c.1291, *Tallunelegan* 13th cent (1332), *Trallwgn Elgan* c.1400, *trallwc elgan*, *Trallwyn Elgan* 1520, (gr.) *Traeth Nelgan* 1540s, *Trathnelgan* 1600, 1633
Referring to the low-lying ground between the rs. Cothi and Annell (CStudies 114). Elgan has been identified with a reputed 6th cent prince of Dyfed (HW II, 641 and n.) recorded in a late medieval pedigree (EWGT 106). This cannot be substantiated but it supports the existence of the pers.n.

Trap (Trapp)   SN 652189
Llandeilo Fawr
'trap (that which attracts and detains)': *trap¹*
(mansion of) *Tuy Watkin or Velin alias Tafarn Trap* 1720, *Tavern y Trap* 1765, *Tafarn y Trap* 1767, *Trapp* 1769, *Trap* 1811, (village) *Trapp*, (br.) *Pont y Trapp*, (ho.) *Park Trapp* 1887
trap¹ is a fairly common pn. el. especially in mid and south Wales, nearly always associated with inns and public hos. Examples include Bwlch-y-trap (SN 415272), Parc-y-trap at Llandyfrïog, co. Cardigan, the former *Tavern Trap* 1852-3, in Gelli/Hay, co. Brecon, Trap, in Aberdâr/Aberdare, co. Glamorgan, and The Trap (SS 592977) at King's Bridge/Pontybrenin, Casllwchwr/Loughor, co. Glamorgan. The furthest known example seems to be the Bridge Inn, Y Waun/Chirk, co. Denbigh, which was known locally as The Trap 'because of its irresistible lure to miners on their way to and from work' (HChirk 143). Not all occurrences of *trap¹* need refer to public hos. Trap Melyn (SN 023387) near Dinas, co. Pembroke, may have applied to a device used to control water leading to a mill (PNPemb 627). The first ref. is for modern sp. Tŷ Watcyn o'r Felin 'house of (man called) Watkin of the mill': *tŷ*, pers.n. **Watcyn**, *Watkin*, *o*, affixed def.art 'r for *yr*, and **melin**.

Travellers Rest   SN 382192
Caerfyrddin/Carmarthen
'(place at) the Travellers' Rest': inn-n.
Travellers Rest 1841, *Travellers' Rest (P.H.)* 1888-9
A resting-place for travellers. A public ho. at what is known as *Bryn-bach* in 1831 and 1841 (**bryn**, **bach¹**) on the main Caerfyrddin/

Carmarthen-Narberth road. The n. is applied to the settlement on OS maps from 1969.

Traws-mawr SN 374242
Llannewydd/Newchurch
?'(property extending over) a large area' or 'large piece of land, large ridge': **traws, mawr**
(cap. mess. etc) *Trawsmawr* 1762, *Troesmawr* 1786, *Traws-mawr* 1831, 1888-9
The n. referred to a ho. and fm. Traws-mawr Hall which was formerly *Traws-mawr-newydd* 1964 (*newydd*) constructed in the 1880s. The former mill Melin Traws-mawr (*melin*) is on the unnamed r. (probably Nant Gwyn) below the ht. *traws* is usually an adj. but here it is a nm. qualified by *mawr*. There are instances too where it is found as a simplex, eg. Traws, in Cil-y-cwm, recorded in 1796, which seem to support this interpretation. The meaning of *traws* is often difficult to determine and pns. which contain this el. should be examined individually and related to topography. B.G. Charles (PNPemb 46) suggests that it may mean 'a feature, such as a piece of land, ridge lying athwart' and that makes good sense here since the ho. bearing this n. lies on the slopes of a promontory between streams.

Treberfedd ?SN 2820
Meidrum
T(re)berneth 1326
'middle township': **tref, perfedd**
Presumably the middle part of the p. around Meidrum ch. The cited form stands for *Treberueth* or *Treberveth* with the common confusion of *n* and *u* in medieval sources. An identical n. in Llansteffan is recorded as (lands called) *Trebervethe* 1592.

Trebifan (Trebevan) see **Twyn**

Trecastell[1] ?SN 2641
Cenarth
Tref Castell, Tref Castel 1303
The castle is probably that at Cenarth. The cited refs. occur in conjunction with an unidentified place *Garth Gwithaul* and *Girch Gwichaul*.

Trecastell[2] SN 5415
Llanarthne
'castle township': **tref, castell**
Tref gastell c.1550, (place) *Tre gastell* 1572, 1639, *Trecastell* 1574, 1851, *Tre Castell* 1811
The southerly part of Llanarthne. The supposed castle may have been in the vicinity of Foelgastell (q.v.) or may be drawn from a small circular earthwork Castell y Garreg (SN 572158) (HER 650) with a nearby ho. Castellygarreg (SN 574156).

Trecastell[3] SN 6319, 6619
Llandeilo Fawr
'castle township': **tref, castell**
Tre Gastell 1606, *Tir Tre Gastell* 1636, *Trecastle* 1723, 1804, *Tre Castle* 1732
Recorded in 1723 with *Melin Arthur* which appears in 1851 as *Felin Arthur* (*melin*, pers.n. **Arthur**) near Glancennen below Trap. The castle is that at Carreg Cennen (q.v.).

Trechgwynnon SN 5419
Llanarthne
Uncertain
(ho. and garden in) *Trechgwynon* 1707, (ht.) *Trech-gwnnion* 1804, *Trêch Gwinnon* 1811, (ht.) *Trechgwynnon* 1851
Historic forms are too late for confident interpretation. The second el. apparently refers to Afon Gwynnon which defines part of the boundaries of the t. It flows from a point (SN 557183) near Penrhiw-goch in a loop to meet Tywi (SN 536204) north of Llanarthne village and is recorded as *River Gwynnon* 1888 apparently containing **gwynnon**, *gwnnon* coll.n. 'fog, long white straw; dry twigs, straw', perhaps a 'stream notable for fog *or* white with foam'. That still leaves Trechgwynnon as a whole unexplained. The first el. is conceivably *trech*[1] 'stronger, mightier, more powerful' as a qualifier but that would demand len. (*gwynnon* > *wynnon*). An unrecorded pers.n **Trechgwyn* combining **trech**[1] and **gwyn** with suffix **-(i) -on** used to describe 'land of (man called) Trechgwyn' is possible but Afon Gwynnon would have to be explained as a back-formation perhaps under the influence of

gwyn, gwynion (locally *gwynon*) and the many rns. containing this el.

Tre-clas SN 531203
Llanarthne
'township at the ?monastic community': *tref, ?clas*
Tre-clas 1804, Tre Clâs 1811, Tre-clâs 1831, Tre-clais 1888
Recorded sps. are too late for certainty with regard to the second el. The OS map 1831 applies the n. to a ho. or hos. a little to the north-east towards Glantowy.

Trefechan[1]**, Trevaughan**[1] SN 401212
Caerfyrddin/Carmarthen
'little settlement': *tref, bechan*
Treveychan 1657, Trefechan Gate, ~ Toll-house 1843, Trefechan 1873, Tre-Vaughan 1889
The n. is transferred from Trevaughan House and ultimately from the fm.ns. Trefechan-fach (SN 395209) and Trefechan-fawr recorded collectively as *Trefechan* 1811 and *Tre-fychan* 1831 and as *Tre-fechan-fâch, Tre-fechan-fawr* 1889-90 (*bach*[1]*, mawr*). The development -fechan > -vaughan is an anglicisation under the influence of the surname Vaughan < W personal cognomen *Fychan* 'the lesser, the younger and cf. Pontneddfechan with anglicised form Pontneathvaughan (PNGlam 179). The ht. developed along the Caerfyrddin/Carmarthen-Bwlchnewydd road between a smithy and the ho. Mile End near quarries.

Trefechan[2]**, Trevaughan**[2] SN 200160
Cyffig
'little settlement': *tref, bechan*
Trefvaghaune, Trevachaun 1531, Trevechan 1739, Trefaughn 1819, Trevaughan 1875, Tre-Vaughan, ~ Mill 1880-9
'little' perhaps in contrast to nearby Hendy-gwyn/Whitland.

Treforis Treforris-fawr SN 382099
St Ishmael
'township of (man called) Moris': *tref*, pers.n. *Moris, Morris*
Trevoris 1676, Trevorrisvach 1736, Trevorice c.1730-40, Tre-forris-fawr, ~ -fach 1831, Treforisfaur 1845
Mor(r)is is presumably the same person as that in Iscoed Morris (q.v.) recorded from 1551-2 in the same p. but he is unidentified.

Trefreuan (Trefroyan) SN 490168
Llanarthne
'township of (man called) ?Breuan': *tref, ?Breuan*
Trevrian 1675, (ht.) Trefreyan 1804, Trêf Rhewin 1811, Tref-royan 1831
The pers.n. seems to be unrecorded and it is possible that the second el. is *breuan*[2] nf. 'carrion crow, raven' signifying an area notable for carrion crows and by implication one which is poor and desolate. The ho. is unnamed and shown as deserted on the OS 1888 map.

Trefynydd SN 444026
Pen-bre
'township on the mountain': *tref, mynydd*
(place) Trevennythe 1555, Trevynydd 1594, tre vynydd 1618, Tre-fynydd 1880, Trefynydd Farm 1918
The n. survives as that of a fm. on the southern side of Mynydd Pen-bre near Cenrhos (*cefn, rhos*[2]).

Tre-garn SN 5013
Llanddarog
'township at a cairn': *tref, carn*
(ht.) Tregorn 1804, (~) Tre Gâr 1811, Garn-ganol, Pentre-garn 1831
carn is an el. in several ho.ns. extending over an area north of Mynydd Cerrig and it is difficult to pinpoint a particular cairn which may have given rise to the pn. It may be that *carn* is employed in a collective sense for 'area characterised by cairns'. The area is notable for limestone quarries and limekilns. *Garn-ganol* and *Pentre-garn* in 1831 refer to hos. called *Ty'r-garn* (SN 505140) and *Garn* (SN 512144) in 1880-7 (*canol, pentref, tŷ*).

Tre-gib SN 633212
Llandeilo Fawr
'township near (r.) Cib': *tref*, rn. *Cib*
(town) *Drefgybe* 1327, *Drefgibe* c.1501, *Tregibe* 1611, *Tregibbe howse* 1660, *Tregibb* 1674, *Tregeeb* 1709, *Tre Gieb* c.1735, *Tregib* 1765-1853
The r. joins Cennen just before this reaches Afon Tywi (SN 633218) near Ffair-fach. The rn. is *Keeb* 1754 and *River Cib* 1886-7 and is also recorded in Cwm-cib as *Cwm Cib* 1773 (*cwm*). The rn. probably has *cib* 'vessel, bowl' perhaps applied figuratively to describe its bed (ETG 112). Best known for its mansion Tre-gib demolished after 1974 (HCH 186).

Tre-glog SN 600352
Llansawel
'settlement at a cliff': *tref*, *clog*[2]
Treglog 1868, 1952-3, *Tre-glôg* 1887, 1906
Located on steep slopes. A ho.n. extended to adjoining hos. built during the late 1960s and 1970s.

Tregynin SN 567228
Llangathen
'township of (man called) Cynin': *tref*, pers.n. *Cynin*
Tregunnin 1776, *Tre-gynin* 1831, 1891, *Tregynin* 1845
Earlier evidence is needed for certainty. For the pers.n., see Llangynin.

Tregynnwr (Tre-gynwr) SN 419194
Llangynnwr
'town of Cynnwr': *tref*, pn. *Cynnwr*
Tre-gynwr 1969
A modern n. for a ho. estate built in the late 1960s. The n. is elliptical for 'town in Llangynnwr'. The OS sp. is irregular.

Treherbert SN 583470
Pencarreg
'town of Herbert': *tref*, surname **Herbert**
Treherbert 1881, 1973, (row of hos.) *Tre-Herbert* 1888-9, 1964
The hos. developed along the A482 midway between Ram and Cwm-ann. The current OS Landranger shows it as part of the latter.

Trehopcyn, Hopkinstown SN 649124
Betws
'settlement named from family called Hopkin': *tref*, surname **Hopcyn, Hopkin**, E *town*
Hopkinstown 1901, 1906
Developed c.1900 along Heol Wernoleu running from Glanaman to Pontaman and named from a local family; cf. Hopkinstown, co. Glamorgan (PNGlamorgan 105). The Hopkin family were farmers in the area recorded in 1861. Trehopcyn appears to be a later adoption.

Tre Ioan, Johnstown SN 399198
Caerfyrddin/Carmarthen
'settlement named from John Jones': *tref*, *Ioan*, pers.n. *John*, *town*
Johns Town 1831, *John's town* 1852-3, *Johnstown* 1871,1969, *John's Town* 1888-9, 1964
Named from John Jones of Ystrad (formerly at SN 39811897), mayor of Caerfyrddin/Carmarthen in 1809 (HCrm II, 60, 470). The n. Tre Ioan appears to be a recent adoption.

Tre-lech SN 283304
Tre-lech a'r Betws
'settlement near a slab of stone': *tref*, *llech*[1]
(dissenting meeting ho.) *Teenewydd Trelech* 1798, *Ty-newydd-trelech* 1831, 1891, *Trelech* 1906 to present
Earlier Tynewydd Tre-lech, 'new house in Tre-lech': *tŷ*, *newydd*. The village developed around crossroads and Rock Independent chp.

Tre-lech a'r Betws SN 309267
'Tre-lech and (chapelry called) Y Betws': pn., *a*, *yr*, *`r*, *betws*
Trenleth c.1291, *Treffelegh* 1361, *Treflegh* 1513, *tre lech* c.1566, *Trelegh* 1580, *Trologh Bettus* 1586, *Trelech & Bettws* 1600, *Trelecharbettws* 1647, *Trelech ar Bettus* 1752, *Trelech-ar-bettws* 1831, *Trelech-a'r-Bettws* 1889-1983
The fuller form (dropped on the current OS Landranger) earlier referred to the p. rather than the village around St Teilo's ch. Y Betws is the former chp. Capel Betws (q.v.).

Tremoilet SN 225094
Eglwys Gymyn
'settlement of ?Moilet': *tref*, ?pers.n. ?*Moilet*
Trefmoillet 1307, *Tremoylad*, *Tremoylet* 1620, *Trefmoylett* 1655, *Tre Moylet* 1809, *Tre-moilet* 1843
Moilet could be a variation of the AN pers.n. Maylard, Mallard though it must be stressed that no person with this n. has been identified here. One might also have expected a regular form with len. **Trefoilet* unless E influence has disturbed this; cf. Llan-maes for Llanfaes and Llanmihangel for Llanfihangel, co. Glamorgan (PNGlam 126-7). Tremoilet is the n. of a fm. and was the n. of a former ho. recorded from the 16th cent and demolished in the late 18th cent (HCH 187).

Tre'rcerrig ?SN 5013
Llanddarog
'house at the stones': *tref, y, `r, carreg* pl. *cerrig*
An unidentified manor according to Melville Richards (WATU) probably centred on the hill Mynydd Cerrig (SN 503138).

Trevaughan¹ (Caerfyrddin/Carmarthen) see **Trefechan¹**

Trevaughan² (Cyffig) see **Trefechan²**

Trichrug (Trychrug) SN 700230
Llangadog
?'prominent hillock': ?*try*- intensifying prefix, *crug*
(*A beacon on the top of*) *Trychryg* c.1700, *Mynydd tre chrig* 1760, (hill) *Tri Chrug, or the Three Hillocks* 1815, (hill) *Trichrug* 1831, *Trichrûg* 1887
We cannot be certain that this is the correct interpretation because historic evidence is late. The 1815 ref. favours *tri* variant *try* 'three' with *crug* (asp.mut. *chrug*) which seems to be confirmed by the OS map in 1831 showing three heaps of stones. The first ref. c.1700, however, applies the n. specifically to the hill and OS plans from 1887 show only a single cairn on it known as *Pen y Bicws*. It is worth comparing Trichrug to Trychrug (SN 542599), co. Cardigan, which is examined by Wmffre in PNCrd 672-4. He argues that *crug* may have been originally a neuter noun and that if *Try-* represents *tri* it would be expected to produce ModW **Trygrug* with len. There is similar ambiguity with regard to the first el. in other ns. such as Tryfan (with *ban¹*) (DPNW 478) and Triffrwd (*ffrwd*), co. Brecon. Pen y Bicws has *pen¹* and the def.art. *y* but the third el. is in doubt. If the 1887 form is reliable, then we may have *bicws* 'type of food consisting of oatbread crumbled into buttermilk'. That might describe scattered stones of a cairn that has been disturbed but it is difficult to think of similar examples; clearly it is a matter for further research.

Trimsaran SN 451048
Pen-bre
?'Sa(e)ran's ridge': *trum*, pers.n. ?*Saeran, Saran*
Trymsarren 1564, *trym saran* c.1560, *Trymsaran* 1570-1766, *Trynsaren* 1578, *Trimsaran* 1615, *Trumsaran* 1686
The pers.n. seems to be very rare. The n. also applies to a ho. Plas Trimsaran (*plas*) ultimately taken from the ridge south of the village. The latter is shown around *Star Inn* 1792 (SN 453046) and *Star T(oll) B(ar)* 1844 which hints that the second el. was locally reinterpreted as *seren* 'a star'.

Trostre SS 521997
Llanelli
?'settlement extending over a large area', *tros, tref*
Dostre (= *Drostre*) 1625, *Trostrey* 1669, *Trostre* 1676
There are similar ns. elsewhere, eg. Trostre (SO 360044), co. Monmouth, recorded as *T(r)ostray* 1296, *Trostri* c.1160 (1330), *Trostre* c.1348, with hos. Trostre Hen (*Trostraxhen* 1314, *Trostre hene* 1465), Trostre Newydd (*Trostrey Newith* 1570, *Trostrey newidd* 1575) (*hen, newydd*). *tros* occurs in other pns. such as Trosnant (Cantref, co. Brecon) recorded in *Trosnant Issa* or *Lower Trosnant, Trosnant Ycha* or *Upper Trosnant* 1763, Trosnant (Trefethin, co. Monmouth) *Trosnant* 1707 (*nant, isaf, uchaf*), Trosymynydd (Llantwit-juxta-Neath), co. Glamorgan, which is *Tros y Mynydd* 1558 (*y, mynydd*) but

the precise sense is often unclear. Cf. ***traws*** identified in Traws-mawr (q.v.).

Tumble see **Tymbl**

Twrch¹   SN 6445
rn. SN 699513 to Cothi SN 654404
'(r.) which behaves like a boar': ***twrch***
(brook called) *Twrch* 1576, 1668, *Turghe flu:* 1578, *the Turche becke* 1587, *Turch R* 1729
In Caeo and Llan-y-crwys. Figurative for a rn. which rushes and roots like a wild pig. The antiquarian Richard Fenton in 1804 describes it as 'a most formidable and dangerous Mountain stream in Winter' (Fenton 13).

Twrch²   SN 7715
rn. SN 821216 to SN 770084
Turch c.1170, *Tourthe* 1203, *Twrch* 1540-1797, *Turch Flu:* 1578, *the Turch or Torch water* 1586
Turch Flu: 1578, *the Turch or Torch water* 1586, *Turch* 1610, (brook) *Twrch* 1576, 1668
In Llanddeusant and Llandeilo Fawr (Cwarter Bach). The r. rises on the southern slopes of the mountains Bannau Sir Gaer and Y Fan Brycheiniog. Tributaries include Nant Gwys, 'stream which behaves like a sow': ***nant, gwŷs***¹.

Twyn   SN 690137
Llandeilo Fawr, Cwmaman
'hillock': ***twyn***
Twyn 1953
Industrial housing which appears on OS maps between 1901 and 1906 above the valley of Cwm Aman, north of Christ Church and adjoining the ho. Twynyboly, from which it takes its shortened n. The ho. is recorded as *Twyn Boley, Twyn y Boley* 1842 and *Twyn-y-boli* 1878 probably meaning 'hillock at the swelling': ***y, bol***¹, *bola* or perhaps a pl. *byly*. Recorded between 1906 and 1938 as *Tre-Bevan* and *Trebifan* (***tref***, surname ***Bifan***, *Bevan*) in GazWPN.

Twynllanan   SN 754244
Llanddeusant
'hillock at a small enclosure', ***twyn, *llanan***
Twyn-llanan 1751, *Twnllaman* 1875, *Twyn Llanan* 1885-7-1964, *Twynllanan* 1979

The location of a ho. *Penlanau* (a printing error for *Penlanan*) 1831, *Pen-llanan* 1891 (***pen***¹). The village developed around a Calvinistic Methodist chp. recorded in 1790 (HMGC 2: 152, 536).

Twynmynydd   SN 664146
Llandeilo Fawr, Llandybïe
'hillock near the mountain': ***twyn, mynydd***
Twyn-mynydd 1907, 1917, *Twynmynydd* 1963-4
The 'mountain' may be Mynydd Isaf (SN 6615) (***isaf***). The ht. developed near the Mountain Inn and Angel Inn near Tynewydd on the road running from Glanaman to Llandyfân.

Tŷ-croes   SN 606107
Llanedi
'house at crossroads': ***tŷ, croes***
Tycross 1716, *Ty Crose* 1735, *Tycroes* 1810, 1903, *Tŷ-croes* 1831, *Ty-croes* 1880
At the meeting-point of five roads. Named from a ho., the village developed from the late 19th cent along the A483, later incorporating Mynydd-bach and the area south of the A483 as far as the lane Heol Ddu and Bancyffynnon (SN 605099)

Tŷ-hen   SN 302241
Tre-lech a'r Betws
'old house': ***tŷ, hen***
Ty-hen 1831, *Ty-hên* 1889
Named from a ho. located next to a Calvinistic Methodist chp. Capel Ty-hen (erected in 1830). The settlement dates mainly after 1945.

Tŷ-isaf   SS 509995
Llanelli
'lower house': ***tŷ, isaf***
(mess.) *Tuy Issa* 1787, *Tir Isha* 1842, *Ty-isaf, Ty-isaf Villa* 1880
Ultimately named from a ho. probably to be identified with *Sythyn ty Isha* 1745 (***syddyn***) and *Tuy Issa* 1787. An area of housing built largely after World War II east of Ty-isha Road. The n. has partly displaced the older development St Pauls built south of St Paul's ch. in the late 19th cent.

Tŷ-mawr SN 542430
Llanybydder
'great house', *tŷ*, *mawr*
(ho.) *Tymaur* 1843, *Ty-mawr* 1889
A ho.n. applied to the ht. before 1905 which developed in the 19th cent along the B4337 near a fork in the road where an unclassified road leads off to Wernant woollen mill and Peithyn (SN554421).

Tymbl, Y, Tumble SN 541120
'steep slope', E *tumble*
Tumble 1817, 1831, 1889-91, *the Tumble* 1843
Probably describing the perilous nature of the road which twists and falls sharply as it runs northwest from Tan-y-graig (SN 543118). The W form is not so well recorded and *Bank y Tumble* 1841 simply reflects E orthography (*banc¹*, *y*). The n. is thought to have applied to a ho. which later became the Tumble Hotel/Inn (SN 541120) with the extended sense of 'inn on a steep slope into which one might tumble'. Cf. 'Dew Drop Inn', a pun on 'do drop in'. The inn is *Tumble Tavarn* 1841 (*tafarn*) and *Tumble Inn* 1866. The n. also appears in a broader sense describing a steep road or hos. on steep slopes; cf. Tumble, in Pontypridd, co. Glamorgan, recorded in 1841. The village developed from the 1880s near Great Mountain Colliery.

Tyn-y-ffordd SN 534437
Llanybydder
'small farm by the road': *tyddyn*, *tyn*, *ffordd*
Tyn y fordd 1704, *Tynyffordd* 1776, 1857, *Ty'n-y-fford* 1905
The n. of two hos., now considered as part of Rhyd-y-bont.

Tŷ'r-frân (Tyrfran) SN 510014
Llanelli
'house inhabited by crows, derelict house': *tŷ*, affixed def.art. `*r*, *brân*
Tir Fran 1831, (hos.) *Ty'r-frân*, (~) *Allt-y-frân Villas* 1880; 1916, *Ty'r-frân* 1938-53, *Tyrfrân* 1972-92
The first recorded form suggests that *Tŷ'r* has displaced *tir* 'land, ground' but len. *f* < *b* would be irregular unless it has lost a def.art. *y*. The n. earlier applied to a small group of hos. but development expanded along Felin-foel Road after 1918.

Tyweli, Afon SN 4238
rn. SN 444362 to SN 412402
?'wild (r.)': ?*dywal*, *-i*
Twely 1794, *Tafwili* 1831, *Twelly* 1840, *Tyweli* 1891
The rn. also occurs in the ns. in Pont-tyweli (q.v.) and several hos. recorded as *Tir glan deweli* 1563, *Glandewely* 1699, and *Glan-tafwili* 1831. The earlier evidence shows that this was Dyweli > Tyweli and that Pont-tyweli < Pont Dyweli. The rn. might indicate a r. which was perceived as 'fierce, cruel' and likely to cause damage by floods though we cannot completely rule out a pers.n. *Dywel* or *Dyfel* with *-i* (DPNW 392). Local pron. is 'Tweli' (HPLlangeler 90). The n. is applied on OS maps for the r. below Afon Talog reaching Teifi near Pont-tyweli (q.v.).

Tywi, Afon
rn.
?'strong river': IE **teu̯ā*, *teu̯-* , *tu*
Tovius c.150, *Tyui*, *tiugui* c.1145, *Tywi* 1129 (c.1175), late 12th cent, *aqua(m) de Tewy* 1326, *Tewy* 1413, *fluvio Tewiensi* c.1191, *Tywy* 1202 (1336), 1602, *Towy* 1303, *the Towie* 1586
The first ref. is a preferred form from Ptolemy probably standing for Brit **Tou̯ii̯ós* > Tywi, ie. -óu̯i- > yw- (PNRB 474; LHEB 474). The root is probably found also in *tyfu* 'to grow, to increase', Ir *téo* 'nerth', Latin *tumeo* 'to swell, to surge' etc., as in *Taw*, co. Devon, but its development is uncertain (DPNW 482-3).

Uwchcoed Morris to Uwch Sawdde

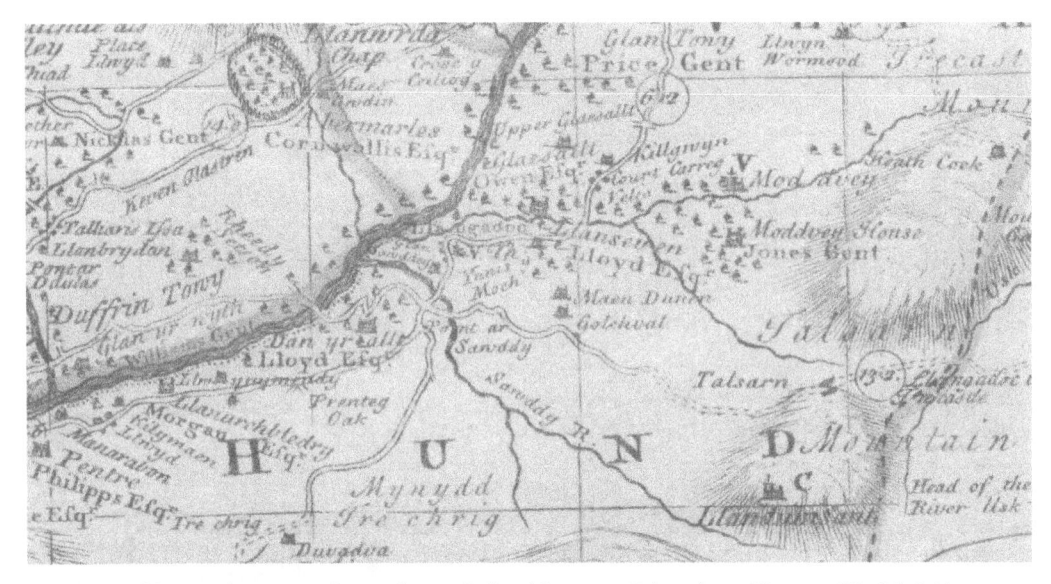

Uwch Sawdde area between Llangadog and Llanddeusant. Taken from Thomas Kitchin's 'Accurate Map of Carmarthen Shire' c.1762.

Upper Brynaman see **Brynaman-uchaf**

Usk see **Wysg**

Uwchcoed Morris
Llanismel/St Ishmael
'(area) above the wood associated with (man called) Morris': *uwch*, *coed*, pers.n. *Moris*, *Morris*
Ughcoide morres 1566, *Uchcoid Morris (and Iscoyd Morris)* 1619
The n. is not current but it presumably made up the southerly part of St Ishmael p. perhaps south of Treforris (q.v.).

Uwch Sawdde
Llangadog
'(area) above (r.) Sawdde': *uwch*, rn. **Sawdde**
Hamlet above the Sawdde 1811, *above Sawthe* 1851
The n. is not current but the n. probably applied to the southerly part of Llangadog adjoining Afon Sawdde or its tributary Sawdde Fechan.

Waun Baglam to Wysg

Waungilwen, Drefach Felindre. Looking northwestwards to houses on the road from Dre-fach towards Pentrecagal c.1910.

Waun Baglam   SN 446039
Pen-bre
'moor at or belonging to Baglam': ***gwaun***, ho.n. *Baglam*
Wainbaglam 1851, *Waunbaglan* 1879, *Waun-Baglam* 1907-1969
Named from a ho. recorded as (land called) *Baglan* 1648, (real estate called) *Baglan* 1788, *Baglan* 1830, *Bagland* 1880, with late development to *Baglam Farm* 1907-8. The n. may mean 'twisted or hooked river' (***bagl*** 'crozier, crook', ***-an***), perhaps a lost n. for the small stream which flows westward to Pinged where it turns northwards to meet Gwendraeth Fawr south of Cydweli. Cf. Baglan, co. Glamorgan (PNGlamorgan 10) which is generally thought to be the n. of a saint but which might rather describe the twisting course of the stream there.

Waunclunda   SN 684319
Llansadwrn
'(the) moor at Clun-da': (***y***), ***gwaun***, and *Clun-da*
Y Waunclynda 1831, *Waunclyndar* 1867, *Waen-clyn-da* 1888, *Waun-clyn-da* 1952-3, *Waunclunda* 1978
Clun-da is 'good, beneficial moor *or* meadow' (***clun²***, ***da***) probably with ref. to its fertility; it seems to have applied to a specific piece of land or area. The moorland is also recalled in the n.

of a former ho. (SN 681319) recorded as *Blaen-y-waen* 1888 (***blaen***).

Waungilwen SN 344392
Penboyr
'moor at white nook': ***gwaun, cil¹, gwen*** len. *wen*
(area) *Waun-gilwen* 1831, (ho.) *Waun* 1887, (hamlet) *Waun-gilwen* 1906, *Waungilwen* 1978-80
A small ht. which developed along the Felindre-Pentrecagal road in the 19th cent near moorland (SN 338397) recorded as *Waungilwen* 1869, *Waun Gilwen* 1887 and ho. Waun.

Waun-y-clun SN 442041
Pen-bre
'moor of the meadow': ***gwaun, y, clun²***
Waun-y-clyn 1880, *Waun-clŷn* 1906-7, 1921-2, *Waun-y-cl n* 1953, 1964
Clun refers to a ho. (SN 450044) north of the ht. recorded as *Clun farm* 1842, *Clyn* 1880, *Clŷn* 1906-7, 1972.

Wen SN 6237
Llansawel
?'white or fair (settlement)': ?lost ***tref, gwyn*** fem. *gwen*
(ht.) *Wen* 1849, *Wen Hamlet* 1851
The north-easterly part of Llansawel extending northwards from the village between Afon Marlais and Afon Cothi. The n. is identifiable with Trewaun-isaf (SN 639371) and Trewaun-uchaf (***tref, isaf, uchaf***) east of the village recorded as *Trewain Issa* 1795 and *Tre-waun-isaf, ~ -uchaf* 1831.

Wern¹ SS 508999
Llanelli
'(the) alder-tree marsh': (*y*), ***gwern***
(cottage and garden on) *the Wern* 1787, *the Wern* 1804, *Wern, Wern Works (Iron)* 1880, *Wern* 1916
Earlier OS maps use it to describe the area between Ann Street and Bigyn Road. The n. is applied on current maps to an area immediately north of St Paul's church (SN 510998) extending towards the site of the former Wern Iron Works. The area developed in the mid 19th cent near coal-pits, brickworks and small iron and tin works.

Wern² SN 590406
Pencarreg
'(the) alder-tree marsh': (*y*), ***gwern***
(hos.) *Wern* 1851, *Wern* 1887
The n. extended from the ho. to a Nonconformist chp. *Capel y Wern* (***capel***) established between 1891 and 1906 and adjoining hos. from the late 1960s. Not named on OS map in 1831.

Wernolau SN 643123
Betws
'bright alder-tree marsh': ***gwern, golau***
(ho.) *Wern-oleu* 1878, 1962, *Wernoleu* 1953
Historic forms have a variant *goleu*, a back-formation from words such as ***goleuni*** nm. 'light, brightness'. An area of housing developed from the late 1970s near a ho., later a hotel. Cf. Maenor Gwernolau (q.v.).

Westfa, Y SN 4901
Llanelli
'(territory paying) food-rent': *y*, ***gwestfa***
(manor) *Westva* 1549, (~) *Y Westva* 1571, *Westva* 1591, *Westfa Hamlet* 1792, *Westfa* 1852-3
That part of Llanelli between Afon Dulais and Afon Lliedi and around Strade. The absence of any qualifier is unexpected. It is worth noting, however, that Westfa is located in an area where ***gwestfa*** is otherwise absent implying that a qualifier was unnecessary.

Whitehill Down SN 2912
Llanddowror
'downland on white or pale hill': E ***white, hill, down***
Whitehull 1437, (hill) *Whitehill down* 1831, (~) *Whitehill Down* 1888, 1973
down is apparently a later addition referring to open pasture-land on the hill. The el. 'white' sometimes describes infertile land.

White Mill see **Felin-wen**

Whitland see **Hendy-gwyn**

Wysg, Usk   SN 8123
rn. SN 819239
?'swift (r.): OW *uisc*
(*flumen*) *Uscha* 1100 (1234), *Osce, Oscę* c.1100 (c.1200), *auon Vysc* c.1136, *Usch* 1146, *huisc* 1129 (c.1170), *uÿsc* c.1170, *Oscha* 1188, (water of) *Husk'* 1326 (1516), *the Usk* 1332, *Osk water* 1479, *6isc* late 15th cent, *Wysg* late 15th cent, *River Usk or Wysg* 1887

Wysg rises on the border of cos. Carmarthen and Brecon at Blaen Wysg recorded as *Blaen or Llygaid Uske* 1754 (**blaen, llygad** pl. *llygaid*) on the north side of Mynydd Du/Black Mountain, taking a northerly course before swinging eastwards at what is now Cronfa Wysg/Usk Reservoir (SN 8128) towards Aberhonddu/Brecon, ultimately reaching the Severn estuary below Casnewydd/Newport, co. Monmouth.

The rn. is thought to be represented in latinised *Isca*, the Roman n. for the legionary fortress at Caerllion/Caerleon, and to be identical in meaning to the r. Exe in Exeter. The latter is recorded as *Iska* c.150 and as recorded as OW *Cairuisc* c.894 (c.1000) (*caer*). The etymology of these ns. is disputed but a summary of the evidence is provided by J.E. Caerwyn Williams (Celtica XXI, 670-8, and refs.). The meaning remains uncertain but most scholars now reject the suggestion 'r. abounding in fish' and prefer a derivation from Brit *Iscā* (DPNW 484). This may have a root in Celt **eiskā* (of disputed origin) meaning 'to move' or to move swiftly'. The latter seems most appropriate and is the meaning favoured for the r. Exe by Watts (CDEPN 221).

Ydw to Ystumgwili

Ystrad Tywi near Tŷ-gwyn Mawr, Llandeilo. Looking northwestwards over Afon Tywi c.1920.

Ydw, Afon SN 8032, 7631
rn. SN 804324 to Brân SN 725291
?'river which shifts': *mud-* or perhaps ?'mute, silent river' ?*mud¹*, and *-wy¹*
River Ydw, Blaen-Ydw 1886
Historic forms are very late and it is possible that the modern form developed from a misrepresentation of *-mudw, -y*. The first suggested el. *mud-* may be the stem in *mudaf¹*, *mudo¹* 'to move, to convey, to bear away' describing a r. which moves and shifts adjoining soil and rocks and by implication one which is strong and forceful. The rn. is otherwise recorded in (ho.) *Blaenudo* and (ho.) *Cwmudo* 1831 (SN 734294) and Cwmydw (SN 783316) (*blaen, cwm*). It can hardly be coincidence that one of Ydw's tributaries is Afon Mudan recorded as *Mudan* 1754, *Mydan Brook* 1886, *Nant Mudan* 1906 (*nant*) which joins Ydw near *Abermydan* 1831, *Aber-Mydan* 1886 (SN 778319) with *-an*. The form with *mydan* may be the development of accented *-u- > i- > -y-* (EANC 77-78).

Ynysdeilo see **Llandeilo'r-ynys**

Ysbyty (Yspitty) SN 557983
Llanelli
'hospice, hospital', *ysbyty*
(place called) *Yspitty* 1605, *Spytty* 1651, *Spitty Copperworks* 1830, *Yspytty Works (Tin Plate)* 1879, *Yspitty* 1915-68

This is almost certainly *locus hospitalis* recorded in conjunction with a fishery *Pencoit* identifiable with Pencoed (SN 558000, SS 561008) between 1223 and 1250. Judging from its location it may have been a hospice used by travellers and perhaps pilgrims on the ancient road running from Caerfyrddin/Carmarthen through Cydweli/Kidwelly and Llanelli to the ferry over the r. Llwchwr to Casllwchwr/Loughor.

Ysgubor-fawr　　　　　　　　　　SN 4112
Llandyfaelog
'great barn': ***ysgubor, mawr***
Iskybor vawre 1590, *Skyborvawr* 1591, *Skybor Hamlet* 1792, (ho., ht.) *Scyborfawr* 1851
Not current. 'Great' in comparison with the former ho. Ysgubor-fach (***bach**[1]*) (SN 413141) recorded as *Ysgubor-fâch* 1890. Cf. Ysgubor Gwempa and Ysgubor y Glyn, in the adjoining p. of Llangyndeyrn, recorded as *Skibor gwempa* and *Skybor y glyn* 1618, (place called) *Skebor gwempa* 1624, *Scibor Gwempa* and *Skybor y glyn* 1629 (**Gwempa, Glyn**).

Ysgwyn　　　　Ysgwyn-fawr SN 585253
Llangathen
Uncertain
Ysgwyn 1731, *Ysgwin* 1745-6, *Ysgwyn* 1747-8, 1848, (ht.) *Usgwyn* 1772, *Ysgwynnfach* 1783, *Ysgwyn-fawr*, ~ *-fâch* 1887
Possibly a misdivided form Ynys Gwyn > *yn Ysgwyn* 'water-meadow of (man called) Gwyn': ***ynys*** and pers.n. ***Gwyn***. See also Maenor Ysgwyn.

Ysterlwyf　　　　　　　　　　　SN 3013
manor, commote
?pers.n.
Ostelof c.1135, (land of) *Oysterlayth'* 1215, *Estrelef* 1257, *Ystlvyf* 1261, *Ostrelof* 1265, 1324, *Osterlawe* 1287, *Oystrelof* 1288, *Ostirlowe* 1352, *Oystrelowe* 1396, *Oysterlowe* 1546, 1596, Comm. *Estholoef* c.1537
Early forms – all from AN sources – and many later ones appear to be attempts at representing *Ysterlwyf* or *Ystrelwyf* (in ModW orthography). Several forms from W sources such as *Ystlvyf* in 1261 and c.1400 and *Ystlwy* 15th cent, however, favour a two-syllable pn. which has prompted the alternative use of the form *Ystlwyf* among historians. The reliability of this form is questionable because a larger sample of 29 historic forms produces just a handful in favour of two syllables. One possible solution in reconciling historic forms is that the pn. was either a three-syllable *Ysterlwyf/Ystrelwyf* or a two-syllable *Ystrlwyf* both stressed on the first syllable. That might account for the variation *-stre-, -ster-* seen in most historic forms. The etymology unfortunately remains obscure and the best that can be suggested is that the pn. bears similarity to pers.ns. such as *Ystrwyth* and *Ysfael*. Some forms appear to have been influenced in E sources by AFr ***oistre*** 'oyster'. The cmt. covered Llanfihangel Abercywyn and Meidrum (Maenor Cefndaufynydd, Maenor Garllegan, Maenor Ddwylan Isaf, Maenor Ddwylan Uchaf, Maenor Lleision, Maenor Rhiwtornor and Maenor Tre-goch).

Ystrad[1]　　　　　　Ystrad-isaf SN 745330
Llandingad
'valley-bottom (of r. Tywi)': ***ystrad***
(ht.) *Ystrad vynis* 1610, (ht.) *Ystrad Vynis* 1685, *Ystrad* 1710-1851
A very common pn. with other examples in Llangadog (*Istret* 1543) and St Ishmaels (*Ystraed* 1638). Early forms qualify the n. with the pers.n. **Mynys** lenited *Fynys*.

Ystrad[2]　　　　　　　　　　　SN 497298
Llanegwad
'valley-bottom': ***ystrad***
(ht.) *Ystrad* 1710, 1842, 1889, *Ustrad* 1715, 1842
The t. was in the northern part of the p. centred on the ho. of this n. which is located in the valley of an unnamed tributary of Afon Pib.

Ystradau see **Strade** (Llanelli)

Ystrad-ffin　　　　　　　　　　SN 788466
Llanfair-ar-y-bryn
'valley-bottom at the boundary': ***ystrad, ffin***

Stratfyn 1202 (1336), *Straffin* c.1291, *Ystrad ffin* c.1560, 1798, *Istrodefyne* 1578, *Ystradfin* c.1600, *Ystradffin* 1739, (ho.) *Ystrad-ffin, Capel Ystrad-ffin* 1834

The n. might also be elliptical for 'valley-bottom of (stream called) Nant y Ffin' a tributary of Tywi. This is *Nant y Ffin* and leaves its n. in that of a ho. *Nant-y-ffin* 1887 (SN 788470) (***nant***). The 'boundary' is clearly that between the ancient historical divisions of Ceredigion and Ystrad Tywi. Ystrad-ffin is also an alternative n. for Capel Peulin (q.v.). The p. of Ystrad-ffin was formed in 1875 taking in parts of Cil-y-cwm and Llanfair-ar-y-bryn and was provided with a ch. dedicated to Barnabas, in Rhandirmwyn, 1878 (TCymm 1911-12: 225-6).

Ystrad Tywi

'valley of (r.) Tywi': ***ystrad***, **Tywi**

Strat Tiui c.1100, *Estratewi* after 1130, *ystrat tywy* 13th cent (early 14th cent), *Estrad Tywy* 1282, *Stratewy*, (land of) *Strateuwy, Stretewy* 1292, (balliva) *Stretthowy, Strethowy* 1289, (deanery) *Strattewy* c.1291, *Estratewy* late 13th cent), (bedelry) *Stratwy* 1339, *Ystrat Tywi* c.1400

Ystrad Tywi was an ancient *gwlad* which included Cantref Bychan, Cantref Eginog and Cantref Mawr, taking up the more than half of the later co. of Carmarthen and including G yr (Gower), co. Glamorgan.

Ystumgwili SN 4127, 4323

Abergwili later Llanpumsaint

'(area within) bend of (r.) Gwili': ***ystum***, rn. **Gwili**

Istym Gwylly 1548, *Ystym-gwili* 1754, *Ystym Gwili* 1811

The n. is not current but clearly referred to the area north of the village of Abergwili on the eastern side of Afon Gwili extending eastwards to Nant Crychiau.

LIST OF SUBSCRIBERS

The author and publisher are very grateful to those listed below for their support for this publication.

Individuals

Antur, Gruffudd
Beek, C. J. H. van
Bevan, Gareth A.
Bird, Margaret
Briggs, John
Briggs, Keith
Broderick, Prof. Dr. George
Coates, Richard
Comeau, Rhiannon
Davies, David Leslie
Davies, Eurig
Davies, Richard
Emlyn-Jones, Stephen
Flowers, Ness
Foster Evans, Dr. Dylan
Fychan, Gwerful Angharad
Griffiths, Alun
Gwyndaf, Eleri & Robin
Headon, Michael
Hines, Prof John
Hughes, Glyn
Huws, Richard
Jankulak, Karen
John, Deric
Jones, Berwyn
Jones, Dennis
Jones, Denzil
Jones, John
Jones, Mary
Jones, Pat & Meurig
Jones, Rhidian H. B.
Jones, Thomas H.
Jones-Jenkins, Chris
Kidwell, Matthew
Kitson, Peter. R.
Lewis, Elfyn
Lloyd, Thomas
Newley, Jane
Morgan, Ian
Padel, Dr. Oliver
Parker, Lesley
Pierce, Gwynedd O. †
Rees, Eiluned
Seaman, Andrew
Skym, Richard
Smith, Peter
Stopp, Peter
Thorne, David
Webb, Cliff
Williams, Dewi
Williams, Will
Wooding, Jonathan

Institutions

Carmarthenshire Archives / Archifau Sir Gaerfyrddin
Comisiynydd y Gymraeg / Welsh Language Commissioner
Cymdeithas Enwau Lleoedd Cymru / Welsh Place-Name Society
Gwasanaeth Archifau Gorllewin Morgannwg / West Glamorgan Archive Service

welsh academic press

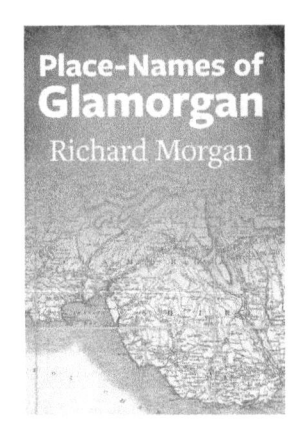

PLACE-NAMES OF GLAMORGAN

Richard Morgan

'An impressive achievement ... a fascinating book to dip into local history and a perfect travelling companion.'
Western Mail

'Handsomely produced ... popular and scholarly'
Cymdeithas Enwau Lleoedd Cymru - Welsh Place-Name Society

'A scholarly work that deftly accomplishes its task ... a comprehensive study ... detailed and informed without ever being overwhelming.'
Guto Rhys, Y Casglwr (The Collector)

Based on many years of detailed research, *Place-Names of Glamorgan* investigates the historical evidence and meanings of more than 1,100 place-names in the historic county of Glamorgan, stretching from Rhossili to Rumney and Rhoose to Rhigos.

The illustrated volume contains a concise introduction to the subject, a bibliography, a glossary of common place-name elements, and a close examination of individual place-names and their historic forms.

Richard Morgan is a former archivist at Glamorgan Archives and co-author of the *Dictionary of the Place-Names of Wales*.

978-1-86057-132-9   304pp   £19.99   PB

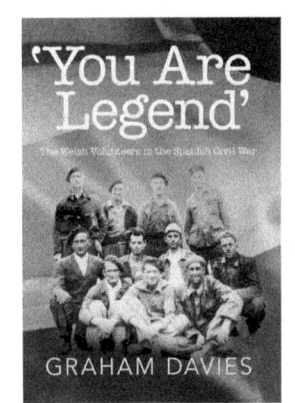

'YOU ARE LEGEND'
The Welsh Volunteers in the Spanish Civil War

Graham Davies

'Excellent. A paean to the working men and women of Wales who went to Spain to fight in defence of the fledgling Spanish Republic.'
Keith Jones, son of volunteer Tom Jones from Rhosllanerchrugog

'Well researched, and using previously unpublished sources, 'You Are Legend' is recommended reading. It is important that the contribution of the large number of Welsh volunteers continues to be recognised.'
Mary Greening, daughter of volunteer Edwin Greening of Aberdare

'A highly readable and comprehensively researched account of the Welsh Brigaders.'
Alan Warren, Spanish Civil War historian

Almost 200 Welshmen and women volunteered to join the International Brigade and travelled to Spain to fight fascism alongside the Republican government during the 1936-1939 Spanish Civil War. While over 150 returned home, at least 35 died during the brutal conflict. *'You Are Legend'* is their remarkable story.

978-1-86057-1305   224pp   £19.99   PB
978-1-86057-1558   224pp   £19.99   EBK

welsh academic press

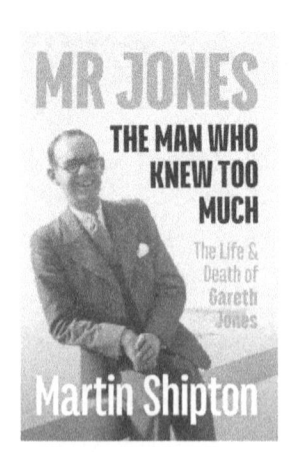

MR JONES
THE MAN WHO KNEW TOO MUCH
The Life and Death of Gareth Jones

'Martin Shipton's biography is a much needed and welcome contribution to our understanding of Jones' experiences and his life'
Mick Antoniw SM

Murdered in Mongolia in 1935 aged only 29, the Welsh investigative journalist Gareth Jones is a national hero in Ukraine for being the first reporter to reveal the truth about the Holodomor – the 1932-33 genocide inflicted on Ukraine by the Soviet Union during which over four million people perished.

Drawing upon Jones' articles, notebooks and private correspondence, Martin Shipton, the highly respected political journalist at Jones' former newspaper, the Western Mail, reveals the remarkable yet tragically short life of this fascinating and determined Welshman who pioneered the role of investigative journalism.

978-1-86057-1435	374pp (80+ images)	£19.99	PB
978-1-86057-1565	374pp (80+ images)	£19.99	EBK

ABERFAN
Government and Disaster
(Second Edition)

Iain McLean & Martin Johnes

'The full truth about Aberfan'
The Guardian

'The research is outstanding...the investigation is substantial, balanced and authoritative... this is certainly the definitive book on the subject...Meticulous.'
John R. Davis, Journal of Contemporary British History

'Excellent...thorough and sympathetic.'
Headway 2000 (Aberfan Community Newspaper)

'Intelligent and moving'
Planet

Aberfan - Government & Disaster is widely recognised as the definitive study of the disaster and, following meticulous research of previously unavailable public records – kept confidential by the UK Government's 30-year rule – the authors explain how and why the disaster happened and why nobody was held responsible.

978-1-86057-1336	224pp (+16pp photo section)	£19.99	PB
978-1-86057-1459	224pp (+16pp photo section)	£19.99	EBK

Lightning Source UK Ltd.
Milton Keynes UK
UKHW031811110822
407176UK00005B/303

9 781860 571572